CW00723049

MAPPING THE WINDMILL

The Ordnance Survey in England

Bill Bignell

Dedicated to the memory of

Major General Sir Charles William Wilson

14 March 1836 - 25 October 1905

Soldier, scholar and Director General
of the Ordnance Survey 1886 - 1894

MAPPING THE WINDMILL

The Ordnance Survey in England

Bill Bignell

London

The Charles Close Society

2013

Published by The Charles Close Society for the Study of Ordnance Survey Maps,
c/o The Map Library, British Library, 96 Euston Road, London NW1 2DB
www.charlesclosesociety.org

A catalogue record for this book is available from the British Library

ISBN 978 1 870598 29 3

Printed and bound by Short Run Press Ltd, Exeter

Contents

Illustrations

Cover picture: Little Hadham Smock Mill, Hertfordshire *circa* 1910

Plates

(between pages 116 and 117)

Figures

Tables

Preface

In keeping with other publications of the Charles Close Society this monograph hopes to shed light on some small part of the historical affairs of the Ordnance Survey. It is different to most other publications however in that it does not take a particular map series or type of mapping and develop a narrative and cartobibliography for it. Instead, it takes as its focus the effectiveness with which one particular landscape feature found itself being mapped by the Survey (as the Ordnance Survey can be shortened to – or even OS if used adjectivally) throughout the nineteenth century and into the twentieth. The choice of feature has been a personal one but, as it turns out, a fortuitous one in that it satisfies the main requirement of presenting a range of traits that might have caused a surveyor to think twice about what he was looking at. It was also a feature whose existence could, for a range of reasons, be short-lived. So recording ground-truth accuracy for such a feature would have constituted a harsh test for the Survey and one that could leave a lot of room for its maps to misinform their users.

If there is any uncertainty over how an Ordnance Survey map might depict a natural or man-made landscape feature, then that vagueness should be dispelled, one would think, by simply understanding the map-making protocols developed over a period of time for each of the major OS scales. More helpfully of course, and readily to hand for modern 1:50,000 maps and their predecessor one-inch maps, are the range of conventional signs shown in their legends. These, one might assume, should be sufficient to ascertain the truth of what is happening on the ground. The experience of using maps can show however that this is not necessarily always the case. This writer's realisation that care needed to be exercised when using one-inch maps in the field came at a formative age, something more than forty years ago, when cycling around the countryside looking for the remains of old windmills. At that time the Survey used two types of symbol mimicking, on the one hand windmills that still had their sails, and on the other, those that had lost them. Not only did it quickly become apparent that this supposedly neat distinction between two types of windmill did not reflect what was being seen on the ground, but that very often mills recognised by the Survey using either of the two types of symbol were discovered to have long since disappeared from the landscape, and certainly before any latest date of correction that a map might carry.

Accordingly, this study takes a selective view of the extended history of the Ordnance Survey to try to understand how this shortcoming leading to an early disenchantment with it might have come about. It quite simply intends to try to gauge how effective the Survey has been, over the long course of its existence, in the mapping of windmills at each of the different scales. In doing this, it hopes as well to act as a primer for those who wish to develop a deeper interest in the story of OS maps and in the ability of map symbols to be throughout both a help and a hindrance. Two points should be made at the outset. First, the span of years that will be considered starts from when the Survey was first tasked to provide maps and it continues up until the current day. But the importance attached to different periods within that time-span – as crudely measured by the numbers of pages devoted to each – varies greatly and reflects the extent to which windmills were working in an 'authentic' everyday capacity during each of those periods. Hence the nineteenth century easily outweighs the twentieth in importance. This is not to suggest that the OS treatment of windmills from, say, *circa* 1920 onwards is not as interesting as the one from the preceding century, but just that the increasing obsolescence of windmills and a severe reduction in their numbers by the early twentieth century allows the handling of them by the Survey to be discussed in a single chapter. This is in sharp contrast to a commentary for earlier years that stretches over five chapters. Maybe for those people whose interests lie more with the colourful and delightfully idiosyncratic OS maps of the last century, this will come as something of a disappointment. But a moment's reflection will be enough to recognise that the most rewarding period for an assessment of how mills were mapped is surely that of their working heyday, rather than some later period when their status was becoming ambiguous and, in any case, when their obsolescence was becoming obvious to all. Second, to anyone with knowledge of the secondary literature that provides a national overview for the windmill, it will readily be understood that this literature places a strong emphasis on those parts of the country either where a modicum of mills still survive, or where they are known to have been a familiar landscape feature as late as 1920. The focus of this study is somewhat different since the windmills that are of particular interest to us are those that the Survey had to recognise, or perhaps not, in the early years of getting to grips with new and innovative map scales. For this reason, the windmills belonging to the northern counties, where the larger scales were first introduced, loom much larger in this account than is customarily the case in the mill literature.

One further point: what this book does not do, and could not properly do in a single volume, is to to chronicle how every windmill that has been built in the country over the past two centuries was shown at each of the Survey's scales – indeed, such a text would quickly become tedious, unmanageable and probably very quickly lose its audience rather than enlighten it. So, rather than try to test the stamina of the keenest devotee of mills or of Ordnance Survey maps, it instead considers in depth sufficient numbers of mills and the nature of their recognition by the Survey such that something can be inferred of the adequacy of depiction of any other mill whose existence we know something of, and can find – or maybe fail to find – on a map. The approach of this study, therefore, has been to embark on area studies for different periods and at different scales, so as to be able to say something legitimate and meaningful for the Survey's overall treatment of windmills. Hopefully in offering insights into its chosen area, this book will be of equal interest to those seeking an introduction to the subject, and who may be prompted then to follow their own related research; to those interested in the mapping of other landscape features, and also to the curious who may initially have thought the topic to be good for just a few pages.

Having been quick to declare what this book does not do, a little more of what it does aim to do might be reassuring. In the nineteenth-century, the range of OS scales was not vast. Those that became mainstream mapping included, in addition to the one-inch, the larger six-inch, 1:2500 and 1:500 scales. Each of these scales had its own beginnings and a later story to tell – all of which will be related here. Part of the remit for each scale was to depict windmills but it turned out to be a different remit for each scale and the degree to which each was successfully met varied to quite a surprising extent. The main task of this book has been to unravel these differing remits and so then to illustrate how successfully and transparently each was met, answering along the way such questions as:

- Could every known windmill expect to be depicted at every scale?
- Were windmills effectively deleted from maps once they no longer existed?
- Was the status of each windmill accurately modified on successive map editions?

The twentieth century saw the introduction of different scales and ultimately, of course, the complete switch to metric ones. As already implied, the story of mill representation by the Survey was also to become mired in the shifting status of those mills that survived long enough for them still to be standing late in the twentieth century, and the treatment of such mills took on a complexity that needs discussion. The roots of this study lie in a doctoral thesis submitted many years ago.[1] This has developed into what is thought to be the first major text whose sole focus is to discuss how windmills have been represented by the Ordnance Survey. It has been able though to draw upon a wealth of information from different types of source. Primarily of course, this includes the published volumes of the Charles Close Society for the Study of Ordnance Survey Maps, along with articles from *Sheetlines* – the society's regular journal. Second, the literature that has accumulated down the years through the efforts of the community of people with an interest in mills has been indispensable. The select bibliography provided at the end brings together for the first time within one set of covers the principal texts from the literature worlds of OS mapping and of mills.

No book such as this is written in isolation. It has taken the help and encouragement from friends over many years for it to reach any kind of fruition. Often, snippets of mill-related information will have happily been exchanged without any awareness on the part of teller or listener that, one day, some would end up in print. Those who may recognise their contribution to these pages and who, with luck, will approve of the way it has been handled include Guy Blythman, Duncan Breckels, Tony Bryan, Peter Filby, Tom Hay, Gareth Hughes, Nick Kelly, Ken and Shirley Kirsopp, Jon Sass. Help of another kind has been readily offered by staff at the British Library and at County Record Offices around the country. Paul Satchell, Hannah Shepherd and Simon Spokes from the University of the West of England at Bristol have helped in the preparation of the illustrations. From the membership of the Charles Close Society, thanks are particularly due to David Archer and Alison Brown who read through the script and made practical suggestions, and to Chris Higley who has done all the final preparation work in readiness for the printers. As well as David and Alison from Kerry in mid-Wales, I would like to thank John Coombes of Dorking and Ann Downes together with Pauline both late of Southsea, all of whom as sources of old OS maps have been free with their advice and hospitality over many years. I especially want to thank my good friend Richard Oliver for his help freely given over a sustained period of time: this book would have been much the poorer had it not been for his comments and suggestions. My thanks are definitely long overdue to Professors Peter Checkland from Lancaster University and Roger Kain, late of Exeter University, both of whom in the past thirty years have offered friendship and guidance. There could well be others who might feel slighted by not receiving a mention here and, inevitably, there will be factual errors in the text or at least events from the past that have been misconstrued,

[1] A thesis that considered *some* of the basic ideas discussed in this book. See W. H. Bignell, 'The cartographic representation of landscape features by the Ordnance Survey: a nineteenth-century perspective', unpublished University of Exeter Ph.D thesis, 2001.

possibly prompting those involved at the time to raise a spiritual eyebrow. Apologies all round. Closer to home, I especially want to thank Olive Morgan and Milica Vernon for their long-term support; and lastly, my gratitude goes to Sue Richards for her particular brand of encouragement over these past few years.

The plates and figures are mainly of two types – extracts from Ordnance Survey maps and photographs of windmills. For all these illustrations – except just two – the original material is of sufficient age to be out of copyright. For allowing me to reproduce the one map extract that is less than 75 years old, I thank the Ordnance Survey for its courteous help. The acknowledgement of those responsible for the photographs of mills included here is a more difficult matter. Most of the windmill images were made by photographers working a century and more ago, and their names are sadly lost to us. So, ownership of an original copy of a mass-produced photograph, often in the form of a picture postcard, is usually deemed sufficient for the publication of a book about mills to be litigation-free. That line is followed here though, as somebody whose own family was responsible for at least one of the early twentieth-century windmill images together with the much later one from 1962, I empathise with, and unreservedly apologise to, anyone related to an original photographer of any of the pictures here who may feel that their sensitivities have been trampled upon. With the four exceptions noted below, I have used a windmill image only if I hold an original copy of it. In a world where electronic copying of images has made them more easily accessible but also aggravated the problems of copyright probity, this approach may be thought a somewhat old-fashioned and needless one to take. This may be so, but it is an indictment that I can live with! This stance has been followed less so in respect of the map images seen here, so further acknowledgements for maps 'brought in' are also provided below. Use of abbreviations and military acronyms has been kept to a minimum: where one that is not self-evident appears, it is accompanied by an explanation for it. The style of military designation used is the one followed at the time to which the text refers. This has therefore resulted in hyphens and full-stops habitually forming part of army titles. For example, Lieutenant-Colonel Colby R.E. would by today's convention condense, for anything but the most formal occasions, to Lt Col Colby RE.

Illustration credits:

The image of Aldbourne Mill (Plate 9) is reproduced by permission of English Heritage; the image of Darn Crook Mill (Plate 30) is reproduced by permission of Newcastle City Library; the extracts from Ordnance Surveyors' Drawing 280 (Plate 33) and the Louth area Revision Drawing (Plate 34) are ©The British Library Board (Maps OSD 280 and Maps 176.o sheet 84); 'The Tyne from Windmill Hills' by T.M.Richardson (snr) 1818 (detail) (Plate 37) appears by permission of the Laing Art Gallery ©Tyne & Wear Archives & Museums; (Plate 45) is from the Cambridge Collection held at Cambridge Central Library; Figures 1.1, 1.2 & 4.2 are from the Charles Close Society; Figure 1.3 is reproduced by permission of the Mills Archive, Reading; Figure 1.4 is by permission of Taylor and Francis – it first appeared in 'The English windmill', Rex Wailes, Routledge Kegan & Paul (1954) page 227; Figure 3.1, drawn by Rodney Fry of the Department of Geography, University of Exeter for The Old Series Ordnance Survey Maps of England and Wales, Volume V, is reproduced by kind permission of the publishers, Harry Margary (*www.harrymargary.com*); in addition a number of 1:2500 map images have been accessed through Digimap Historic – these remain '© and database right Crown Copyright and Landmark Information Group Ltd *(All rights reserved 2012)*'.

Part One

Limits to Representation

1

Setting the scene

As suggested in the preface, if one single landscape feature has to be chosen to explore the competence of Ordnance Survey maps, then a feature that offers a variety of visual traits becomes a good choice. The windmill matches this criterion and, with all the vagaries of a fragile yet often well-documented existence, it offers a way of being able to either endorse or criticise the 'correctness' of an OS map. How the range of circumstances presented by windmills for any observer to see was then mirrored on the maps would, essentially one might think, have been dependent on a sequence of OS directives guiding its field staff in this respect. Certainly some guidelines from the later years are known to us, but they tend to be fairly cryptic in content and offer no insight into the rationale that lay behind them. What is more, there is only scant documentary evidence for the thinking that led to these – and presumably earlier – directives for achieving map coherency in the recording of all features, windmills included. Whatever else of a documentary nature that may once have existed, and which may have helped to explain why the maps show some things to have been done but not others, has simply not survived, so echoing a story bleakly familiar to anyone who has experience of delving into the Survey's history. It is with the idea of reconstructing one part of this history that this study, ostensibly of the nineteenth-century windmill, is concerned. By examining mainly nineteenth-century OS maps of different scales for different styles of windmill depiction, our understanding of the remit and function of each scale could be enhanced. For those whose interest lies as much with windmills as it does with maps, the extent of any incorrect depiction (judged by other types of knowledge for individual mills) can serve to colour our sense of the importance of these mills in their declining years. To develop our thinking around these issues it will be necessary to include a basic discussion of what it means to use map symbols, but before that the reader is offered concise narratives both for the pre-1914 history of the Ordnance Survey and for the specific landscape feature of interest – the traditional windmill. This may test the patience of those approaching this study already well-versed in the story of the Survey or of windmills, or perhaps both. But for others, these preambles might ensure that what then follows is more easily digestible. [1]

[1] This study is mainly concerned with OS map series of the nineteenth century – both large-scale ones and, more especially, those at the smaller one-inch scale – but it limits itself to coverage of England. This is not to suggest that windmills could not be found elsewhere in the British Isles, or that such mills were not mapped by an equivalent map series (or, in the case of Wales, the same ones) or that they were not of interest. It is simply that, as demonstrated by the windmills of Wales (with the exception of Anglesey), they were very thin on the ground compared to their English counterparts and their story adds nothing that a review of English mills alone cannot say. In addition, such windmills as there were beyond England's borders tended to be fairly

Throughout the nineteenth century the English windmill continued, as in earlier years, to be in widespread use within both rural and urban communities, but by 1900 their numbers were in decline with losses particularly significant in built-up areas. Across the country their time was certainly coming to an end by the outbreak of war in 1914. Up until this point, the general practice of milling will have made some sort of mark on the consciousness of the population. The prevailing view of windmills during the nineteenth century may have been one of association with the supply of food, or maybe as a familiar and accepted technology in an age of more progressive ones, or just as an easily recognisable feature in the landscape. What can be said with certainty is that the status of the windmill as a technological icon was severely weakened between the earlier and the later years of a century that experienced huge industrial and cultural change. Developing an insight into quite how the Ordnance Survey perceived windmills during these fast-changing times; how it considered the representation of mills to be still part of its remit; how that perception and remit may have changed during the passage of the nineteenth century and into the early part of the next, all lie at the core of what this book is about. So, before discussing the story of the windmill any further, there follows a short narrative for the history for the Survey through the years of key interest.

The Ordnance Survey

Any such narrative would certainly acknowledge that the end of the 1850s can be seen as representing a coming of age for the Ordnance Survey. A period of intense uncertainty lasting very nearly a decade, due to vacillation over the prime survey scale at which mapping for Britain should be made, came to an end in 1858. This ending to the so-called 'Battle of the Scales' coincided with technological advances that dramatically changed the potential for providing maps at a fraction of their earlier costs. Consequently, public awareness of the Ordnance Survey was set to change from the late 1850s. Along with the new provision of larger-scale maps, the activities that had been the mainstay of OS work in England up to this point – those needed for producing maps at the one-inch scale – were coming to an end. So, by 1860, mapping for nearly all of England and Wales was available at a scale of one inch to one mile and, for just a few counties in the north, maps at the newly-introduced scale of 1:2500 (approximately 25 inches to one mile) were becoming a reality by this time. This then quickly led to maps at a scale of six inches to one mile also being issued for these northern counties, complementing the earlier mapping done at this scale for Lancashire and Yorkshire.

But to go back to the beginning. If one accepts the founding moment for the Survey as 1791, any narrative outlining its major operational achievements, starting with its pre-history and work in Scotland in the aftermath of Culloden, would inevitably proceed to the genesis of its one-inch mapping amid fears of Napoleonic invasion.[2] The decision to provide maps

small and rudimentary in their construction, though there were exceptions to this, as shown by Llancayo Mill near Usk in Monmouthshire.

[2] There are not too many accounts of the Survey that go into overwhelming detail for all the fundamentals of history that will be discussed here, but see Tim Owen and Elaine Pilbeam, *Ordnance Survey: map makers to Britain since 1791,* London: HMSO and Southampton: Ordnance Survey, 1992. This semi-official volume, published to mark the bicentenary of the founding of the Ordnance Survey, provides a broadbrush yet adroit review of OS history. See also W.A. Seymour (ed), *A history of the Ordnance Survey,* Folkestone: Dawson, 1980: the compilation of this large and informative volume had a protracted history reminiscent of some of the events it describes. Also, a

at this scale was in keeping with tradition since it had been the customary scale of choice for private cartographers when mapping individual counties in the second half of the eighteenth century. At this time under the control of the Board of Ordnance, and with the freedom to liaise with the map trade and exploit some of its expertise, the Ordnance Survey was capable at the turn of the nineteenth century of meeting the military need for maps of the exposed southern coastline.[3] Thus, the earliest OS maps, those of Kent and Essex, were published between 1801 and 1805 with more coastal sheets following shortly afterwards. By 1819 that part of the country south of an extended line stretching between Bristol and London had been surveyed and published. But, with the end of the Napoleonic threat in 1815, military considerations receded though work for the military map, drawn to the one-inch scale and eventually to become known as the Old Series, was extended further and further towards the north of the country, but now under somewhat less of an imperative than before.[4] With the arrival of Major Thomas Colby as Superintendent of the Survey in 1820, strenuous efforts were made, and continued to be made for the next twenty years, to comply with his enthusiasm for high standards and for a consistency in the published Old Series sheets. But evolving ideals and practices as well as changing fortunes for the Survey, inevitably resulted in the style of later sheets moving away from that of earlier ones.

Meanwhile other tasks had come the way of the Survey. Field Survey Companies from the Royal Sappers and Miners, officered by subalterns seconded from the Royal Artillery and the Royal Engineers, were sent to Ireland in 1824-5 to provide maps for land taxation reform.[5] These maps had to be at a scale larger than the one-inch and the experience gained from being in Ireland working with the new scale – the six-inch – inevitably raised the issue

---

detailed study by Richard Oliver, *The Ordnance Survey in the nineteenth century: maps, money and the growth of goverment*, is at an advanced stage for publication by the Charles Close Society.

[3] The Board of Ordnance was a department with civilian responsibilities of disbursement and supply to both the army and the navy. But it held as well the military command function for the 'Scientific Corps' of the army – the Royal Artillery and the Royal Engineers. It was presided over by the Master-General of Ordnance who, though a soldier, was a political appointee and sat in the Cabinet to provide military advice. This was in contradistinction to the departments of the Adjutant-General and the Quartermaster-General. The Ordnance Survey therefore had a military as well as a civilian pedigree and it was commanded by a military officer who was accorded the neutral title of 'Superintendent'. This would later change to 'Director-General': see Hew Strachan, *The politics of the British Army,* Oxford: Clarendon Press, 1997.

[4] For a capable description of the earlier years of the one-inch scale see Harry Margary, *The Old Series Ordnance Survey maps of England and Wales,* volume I, *Kent, Essex, E. Sussex and S. Suffolk,* Lympne Castle: Harry Margary, 1975, with *Introduction* by J.B. Harley and Yolande O'Donoghue, *passim.* This is the first of eight volumes that each have an introductory essay and a cartobibliography, and that collectively provide a history for the one-inch scale Old Series coverage of the country. When published, they were considered quite exceptional in their rigour and are still, conventionally and deferentially, referred to as the Margary volumes: Harry Margary, *The Old Series Ordnance Survey maps of England and Wales,* volumes I to VIII, Lympne Castle: Harry Margary, 1975-92.

[5] The core involvement of the military establishment in the affairs of the Survey has been insinuated, as has the separation at this time of the 'Scientific Corps' – essentially a combination of the Royal Artillery and the Royal Engineers – from the rest of the army. The former initially provided its share of seconded officers but was to fade from the scene, leaving the Survey as the exclusive preserve of the latter. The engineering arm of the army at this stage was itself split confusingly into two Corps, one for the officers (the Royal Engineers) and one for the other ranks (the Royal Sappers and Miners). As part of the army reforms made in the light of experience gained in the Crimea these two were later amalgamated to become the Corps of Royal Engineers, at which time also the Board of Ordnance disappeared and the Survey came under control of the War Office. Gwyn Harries-Jenkins, *The army in Victorian society,* London: Routledge & Kegan Paul, 1977, 113 ff. See also R.A. Buchanan, *The engineers: a history of the engineering profession in Britain 1750-1914,* London: Jessica Kingsley, 1989; and M.D. Cooper (*et al*), *A short history of the Corps of Royal Engineers,* Chatham: Institution of Royal Engineers, 2006.

of whether the steady continuation of surveying for the Old Series, which by this time had got as far north as Lancashire and Yorkshire, should be discontinued in favour of surveying at, and for, this larger scale. A Treasury decision of 1840 permitted the more costly option with the result that these two counties were surveyed between 1841 and 1854 for maps to be published at this larger scale. This decision can retrospectively be seen as the prelude to the 'Battle of the Scales' when in 1851 it was revoked for further counties on the grounds of cost. For the next few years argument and counter-argument were put forward, not just for retention of the six-inch as the larger scale, but even for its discontinuation in favour of a still larger scale – the 1:2500.[6] This political indecision was seemingly brought to a close in 1855 when, to the huge satisfaction of the newly-appointed Superintendent, Major Henry James, the much larger scale was sanctioned for all but uncultivated upland areas. But this decision was overturned however in 1857 in renewed favour of the six-inch scale, leading James to remark bitterly that the position was now exactly the same as that reached in 1840. But the following year, in 1858, the 1:2500 was reinstated, leaving the Ordnance Survey able to draw breath and, had it but known it, look ahead to a period of stability providing maps at this large scale together with derivative six-inch and one-inch maps. This would be done for each county in turn, continuing with those in the north already started before returning to re-survey the southern parts of the country.

In a development linked to its recent interest in the six-inch scale, the Survey turned its attention also to the mapping of towns. This was partly a legacy of its activities in Ireland, but it also proved fortuitous in the pending concerns over sanitary welfare. Consequently, many towns in Lancashire and Yorkshire, together with London, found themselves being mapped from 1843 onwards at the scale of 60 inches to one mile (1:1056) but, following the Public Health Act of 1848, this scale was thought to be inadequate and a number of local boards of health were encouraged to finance OS town plans at the larger scale of ten feet to one mile (1:528). But these were to prove false starts since both scales were superseded in 1855 by the 1:500 scale. This then remained the standard scale for town plans at which in excess of 400 towns were mapped by 1894, this taking place in parallel with the (re-)surveys undertaken county by county for the 1:2500.[7]

From this short narrative for the major events that led to the position the Survey found itself occupying *circa* 1860, it will be appreciated that the availability and datedness of sheets for each of the different scales varied, and that this would also depend upon whereabouts in the country one was. Crucially by this date, though, the Old Series mapping was less than a

---

[6] This was known colloquially as the 25-inch scale since, as chance would have it, the difference in percentage terms between these descriptions is minimal. Whereas 'one inch to one mile' and 'six inches to one mile' reflect the scaled reductions from landscape to map precisely (1:63,360 and 1:10,560 respectively), so leading to formal labels of the 'one-inch scale' and the 'six-inch scale' (often then abbreviated to the 'one-inch' and the 'six-inch'), the larger-scale descriptions '25 inches to one mile' and '25-inch scale' are informal. They are synonymous with '1:2500 scale' or '1:2500' despite the very slight inaccuracy involved – 1:2500 represents 25.344 inches to one mile.

[7] This is a convenient place to cite three important texts each giving itself the remit of explaining the different OS map series and their scales. The first is J.B. Harley and C.W. Phillips, *The historian's guide to Ordnance Survey maps,* London: The Standing Conference for Local History, 1964. This is now very dated but still recognised for its importance over many years as really having been the only reference of its kind. It has been superseded by the collected efforts of authors contributing to *Sheetlines* and, importantly, by Richard Oliver, *Ordnance Survey maps: a concise guide for historians,* second edition, London: Charles Close Society, 2005. A more recent and very readable text that sets its sights on making the history of the Survey and its products accessible to all is Chris Higley, *Old Series to Explorer: a field guide to the Ordnance map,* London: Charles Close Society, 2011.

decade away from completion – if the Isle of Man is discounted – and revised states were continuously being issued for the majority of the one hundred or so of the sheets published by this time.[8] Though these revised states were keeping abreast of new railway construction, mounting criticism of the datedness of many of the early southern sheets was justified, as James himself was forced to admit when defending the Old Series in a letter to *The Times* in September 1862.[9] Part of the problem lay with the original copper plates. After years of use and constant revision, these were wearing down and no longer able to provide clear copies. The duplication of these plates by a process known as electrotyping – where, by depositing copper on the original, a copy is made through galvanic action – was a move sanctioned for use in Britain in 1847 following development of the process in Dublin. By the mid-1850s most Old Series sheets were being printed from electotypes and, from this time, both new sheets and revised states of earlier sheets began to include, in the marginalia, the annotation *Printed from an Electrotype*. From 1862 this annotation included a date.[10]

As a chronicle for the historian, these maps are supplemented by the manuscript records drawn for their compilation, namely the 'Ordnance Surveyors Drawings' (OSDs) and, from 1836, their later more systematically-drawn and regularly-shaped counterparts.[11] In addition, the records sometimes include the 'Hill Sketches' and 'Revision Drawings' that were needed for a broad swathe across central England where the delay in time between surveying for an OSD and publication of the maps could extend, in the worst cases, to more than twenty years.[12] These are all eminently useful records in that the information appearing on the two-inch (1:31,680) manuscripts did not always make it onto the final Old Series sheets. So, by 1860, with an availability that can be gathered from knowing the initial dates of publication for each of its 110 sheets, the Old Series was meeting many needs (Fig. 1.1).[13] If, however, the requirement was for mapping at a larger scale, then in 1860, as already seen, it was only just beginning to be met. In addition to the earlier issue of six-inch sheets of Lancashire and Yorkshire, the Survey could now provide derivative six-inch sheets from its 1:2500 sheets of Durham, Northumberland and Westmorland. This it did, but not immediately. But the issue of 1:2500 sheets was gathering pace. This was due in part to the survey of the northernmost counties being completed reasonably quickly, but in part to some areas from the southern

---

[8] The cartobibliographies of the Margary volumes identify for each numbered map sheet a sequence of states as determined by additions and deletions to the topographic detail or to the sheet marginalia. New states are being uncovered all the time so, from a current perspective, some volumes are now more authoritative than others. A comprehensive revision of the Margary cartobibliographies is being prepared by Roger Hellyer for publication by the Charles Close Society.

[9] *The Times*, 22 September 1862, p.7, c.3.

[10] For a contemporary account of the technical processes used by the Survey during the nineteenth century, see the sequence of articles by Riall Sankey in *Engineering*, volume 45 (1888). These have been collected and reprinted as H. Riall Sankey, *The maps of the Ordnance Survey: a mid-Victorian view*, London: Charles Close Society, 1995. For a more lucid account of these methods, and written as a retrospective prelude for her history of a later genre of OS maps, see Yolande Hodson, *Popular Maps: The Ordnance Survey Popular Edition one-inch map of England and Wales 1919 - 1926*, London: Charles Close Society, 1999, chapter 2.

[11] The manuscript OSDs are held in the British Library Map Room under the pressmark 'OSD'.

[12] Held in the British Library Map Room, this time as 'Hill Sketches and Revision Drawings', they are collated in groups according to Old Series sheet number under the pressmark 'Maps 176.0'. See Oliver, *Ordnance Survey maps: a concise guide for historians*, 40.

[13] Some of the sheets shown in Figure 1.1 were issued as full sheets. But from 1831 it became more expedient to issue quarter-sheets, in which case a suffix (*eg* NE for North-East) was added to the sheet number.

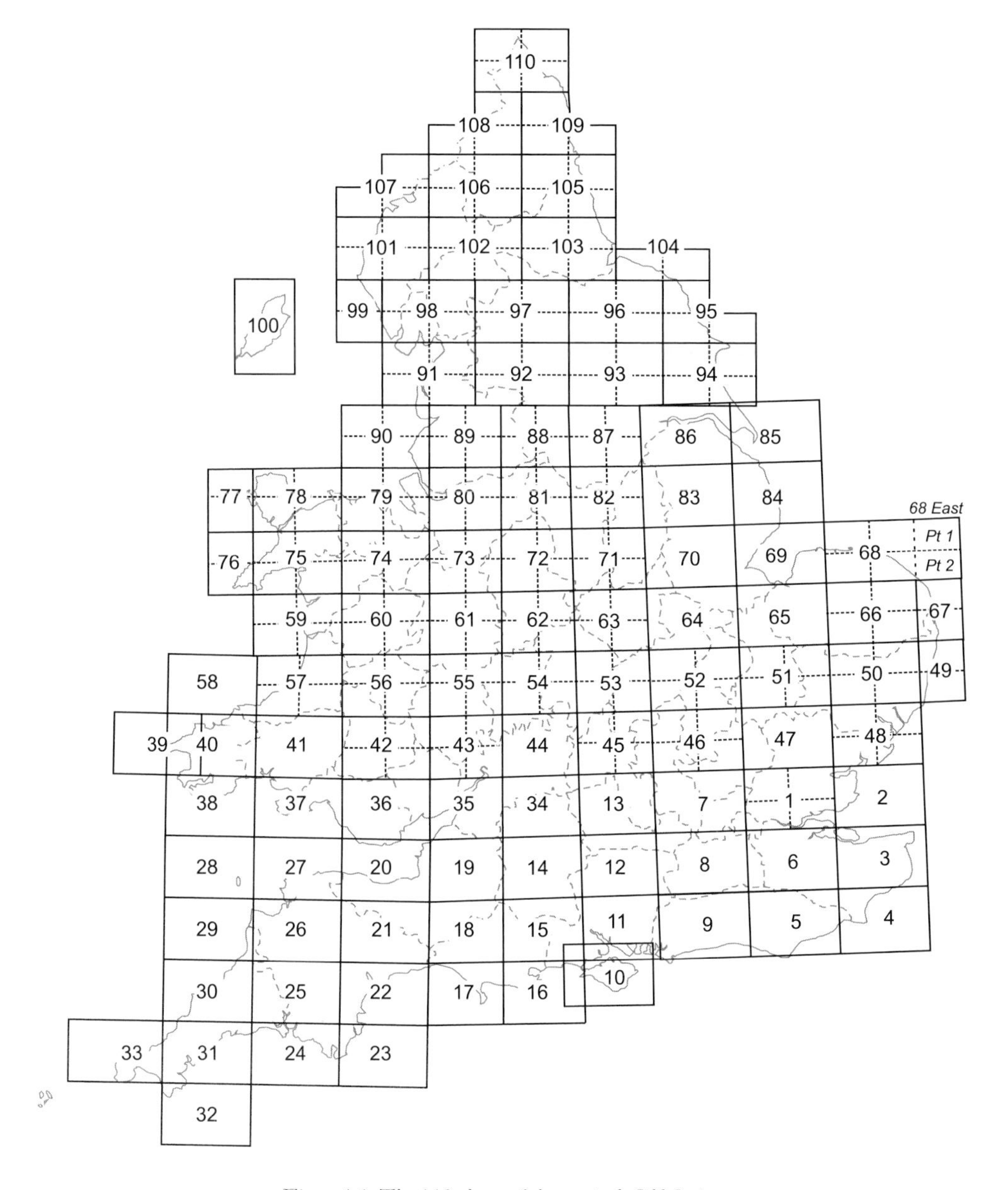

Figure 1.1. The 110 sheets of the one-inch Old Series.

coastal counties again supposedly facing the threat of invasion, thus prompting re-survey to take place there without delay.

For these large scales too, technical improvements were not long in coming. The limited use made by the Survey of lithography gave way in 1855 to zincography. This process could not match the finesse that came from working with copper plates, but it was appropriate to large-scale and town-plan work. Not long afterwards, in 1859, amalgamating zincography with photography led to the invention of photozincography which, though enthusiastically hailed by James, was not to have a significant effect on the work of the Survey until some twenty years later. As had been decided in 1858, and confirmed in 1863, the new mainstay of OS work was to be the 1:2500 scale but, like the one-inch Old Series, this too lapsed into progressing extremely slowly. It was finally completed for England in 1893 at which point the maps for some counties were anything up to 40 years old.[14] In contrast, using the initial edition as a template, revision to provide a second edition was a much easier task, and this was achieved by 1914 with most counties completed a decade earlier. The one-inch scale, which had been displaced from its position of central importance in the 1850s – though the writing for this had been on the wall by 1840 – had hardly fared any better. Disgruntlement over the value of the earlier Old Series maps had led to a proposal that a replacement series be provided: this was obviously considered to be a good idea since the authority to proceed with a New Series was given in 1872. Soon, however, this New Series also attracted adverse comments, and this contributed to a decision that the Ordnance Survey be investigated by a Departmental Committee of the Board of Agriculture. One resulting recommendation of this investigation was that the New Series be revised. This was quickly acted upon, resulting in a second edition for the New Series that was officially labelled as the Revised New Series. As with the initial edition, which can now best be referred to as the New Series, this revised edition appeared in a variety of formats. A second national revision led to the Third Edition in the early years of the new century.[15]

These last four simple sentences, though not wrong in summarising how the first three editions of the New Series came to replace first, the Old Series and then one another, totally belie the complexity of OS one-inch mapping at this time. In the context of the Old Series, mention has been made of new states being regularly issued, implying that it was normal to expect a succession of updated states for each sheet. Before the advent of a computer-aided capability that allowed maps to be updated ever more incrementally, a concern to make full use of its survey data had always been a feature of OS map production. So it was that many of the 360 sheets of these early New Series editions were published, not just in the normal outline format, but also in two formats that each emphasised relief using a different colour. All of these could be updated with successive states for each edition.[16] The resulting large number of variant maps was later increased by the introduction of a more colourful version just before 1900. In contrast to Old Series sheets, those of the New Series were numbered from the north of the country southwards, and the arrangement of them has a more regular look to it than does that of the Old Series (Fig. 1.2).

It will be appreciated that this family of New Series editions complete with all variants in which each was issued generates a complexity for which a descriptive shorthand would be most welcome. Accordingly, mention should be made of a notation developed by Richard Oliver whereby sheets of the New Series and any subsequent states for them – all of which would later be seen as the first edition of the New Series – are described as NS-1. Sheets of the Revised New Series and their subsequent states – all later viewed as a second edition – are described as NS-2. Sheets and states of the Third Edition are obviously then described as NS-3. The notation can be extended using single-letter suffixes to distinguish between the different variants of an edition, and extended also to incorporate the number of the sheet.

---

[14] Seymour, *A history of the Ordnance Survey,* 133.

[15] For further explanation of the genesis of the New Series (and much more besides) see Roger Hellyer and Richard Oliver, *One-inch engraved maps of the Ordnance Survey from 1847,* London: Charles Close Society, 2009.

[16] Hellyer and Oliver, *One-inch engraved maps of the Ordnance Survey from 1847,* chapters 4 and 5.

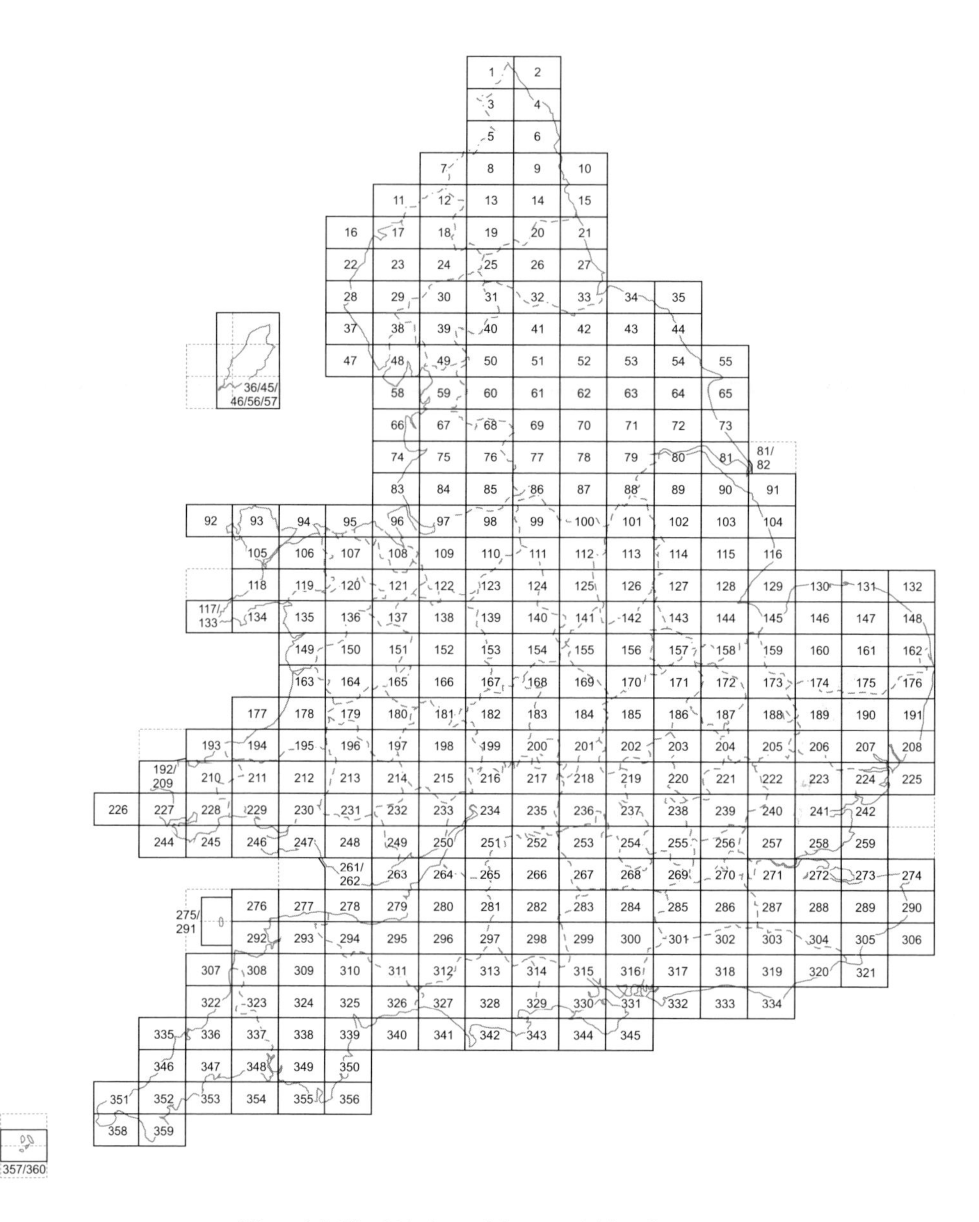

Figure 1.2. The 360 sheets of the one-inch New Series.

For the plethora of maps belonging to these three New Series editions, this is a convenient way of classifying them.[17] Initial states of each of these New Series editions can, in theory,

[17] Not unnaturally, the New Series sheets carry no recognition of the fact that in time they would be seen as a first edition. Neither did the revised sheets of the New Series bear any mention of being a next edition, although they were tacitly accepted by the Survey as providing a *de facto* second edition as sheets of the subsequent edition were labelled 'Third Edition'. The Oliver notation has not found wide usage but, then again, there have not been too many studies that could have been expected to use it. Later chapters of this volume make limited use of it. A broader explanation for it appears in Richard Oliver, 'What's what with the New Series', *Sheetlines* 5 (1982), 3-8. A complication with all of this, and one that can easily cause confusion, is that the expression 'New Series' is used as a label both for the entire family of New Series editions – which spanned many decades, but initially had to be differentiated from the Old Series – and, more specifically and understandably, for sheets of the first edition that

be found in the legal-deposit libraries but later states, where issued, need to be sought for elsewhere.[18] The format in which later states often now exist is the one in which they were sold to the public by the map-selling agencies. These agents, as a matter of course, would unhelpfully remove the margins which carried the explicit means of dating any sheet, and simply paste the legend – not always the one appropriate to that edition – and a scale onto the back of the linen-mounted map.[19] Later states found with their margins intact will often have electrotyping dates or the dates to which railways had been corrected in the same way as this sort of information appeared on successive states of Old Series sheets. An extensive cartobibliography for the New Series, more than mirroring that achieved for the Old Series by the Margary volumes, is a major part of the outstanding study by Hellyer and Oliver for the whole range of OS one-inch engraved maps.[20] To précis drastically: after the protracted preparation and publication of the first one-inch non-county-based maps – the Old Series – initial states of New Series sheets were prepared in a very much shorter space of time using data from the large-scale survey. Later, sheets of the Revised New Series were also quickly provided followed by sheets of the Third Edition.

As anticipated, the changing practices adopted by the Old Series, and later by three early editions of the New Series as well as by the larger scales, all for the depiction of one specific landscape feature – the windmill – are going to be our focus of attention. As this particular feature potentially fulfils the need for one that was difficult to gauge in a consistent manner, it will be interesting to try to measure the possibility that the language of symbols and signs adopted by the one-inch was unable to handle the variety of circumstances that windmills found themselves in. To affirm the appropriateness of this choice of feature when assessing nineteenth-century OS maps, the next section of this chapter will now examine the windmill – for which there is a diversity of documentary sources – and its role in the landscape.

The windmill

As a feature in the landscape, the windmill, both in what is held to be its customary role of grinding cereal and working at the many other uses to which it has been put, is visually very emphatic. For centuries it was a common enough sight over large parts of England, initially being seen in the historical record at a point late in the twelfth century, before developing through into early modern times in increasingly larger numbers while undergoing changes in

---

were the immediate replacements for the Old Series sheets. The context in which 'New Series' is used should let the reader know which of the two meanings is applicable.

[18] The situation for New Series sheets is not quite as straightforward as for sheets from the Revised New Series and the Third Edition. A delay in the British Library receiving its first batch of New Series sheets led to some of its holdings not being initial states.

[19] Each sheet would then be folded and have attached to it a label which, as well as identifying the sheet, carried the agent's address. Stanfords, the most important of these agents, by twice changing their address at times (1888 and 1901) that were watersheds for the depicting of topographic detail (and in particular windmills) unwittingly provided, as a *quid pro quo* for their eager use of the scissors, some means for dating later-state New Series sheets.

[20] Hellyer and Oliver, *One-inch engraved maps of the Ordnance Survey from 1847*. This corrects a situation where little of this kind had been done for either the New Series or the Revised New Series. In 1988 Guy Messenger wrote about a later coloured version for the Third Edition and in 2002 Tim Nicholson wrote a study on the coloured maps of the Revised New Series. But, with these exceptions, the only acknowledgement that sheets of these editions may have been issued in more than one state, before Hellyer and Oliver, had come from Harley and Phillips, *The historian's guide to Ordnance Survey maps,* 12 and Oliver, *Ordnance Survey maps: a concise guide for historians,* 36.

morphology. The precise genesis of the windmill is obscure and the subject of some debate, but what *is* accepted is the fundamental division of seeing it as basically a wooden structure – in which form it came to England in early medieval times – or as a stone-built edifice of the type common to the Mediterranean shores. The so-called wood and stone trails both led to England but arrived at different times, with the wooden post mill preceding the stone tower mill by a couple of centuries or so. This basic differentiation in windmill typology is important because the visual, cultural and mechanical contrasts have continued down the centuries to the point that mapmakers have often distinguished between the two types of mill. Making this distinction, though it has been done only intermittently, is a first indication of a complexity in the history of windmill representation on maps that needs unravelling.

Acknowledging the ground presence of a recognisable and salient landscape feature by using an admixture of representational means that includes the use of symbols is a basic OS craft. Invoking the windmill as a vehicle for exploring the ability of the Survey to do this in a satisfactory way suggests the need to explore further this idea of two types of windmill. The earlier wooden post mill essentially comprises a wooden-framed and weather-boarded box structure into which all of the machinery is fitted, and to which the sails are attached at what can be considered the front of the mill. This box structure is supported by a vertical post around which it is free to rotate, so allowing the sails to face the wind from whichever direction it may blow. Access to the mill is by means of a ladder at the back, while the post is maintained in its upright position by a substantial timber trestle that conveys the weight of the mill through brick piers to the ground. Both during the nineteenth century and later, the simplicity of this arrangement could be seen in many windmills, including the one that stood at Mumby in Lincolnshire (Plate 1).

Gradual upgrading of the machinery of a post mill occurred, as would be expected over a period of several centuries, but visually, from the outside anyway, little altered in 700 years, except that the supporting trestle might be enclosed by a protective but non-structural stone or brick roundhouse, or that cloth-based sails might be superseded by shuttered sails. In the case of Mumby Mill, the sole concession to anything approaching modernity can be seen to have been the replacement of two cloth sails by shuttered ones. In contrast, each of the two mills at Brill in Buckinghamshire can be seen to have kept all four of their cloth or common sails but the nearer – Nixey's Mill – has had a brick-and-tile roundhouse added, unlike the more distant Parson's Mill (Plate 2).[21] Many post mills similar to these at Mumby and Brill continued to work into the twentieth century using only the most rudimentary machinery. The other type of windmill, the stone- or brick-built tower mill, arrived in England much later than the post mill had done and did not find widespread use until the early years of the eighteenth century. In appearance, a solid tower now houses most of the machinery with just the topmost wooden cap (enclosing the shaft supporting the sails) able to turn, thus allowing the sails to face the wind. This is altogether a far more robust windmill design that was better suited for the economies of scale aspired to in Georgian and Victorian England. Built at a greater cost, these mills could accommodate more pairs of stones and better resist extremes of weather, but to suggest that post mills were eclipsed by the newer tower mills

[21] The custom of prefixing the locational name of a windmill by that of the owner or tenant miller was used by the milling fraternity in those instances where many windmills operated in close proximity, and they needed to be distinguished one from the other. This practice has been adopted as a convention across the broad spectrum of secondary mill literature, and is therefore used here.

would be wrong. Just as there were modest examples of post mills, so there were of tower mills, as at Barrington in Cambridgeshire (Plate 3). This mill was a simple one from an area otherwise populated by more sophisticated tower mills but in those parts of the country where mills were less numerous, such ones as there were tended to be tower mills of this simpler nature.

Technological and economic aspirations led naturally to more powerful tower mills that used refinements such as automatic turning to wind by means of a fantail and the automatic reefing of sails. These two developments alone changed the visual impression to be gained from looking at a mill. Greatham Mill in County Durham was not exceptional in appearing elegant with a finial adornment on top of the cap, or in the technology it made use of (Plate 4). Mounted at the rear of the cap is the fantail which will turn if the wind strikes it from any direction other than directly through the sails and, through gearing, rotate the cap, thus bringing the sails to face the wind again. Also shown are the roller-blind arrangements set within each sail bay, which can be furled or unfurled according to the strength of the wind without the sails even needing to be stopped. Windmills such as the one at Greatham were built in large numbers during the early decades of the nineteenth century. Often they were replacements for humbler post mills, but post mills were themselves also capable of being refurbished to incorporate the same technologies developed for tower mills. The peak of refinement for the post mill, and arguably for windmills of any kind, was reached in East Suffolk where mills such as the one at Friston had been raised so that their trestles were now supported on high brickwork piers within substantial roundhouses (Plate 5). Inside these mills, the sophistication of the machinery had kept pace with that of the tower mill.[22]

In the face, however, of increasing efficiencies to be had from steam power, the future of the windmill had already been curtailed. The powering of a flour mill by steam is usually accepted as dating from as early as 1784 with the building of the Albion Mill, Blackfriars.[23] But the best part of another century was to pass before a combination of factors, each one in itself a threat to the continuing utility of the windmill, caused the decline in their numbers to become more than gradual. These factors included rural use of the portable steam engine and improved access to rural areas from the ports that were importing an increasingly large quantity of grain but then, starting in the 1870s, using their own steam-driven roller mills to grind it. The span of years that passed between the high point of constructing tower mills in the 1820s and the onset of this decline would not be long. The number of windmills known to be at work in Essex was on the wane by 1850 and in counties further from London, both those with a large windmill-dependency culture and those with a much smaller one, it can be estimated that the same position was reached by the 1860s, or by the 1870s at the latest. A perception that the windmill no longer stood as cutting-edge technology in mid-Victorian England is provided by one contemporary commentator who somewhat sternly declared:

---

[22] For a general introduction to the windmill and further explanation of the technologies touched upon here the reader is directed to Rex Wailes, *The English windmill,* London: Routledge & Kegan Paul, 1954. A more recent text is offered by R.J. De Little, *The windmills of England,* third edition, West Sussex: Colwood Press, 1997. Later in this book, a concise consideration of the historiography of the windmill will cite other texts but will also acknowledge Wailes as the classic writer of the genre.

[23] Aubrey F. Burstall, *A history of mechanical engineering,* London: Faber and Faber, 1963, 255; Marilyn Palmer and Peter Neaverson, *Industry in the landscape, 1700 - 1900,* London: Routledge, 1994, 23. It can also be noted that a claim for an earlier (steam) atmospheric-engine-powered flour mill (by Wasbrough in Bristol) was made in H.W. Dickinson and Rhys Jenkins, *James Watt and the steam engine,* London, 1927.

> 'For giving motion to machinery, windmills have been and still are very extensively used. Engineers of the last generation devoted great attention to the construction of windmills, and brought them to great perfection. The introduction of steam power – a power economical, manageable, and always to be depended on – has, in a great measure, superseded that of wind as a mover of machinery. It is true that after the first cost of a windmill, the power is comparatively inexpensive; but it is so variable in intensity – sometimes, when it is not required, exerting great force, and sometimes, when it may be most wanted, totally ineffective – that it is generally preferable to apply a force, perhaps considerably more expensive in its production, but constant, steady, and completely under control.' [24]

Clearly a lack of control in these matters was iniquitous to the Victorian mentality, and even though the windmill was nowhere near approaching obsolescence, its status at this time was changing fast. Indeed, the role of the windmill as a technological icon went through a huge transformation in the fifty years between 1830 and 1880. At the beginning of the nineteenth century, despite the onset of industrialisation, the mechanical clock was still the universal metaphor for technological capabilities and, some might argue, an analogy or even cause for the authoritarian concept of social order prevalent throughout early modern Europe. The move towards a more liberal concept of social order is linked, in one writer's mind at least, with the progress of machinery towards self-regulation. Citing as examples the use of the fantail by windmills (patented by Edmund Lee in 1745) and the centrifugal governor – that had been adapted for steam engines from its earlier use in windmills – Otto Mayr concedes that any notion of causation between a prevailing model of social order and technological control can only be speculative, but he makes a persuasive case for accepting that an affinity exists between the two:

> 'If a technological innovation displays in structure and functioning an unmistakable analogy to the structure that a society prefers to give its various practical and theoretical systems, if it reflects the various mentalities and attitudes that shape public life, in short, if it matches and reinforces the prevailing conception of social order, it will be received more warmly, regardless of its technical merits, than other inventions.' [25]

It may be stretching credulity to imagine that the English collective state of mind at the time of Waterloo was overly sympathetic towards, or even cognisant of, the latest aspects of windmill machinery. But, given the large numbers of mills then operating, their high dispersion, visibility and obvious utility to all, it may not be too fanciful to consider that even compared to the use of steam power, the windmill was held to be at the forefront of technologies at this time, and accorded a corresponding status. This, as already seen, was to change both rapidly and radically. By *circa* 1860 the status of the windmill was far more

---

[24] James Wylde (ed), *The circle of the sciences,* The London Printing and Publishing Company Limited, Paternoster Row, volume 1, 1868, 822 ff. This was a monumental two-volume work embracing, for its time, all science-based knowledge.

[25] Otto Mayr, *Authority, liberty & automatic machinery in early modern Europe,* Baltimore: John Hopkins Press, 1986, 190 ff. For a broader insight into the arguments for technological determinism see Merritt Roe Smith and Leo Marx (eds), *Does technology drive history? The dilemma of technological determinism,* Massachusetts Institute of Technology Press, 1996.

ambiguous. Without doubt, in the years after the Great Exhibition of 1851 it was no longer the technological icon that it had once been, but this was still a prosperous time for English agriculture, and millwrighting practices in rural areas were able to trade on their ability to be innovative in mill repair, if not new mill construction. This changed with the depression that affected agriculture from the mid-1870s onwards, when many mills went out of business and others were forced to rethink any ideas they might have had of regenerative technology. But, paradoxically, from this low point came some salvation for the survival of the windmill as an icon. The depopulation of the countryside and a significant shift in the economy from an agricultural to an industrial base was eventually to result in a compensatory swing back to what can be described as a rush of nostalgia for the rusticity of the countryside. It became fashionable, increasingly from about 1885 onwards, for those who could afford it, to escape the reality of urban fuss and squalor and return to what were seen as wholesome, traditional values. The values espoused by Ruskin and Morris began to resonate with legions of people who could decry the benefits that urban life was bringing them, and they found an outlet in myriad back-to-the-land activities.[26]

These sentiments allowed writers such as Thomas Hardy to find widespread favour for their portrayal of a countryside rooted in backwardness and a sense of bucolic picturesque. The image and condition of the country mill were, of course, caught up in this enthusiasm. Thus Hardy felt able to be suitably mellifluous when setting his watermill from *The Trumpet-Major* in context:

> 'In this dwelling Mrs Garland's and Anne's ears were soothed morning, noon, and night by the music of the mill, the wheels and cogs of which, being of wood, produced notes that might have borne in their minds a remote semblance to the wooden tones of the stopped diapason in an organ. Occasionally, when the miller was bolting, there was added to these continuous sounds the cheerful clicking of the hopper, which did not deprive them of rest except when it was kept going all night; and over and above all this they had the pleasure of knowing that there crept in through every crevice, door, and window of their dwelling, however tightly closed, a subtle mist of superfine flour from the grinding-room...' [27]

As a result of this kind of sentiment, windmills and watermills were not doomed to oblivion as outmoded forms of technology. Instead, they changed – the windmill particularly – from being technological icons to becoming architectural ones; a position that they have held ever since. For this reason, though continuing to provide a service to the communities in which they were placed, windmills have, from the beginning of the twentieth century, been viewed as a metaphor for a lost way of country life and eulogised as such. Whether through using a painter's brush or early and fairly rudimentary camera equipment, the mill became a popular subject for artists. Inevitably, an effort was made to preserve some windmills as emblematic of a bygone age, and this initially manifested itself as the desire to preserve some of them as landmarks. This move is popularly considered to date from as early as 1899 when the soap-maker industrialist Mr Hudson saved a tower mill at Bidston on the Wirral though, as a later

[26] For a good account of this movement see Jan Marsh, *Back to the land: the pastoral impulse in England from 1880 to 1914*, London: Quartet Books, 1982.

[27] Thomas Hardy, *The Trumpet-Major*, 1880, chapter 1.

observation will make clear, the interests of the Admiralty may have played a hand here.[28]

The narrative offered here for the fortunes of the windmill throughout the nineteenth century inevitably comes with some complications. Although the decline of the mill had set in by the middle years of the century, and they had all but disappeared in some parts of the country by then, elsewhere they were still being constructed *de novo* as late as the 1880s. The last corn-grinding tower mill to be built in England was constructed on a new site at Much Hadham in Hertfordshire in 1892, although it was only to survive until 1910 before being pulled down.[29] The post mill had continued to hold its own against the tower mill with one of the last to be built erected at Chillenden in Kent in 1868, and other post mills underwent major repairs in order to continue working.[30] As late as 1882 it was thought worthwhile to totally rebuild the post mill at Wetheringsett in East Suffolk after it had been blown down in a storm and, furthermore, to fashion a new main post for it rather than use a redundant one from elsewhere.[31] So, some mills were destined to continue working in an authentic capacity well into the next century, long after their kind were generally considered obsolete if not archaic.[32] This was a reality that no doubt blurred any perception of the windmill as a rustic icon. There were obvious implications for mapmakers: were individual windmills to be perceived simply as still functioning or had they become iconic, or were they collectively now to be viewed as emblematic? Towards the end of the nineteenth century considerations such as these will undoubtedly have influenced the Survey when it was formulating policy for windmill depiction. But other complications were also present that served to confound that depiction, including the wide variety of structures functioning as windmills. This is a theme that will be explored later but first, some small expansion of the bipartite post-mill and tower-mill typology is necessary.

Not unlike the tower mill in appearance, a third type of mill known as a smock mill also houses most of its machinery within a static structure that supports a cap capable of turning

---

28 Wailes, *The English windmill,* 181.

29 Donald Smith, *English windmills,* volume 2, London: The Architectural Press, 1932, 121. See also Wailes, *The English windmill,* 69. This statement does not take into account the many latter-day mill (re)constructions (whose onset many consider to date from the erection of a new mill at St. Margarets Bay in Kent *circa* 1928) where the incentive for building has been rooted more in the modern 'rush of nostalgia' crusade – heir to the back-to-the-land movement – rather than an 'authentic' utilitarian need for a traditionally-built windmill.

30 William Coles Finch, *Watermills and windmills,* London: C.W. Daniel, 1933, 186.

31 Rex Wailes, 'Suffolk windmills: part 1, post mills', in *Transactions of the Newcomen Society* 22 (1941), 59. But then again, this position of the post mill as holding its own against the tower mill did differ across England. In many parts of the country, and across Europe generally, the post mill at this time was seen by many as second-rate in comparison to the tower mill. Accordingly, Theodor Fontane, the German equivalent of Hardy, had his heroine Effi Briest exclaim on arriving at her matrimonial home 'das wäre so was für die Mama ... vorn hätte sie die Stadt und die Holländische Windmühle. In Hohen-Cremmen haben wir noch immer bloß eine Bockmühle'. Theodor Fontane, *Effi Briest,* 1895, chapter 8. '[this room] would suit my mother, ... in front she would have the town and the Dutch [tower] windmill. In Hohen-Cremmen we've only got a trestle [post] mill'.

32 Any discussion invoking the notion of authenticity requires careful use of language. While, for example, an expression such as 'everyday utility' is understandable and attracts broad consensus, the word 'authentic' is open to interpretation and can mean different things to different people. One highly-influential school of philosophy, for example, views instrumental technology as something that suppresses all other possible types of existence – thus rendering it up as 'inauthentic'. Most people would see this as taking a rather arcane perspective on life and, what is more, the literature promoting it can be fairly opaque but the fact that it has been written demonstrates a need to be careful when discussing mapmakers' perceptions of the changing status of mills, and in particular this idea of 'authenticity' – or what at any point in history has been 'real'. Pragmatically, we can use 'authentic' in this study to mean routine, everyday 'invisible' existence, so the use of it here in the text is not inappropriate.

the sails into the wind. Instead of stone or brick, the body of the smock mill is of wooden construction, normally eight-sided and reducing in size the higher one climbs. This body is supported either on a substantial brick base or just clear of the ground on simply just a few courses of brickwork. In the south-eastern counties particularly, many windmills of this sort were built such as the one at Earnley in West Sussex (Plate 6). There were cases too of later smock mill construction; one of the windmills at Shipley in Sussex was built *de novo* in 1879 for a Mr Martin and is preserved today as a memorial to Hilaire Belloc who owned it from 1906 until his death in 1953.[33]

The three types of windmill so far described account for the overwhelming majority of those built and operated in England. However, the differences between them included their patterns of distribution. This will possibly not be seen as much of an issue for the Survey to have had to deal with, nor would any variation in the rates of attrition for windmills seen in different parts of the country have been of real concern to it. A surveyor would have been intent on just recording anything that came into view during the course of his work without asking too many questions but, as issues faced by historians of windmills, these patterns are of considerable interest and clearly underpin our ability to provide a simple picture for the national distribution of windmills at any one particular time.

So, some idea of the changing distribution patterns for windmills in England throughout the nineteenth century is needed if we are to assess the Survey in its attempts to provide a coherent countrywide representation of windmills. Unfortunately, any evidence taken from the secondary mill literature for the geographic spread of windmills is likely to be based on a knowledge of individual sites that has been derived from OS maps. If there is any suspicion that this might be the case, such evidence cannot be used for fear of it corrupting our study. At first sight, for example, the graph illustrating the decline of windmills in Essex shown in Figure 1.3 appears to endorse a natural suspicion that its author has used maps as *prima facie* evidence for his research into the survival patterns of Essex mills.[34] But, on this particular occasion, the accuracy of the graph has been verified by rigorous complementary research using other sources or forms of evidence to establish ground-truth accuracy for individual windmills. Companion maps drawn to accompany this graph in the Farries volumes, which show the distribution of windmills for Essex, are therefore valid to use. But unhappily, not much material of this high quality, which can safely provide non-map-derivative accounts of where windmills could once be found, exists in the mill literature.

This is perhaps less of a concern than one would think. The near-reliability of the Survey with respect to Essex windmills as suggested by Figure 1.3 is an encouraging first indication that the scale of any potential problem the Survey may have had with windmills at the one-inch could be small. The gathering of empirical data on windmills over many years by early workers in the field such as Rex Wailes would suggest that, even if their initial reckonings of windmill distributions had indeed been based on OS maps and possibly biased, then those

---

[33] Anon. *Shipley mill,* West Sussex County Council, 1966. See also A.N. Wilson, *Hilaire Belloc,* London: Hamish Hamilton, 1984, *passim.* Whatever unconventional views on other matters Belloc may have held, it seems that he was happy to acquiesce to the pastoral indulgence of caring for his windmill. There is even a story told of Belloc being protective of his privacy to the point of insisting that the Survey did not record his mill. Apocryphal or not, and whatever the truth of Belloc trying to influence the Survey, the recording of this mill was mishandled, but this had happened before 1906.

[34] Kenneth G. Farries, *Essex windmills, millers & millwrights,* London: Charles Skilton, volume 1, 1981, 30.

accounts would have altered in the light of field experience and come closer to ground-truth accuracy. As hinted at above, the usefulness of distribution maps of windmills found in the literature is variable, certainly in texts dealing with windmills on a countrywide basis. Putting to one side the maps compiled by Farries for *Essex windmills, millers & millwrights*, those that accompanied Rex Wailes' *The English windmill* are as good as any, even if rather rudimentary. All these do is show those mills that happened to be mentioned in his text, and a simplified form of the main map for the windmill-rich areas of eastern England (without the numbers accorded each mill) is reproduced here (Fig. 1.4). Each dot represents a single mill, and they serve *in toto* to demonstrate that windmills were widespread across the eastern counties with the greatest emphasis in East Anglia and Lincolnshire. The second map drawn by Wailes showing the remainder of the country is content to acknowledge only another 37 windmills, meaning that some counties such as Hampshire, Leicestershire and Northumberland only have one windmill apiece, and others such as Durham have none at all. In needing to select specimen mills for his book, Wailes would have been the first to acknowledge that his maps severely understated the true numbers of mills, but would presumably have argued for the pattern of distribution that they show. As a statement for the pattern of mill distribution, it can be used to further our understanding of the nineteenth-century windmill, provided one consideration already alluded to, which must quickly be mentioned again, is not forgotten.

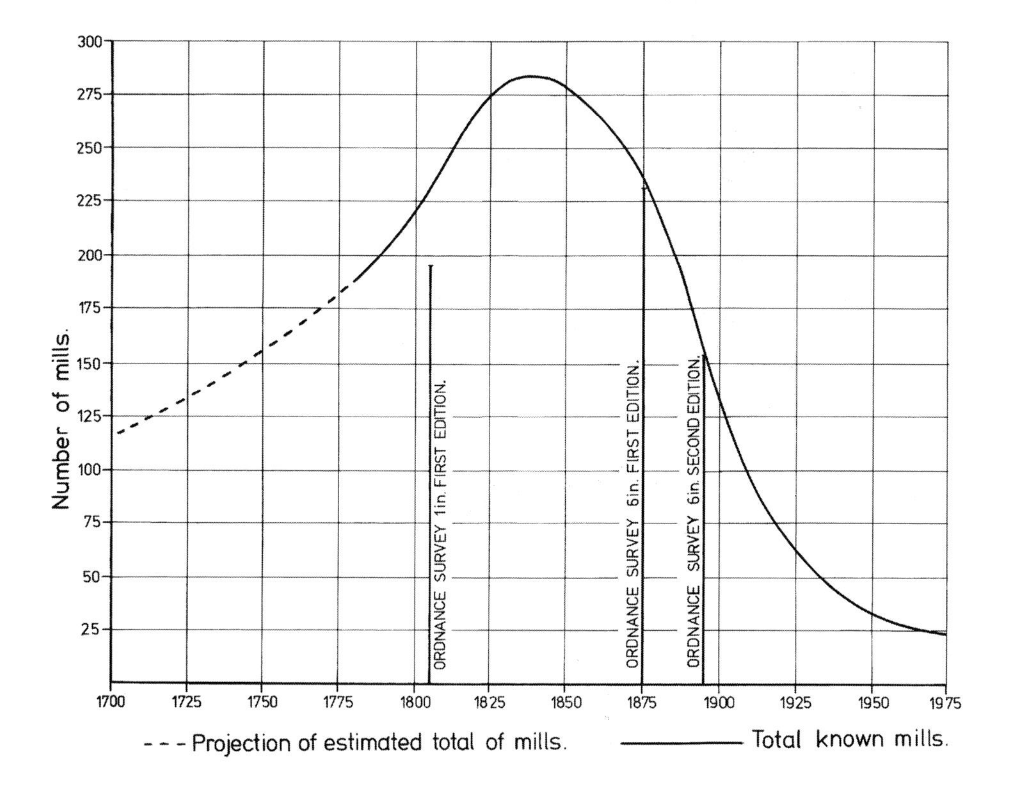

Figure 1.3. The decline of the windmill in Essex (after Farries).

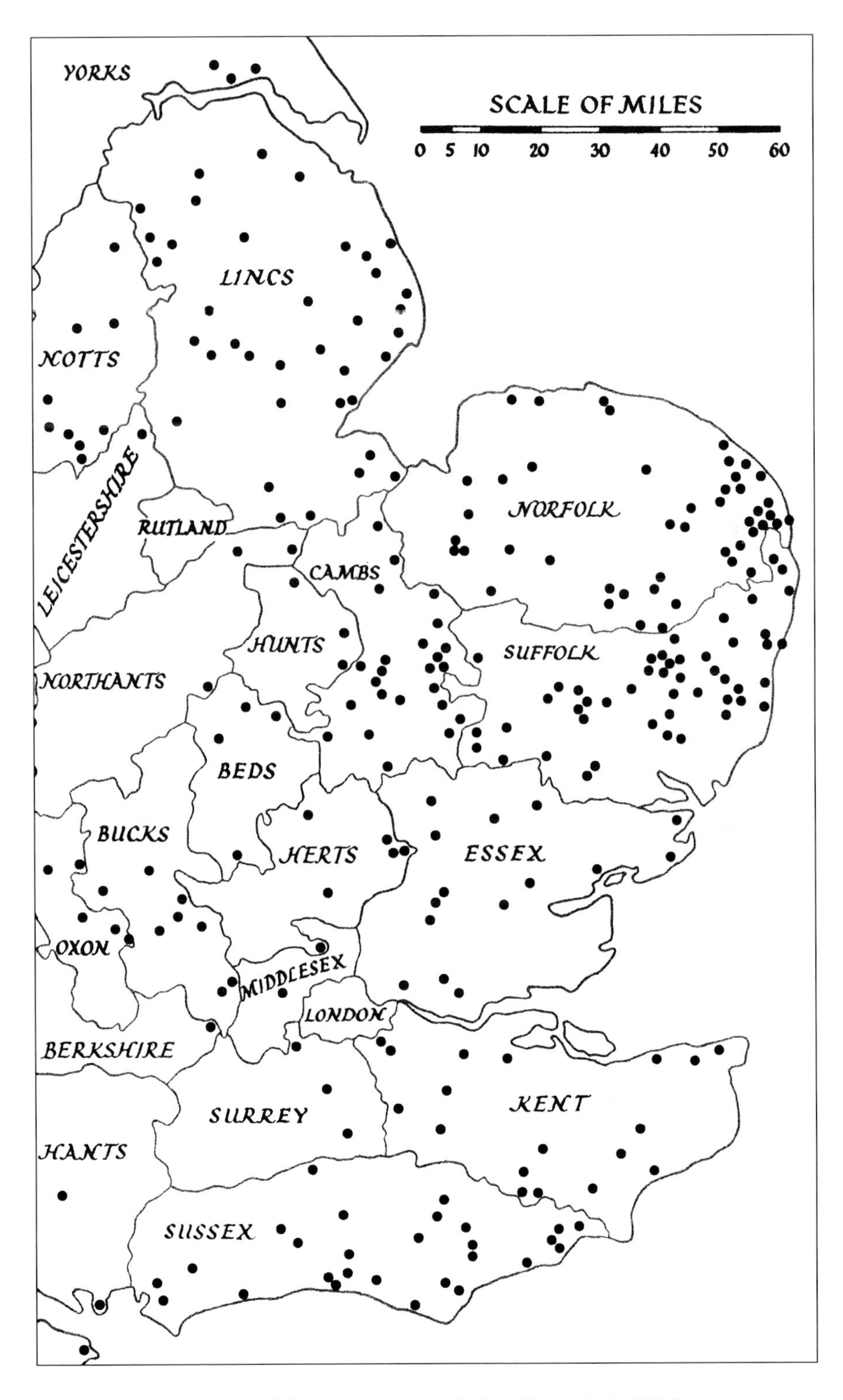

Figure 1.4. The distribution of windmills in England (after Wailes).

It has been suggested that the decline of the windmill was staggered across the country. While mills such as Belloc's Mill at Shipley were newly constructed in 1880, wide areas of the countryside had lost their windmills by that time. The distribution of mills revealed by Wailes represents not so much a centuries-old placement pattern for the English windmill, but rather the pattern of post-1880 windmill survival and a template for the emerging age of the windmill as an architectural icon. Thus Essex, a county where windmills survived until fairly recently, looms large in Wailes' book whereas counties such as Durham rarely feature in such accounts. Yet at the beginning of the nineteenth century, Durham possessed over one hundred windmills with large clusters situated at the seaports and in Gateshead.[35] But, by the middle of the century only 45 are thought to have been active by wind with barely a small handful of these surviving into the next century, and even then not necessarily solely reliant on wind power. Greatham Mill was one of the very last to survive working by wind. As a result, a common misconception is that Durham is a county that has no real history of windmills.[36] Furthermore, the position assumed by Wailes of focusing on the post-1880 set of windmill circumstances has been adhered to by most later authors, so giving to the mill literature an emphasis that obscures the standing of windmills much earlier in the century. Consequently, any study that requires an understanding of English windmills throughout the nineteenth century has, of necessity, not only to be mindful of the varying pattern of decline across the country, but also a tendency for the literature to overlook this fact.

This introduction to the windmill has revealed some of the issues and concerns affecting its visibility in Georgian and Victorian landscapes. The different factors that contribute to a knowledge of windmills, the nature of the literature generated by an interest in mills and the types of primary evidence available to an historian of mills are all relevant to this study and each will be explored further. In the meantime, these issues seem suitably complex for us to maintain our suspicion that the windmill could well be useful for examining the capacity of the Ordnance Survey to provide accurate and comprehensive maps.

Mapping the windmill: an introduction

The short narrative for windmills just offered offers some insight into the drawn-out period of time that marked the end of most mills' authentic working lives. Throughout, instead of all mills surviving in their longstanding role of grinding cereals or other produce within local communities, one by one they became uneconomic to work. Over time, a majority of their number disappeared, leaving a minority of mills to discover that, having survived a time of transition, their status had altered. Those windmills that, quite by chance, maintained their presence into modern times became a focus for the newly-labelled discipline of 'industrial archaeology'. By the late 1960s, measures were in place to help preserve what few of them were left, though in the half-century since then commercial interests have not always helped

---

[35] Texts devoted to Durham windmills are almost non-existent, but see Vera Chapman, *Rural Durham,* chapter 1 'Into the eye of the wind', Durham: County Council Publications, 1977, 1-11.

[36] W.R. Wiggen, *Esh leaves: being drafts upon the memory of an old parishioner,* Durham: Thomas Caldcleugh, 1914, 61-3. The writer, who died in 1917 aged 82, details his reminiscences on many things, including windmills, for which he lists the 40 active examples that he had known in County Durham. At the time of publication in 1914, all mills are declared derelict or destroyed except three or four struggling to stay working, and of these only one – Greatham Mill – is noted as still having all of its sails.

in that process. More recently, surviving windmills have become significant participants in a heritage industry whose *modus operandi* sometimes appears overly populist and not averse to appropriating and traducing a mill's history. One might think that this fundamental change of circumstances for a surviving windmill that had seen it transformed from a situation of everyday ordinariness to one of sanitised icon would have been enough in the intervening years to promote interest among mapmakers. But this is granting the benefit of hindsight to the Survey and capturing this change, as it transpired, was to prove problematic. If, more mundanely, though, the Survey had been asked at the end of the nineteenth century 'what qualities or traits of a windmill's presence might pose a dilemma for its depiction, or even the need for different types of depiction?', then attributes that the Survey might have been expected to consider come readily to mind. For a start: if a surveyor was having to decide whether or not a windmill was capable of work, then this might have been easily determined by looking at the state of its sails, as they are normally the first parts to rot when a mill falls into disuse. Perhaps the size and type of mill and thus its likely output or numbers of pairs of grinding stones would have been a consideration. Whether a mill was in open ground or situated in a built-up area could have been another one simply because there might have been limited space on the map for any type of recognition. Plausibly there are others as well. On top of all of this, the need to recognise the ground presence of a windmill in whatever way would also have depended upon the remit that the Survey was giving itself for each of its scales. It does not help our inquiry that, with the passage of the years, those remits have become obscured. Today, much further on in time, the one clear sign of something being amiss – the recognition of a windmill on a map of one scale but not on a contemporary map of another scale – provides an opening for searching dialogue. Perhaps now it can be better appreciated that ground-truth accuracy for windmills, far from being competently expressed by a dual categorisation of mills acknowledging those that still had their sails or those that had lost them as in the 1960s, is a more complex business that can easily result in the representation of these mills by a mapmaker going badly awry.

Staying mindful of all these points, the story of the Victorian windmill can be expanded to include some demonstration of the sources of nineteenth-century documentary evidence relating to them. If an assessment is to be made of the way in which the Survey recognised windmills, then some independent (*ie* other than map-based) evidence is needed to indicate the context in which individual mills survived, what each one looked like and, if we are very lucky, exactly when each one vanished from the landscape. This latter point is crucial since, for a windmill that is shown by a map of one scale but not by another of a different scale, knowing the precise date of that windmill's demise, and also the probably slightly different survey dates for each of the scales, can mean the difference between denouncing a map due to its record of that mill – so inviting a deeper look into the cause – or simply deciding that this mill was correctly recorded by both maps, and so holds no further interest for us.

There are many different types of primary source available to date precisely the demise of a windmill. At best, perhaps in the event of swift or calamitous loss, this date might be known to within a few days. Hopefully, it will be to the nearest month or at worst to within a couple of years. The sources that can be assumed to be standard for most millers include trade directories, census returns, land-tax returns and electoral registers. All theoretically have the advantage of good coverage and availability yet can be unrewarding in their use. It is not normal for trade directories, for example, to distinguish between wind, water or (later) steam corn mills in their entries. Of potentially greater value, but owing far more to chance for their existence and laborious to locate, are newspaper reports, auction and sale notices, excerpts from antiquarian topographies and, above all from *circa* 1875, dated photographs.[37] The use made of these sources by the secondary mill literature will be reviewed later when further mention is made of Farries who used all these types of primary source to good effect in his study of Essex windmills. In the meantime, a glimpse at the circumstances of two specific mills and then those of a small town – Boston in Lincolnshire – will illustrate the usefulness of these sorts of sources and help confirm their importance to this study.

Belper Mill in Derbyshire is shown and described as *Windmill (Corn, Disused)* on the first edition of the 1:2500 using survey data of 1877-8 (Plate 7). These details are then replicated – the difference in scale permitting – at the six-inch (Plate 8).[38] The one-inch New Series sheet, published a few years later in 1890, depicts many other windmills of the area by use of symbols but, despite adequate space on the map, it ignores Belper Mill – as do the first revisions of all scales in the mid- to late 1890s. This is a typical enough sequence of mapped information for a mill even though the New Series was supposedly derived from the larger scales. Letting this apparent inadequacy of a New Series sheet pass, realising exactly how well this mill was served by the Survey depends upon knowing the precise date and nature of its demise. Here, Belper Mill is better served by the timely portrayal of it written by late Victorian topographer Charles Willott. In 1894 he spoke of the mill as being conspicuous and also 'thoroughly repaired a few years ago', possibly implying that this mill worked later than 1878. However, more conclusively, he relates too that 'the old windmill was dismantled and extensive alterations made to it ... in December 1891', this being the time of the mill's conversion into a house, which has survived.[39] Hard evidence such as this for the demise of a windmill is not so conveniently found even for a minority of mills, but what increasingly

---

[37] Photographs of mills were taken from as early as *circa* 1853 though, understandably, examples of this vintage tend to be very few in number, but they are nonetheless fascinating for what some can tell us of the era before the windmill went into decline. One example of this is the well-known (to historians of photography at least) image of Kempsey Mill in Worcestershire taken by Benjamin Brecknell Turner *circa* 1853 – he actually made four different calotype negatives on the one visit. The image shows a stone tower mill carrying rather worn-looking cloth sails with a rudimentary alternative power source in place. The miller is also present. The point of particular interest here is that Worcestershire simply does not feature in the windmill literature and the number of windmills that survived here up to 1900, derelict or otherwise, was in the order of a handful. The image of one that was working (it was pulled down in 1875) in this county comes as an agreeable surprise and a reminder that their distribution *circa* 1850 was very different from that of only half a century later. There can be few, if any, windmills that were photographed before Kempsey Mill, though Grimsby (Cartergate) Mill would qualify if the story of its demolition in 1848 to make way for a new railway line is correct. The appearance of this old post mill in a photograph that has stovepipe-hatted gentlemen in the foreground, certainly lends credence to an exceptionally early image. But the appearance of both windmill and railway on Micklethwaite's 1848 town map of Grimsby negates the idea of enforced demolition, at least in the late 1840s, and points instead to this photograph having been taken in the 1850s. But, regardless of Cartergate Mill surviving rather longer than generally appreciated, it is probably still the windmill whose date of being fully swept away predates that of any other for which a photograph has survived. From about 1880 a fair number of mill photographs exist, but it is not until about 1905 that one can reasonably expect to find a photograph of every windmill, and even then some persistently eluded the photographer.

[38] Quite abnormally so: rarely where a mill is described as 'disused' is any other description for it given. The lack of clarity seen in Plate 8 reflects how much this – or any six-inch – sheet has to be magnified in order that a mill depiction from it matches the 1:2500 depiction for the same mill – in this case, as seen in Plate 7. The annotation of *Windmill (Corn, Disused)* appears also on the 1:500 town plan that carries a survey date of 1879.

[39] See Charles Willott, *Belper and its people: with historical notes, incidents, reminiscences and recollections of past and present,* Belper, 1894.

does become available from *circa* 1890 onwards is a reasonably comprehensive photographic record.

Any photograph of a windmill is obviously more than useful for assessing its condition and cumulatively they demonstrate the range of visual situations that ageing mills could find themselves in. Moreover, enough photographs have survived from before the turn of the century to offer fair coverage for mills, certainly for the 1890s, which is looking more and more to have been a pivotal decade for the fortunes of the English windmill. Precise dating for many of these photographs is sadly unavailable but where mills were left to deteriorate before suddenly collapsing, or they succumbed to fire or storm, photographers were often soon at the scene to record what had by this time come to be regarded as a newsworthy event. As one example, the sense of occasion surrounding the demolition of Aldbourne Mill in Wiltshire is very apparent in the recording of it (Plate 9). As it happens, the provenance of this photograph includes no suggestion of the date it was taken but enough evidence, it can be argued, lies within it to date it fairly well. Country dress habits may not have changed all that often, but the South African (Boer) War slouch hat would not have been available as a fashion accessory for small boys to copy – in particular, the individual seen sitting atop the remains of the sails – much before the turn of the century. In fact, local sources are able to provide a date for the mill's demolition as 1900 together – thus probably authenticating the information – with the name of the person who oversaw it.[40]

If one was looking to be more ambitious by seeking out a group of mills, not necessarily all co-existing at any one time, and exploring the range of sources of evidence for the losses of those mills, any area sufficiently compact to be suitable for this opening chapter can be chosen. The windmills once to be found in what now comprises Boston in Lincolnshire are the subject of a small booklet by Hanson and Waterfield who state that in the 1820s fifteen windmills were active here, but that 'the 1880s saw the end of most of them and by 1910 in Boston itself only the Maud Foster Mill was still working by wind'.[41] Only a dozen of these mills can be construed as working after *circa* 1860. For the nine of them that subsequently did not survive into the twentieth century in one form or another, the evidence gathered by Hanson and Waterfield for the disappearance of each is more than reasonably impressive. Sale-by-auction notices for demolition and clearance of their sites appeared in the *Boston Guardian* for three of these mills between 1885 and 1891 – including the eight-sailer whose machinery found its way to Heckington Mill, also in Lincolnshire. Another two were very publicly pulled down to enable the construction of new docks in 1882, amid controversy over compensation for their owners. The assumption that two of the remaining four mills had gone by 1870 owing to their being replaced by new buildings is a safe one, and for a further mill similarly so by 1890. Only in the case of one of these nine windmills – a tower mill at Burton Corner – are the authors apparently unable to decide between conflicting evidence for its disappearance, which occurred sometime between 1878 and 1886.[42] All the

---

[40] M.A. Crane, *The Aldbourne Chronicle*, Aldbourne Civic Society, 1974, 21. More information for this mill appears in Martin Watts, *Wiltshire windmills,* Wiltshire County Council Library & Museum Service, 1980, 17.

[41] Martin Hanson and James Waterfield, *Boston windmills,* Boston: privately published, (nd), 3. This publication is typical of a lot of the secondary windmill literature in being useful, but also modest in its use of format. It is less than sixty pages long but is atypical in its thorough use of the different forms of evidence available.

[42] Hanson and Waterfield, *Boston windmills, passim.* Of the three mills which did survive beyond 1900, one is still a working windmill. Another lost its sails in 1905 but continued to work with a gas engine, and the third had been

evidence collated for these nine windmills as well as for the three that survived into the next century is summarised below (see Table 1.1). Turning to the map record, the recognition, or lack of it, of these nineteenth-century mills on the one-inch provides an example of what one may be beginning to suspect will become a familiar story. Sheet 69 of the Old Series

Name of Mill	Evidence for Mill Disappearance
West Street Tower Mill	Named millers from 1859 and, for 1869, the building of a new steam mill on the site.
Black Sluice Post Mill	Named millers for period 1851-61. In 1870 the site was built over with all previous buildings gone, except one – a granary.
Burton Corner Tower Mill	Two conflicting reports of demolition – 1878 and *circa* 1886.
Gallows Post Mill	Demolished 1882. Dated photos. Sketch showing dismantling and report of the Boston Harbour Commissioners' costs of compensation – *Boston Guardian*, April 1882.
Gallows Tower Mill	Same as for Gallows Post Mill.
Mount Bridge Tower Mill	Sale by auction for demolition and clearance – *Boston Guardian*, September 1885.
Brothertoft Road Mill (mill of unknown type)	Features in a sale notice of June 1888, and presumably sold that year.
Brothertoft Road Tower Mill	Features with Brothertoft Road Mill (see above) in the same sale notice of June 1888, but also appears in a sale notice for demolition and clearance the following year in August 1889.
Tuxford's Tower Mill	Sale by auction for demolition and clearance – *Boston Guardian*, November 1891.
Thompson's Tower Mill	Worked by sail to 1882 (photo of 1880), afterwards by steam to 1892, later idle until 1895, then converted to grain silo for adjacent roller mill (photo of 1928); mill demolished in 1960s.
Sleaford Road Tower Mill	Working windmill until 1905 when sails were removed (photo of 1888), then continued to work by gas engine; stump of mill remained until 1960s.
Maud Foster Tower Mill	A windmill that has survived and that still works by sail.

Table 1.1. Evidence for the disappearance of windmills in Boston (after Hanson and Waterfield).

depicted Hanson and Waterfield's fifteen mills when it was first issued, but then resolutely continued to depict all these mills on each successive state of this sheet up to the last one (Margary state 16), which was still being printed as late as 1909. The New Series, based on a survey of 1886-8, perversely shows just one windmill, but the Revised New Series of 1897 shows the correct number of two windmills – so long as one accepts as correct the decision not to recognise Thompson's Mill, which by 1897 had become a grain silo.

stripped of all its innards and used as a grain silo from 1895.

So, we might be encouraged to think that this process of discovering when each of the windmills of Boston disappeared augurs well for the efficacy and worth of the first national revision of the one-inch New Series. We would not be wrong. The Revised New Series, and the Third Edition that followed it, were both better in their recording of windmills than the earlier New Series and, in its later years, the Old Series, had been. The Revised New Series is especially useful for its depiction of windmills simply because it predates the Third Edition and came at a crucial time for the recording of windmills. The gradual decline in numbers of windmills had started many years before the end of the nineteenth century as we have seen. One can speculate that a century or so earlier, windmills had been cruder, much smaller and less permanent as structures, but also that they had been more numerous. The improved working capacities of later, more permanent, mills provided better commercial success and led to their numbers remaining steady for a time but, inevitably, their decline set in. In some areas, the process of attrition and non-replacement of windmills was well underway by the 1860s. It then accelerated until, by the beginning of the First World War, a great many mills had gone. The serious decline of the 1890s was of the last generation of mills, those that had been built or much enhanced in the early years of the nineteenth century. Many would still have been at work in the 1870s and 1880s, with a sizeable proportion of the remainder lying derelict. The last windmills to be built along traditional, though outmoded, lines were taking place, as seen, as late as *circa* 1880, but such cases are rare. We can now appreciate that the Revised New Series appearing as it did within the span of one decade – the 1890s – provides a snapshot view of events. It would be hard to select a more appropriate decade in which to try to capture a coherent picture of the last working days of the English windmill, notwithstanding that the decline of mills has been shown to have occurred in different areas at different times. For mills in the broad rural swathes of southern and eastern England, the pattern was very much as just described. For urban areas and some parts of the country not commonly associated with windmills such as, for example, the Cumbrian seaboard, the decline and near-passing of windmills had happened before the middle of the nineteenth century. However, despite this and other caveats, the Revised New Series provides a useful account for the areas of the country conventionally thought of as being 'windmill regions'.

The suitability of windmills for examining the competence of the Survey in the second half of the nineteenth century stems in part from this watershed in their fortunes and from the corresponding diverse and fluctuating states of dereliction in which they could then be found. Vulnerable both to fire and strong winds, a mill's existence could be a precarious one, especially in an economic climate when repair after some setback could not be justified. These factors militated against the long-term survival of windmills and could be the reason for the sudden loss of one of their number. The result of this is that maps can misinform the user on many accounts; they may depict a reality that is simply defective in its windmill count or they may indicate a situation showing mills as co-existing where such was never the case. This febrile landscape as far as our windmills were concerned was made worse for the Survey by the morphology of those mills still at work. Mills could offer varying sets of circumstances where each was a blend of visual and functional factors. The 'look' of a mill might include its sense of permanency, or of transience in the landscape, and this could be irrespective of any impending removal through dereliction or destruction. Almost at the end of this chapter, it may be salutary to consider the differences between two 'normal' mills. The post mill in the parish of Sullington in West Sussex seems rootless in a landscape that is otherwise empty; in one photograph taken of it, it seems disconnected to the ground in the sense that no firm pier foundations are apparent for the ends of the crosstrees of its trestle (Plate 10). The mill appears capable of easy mobility. This perception of Sullington Mill is deceptive of course, but contrasting this mill with a much more robustly built and seemingly permanent tower-mill type of structure, of the sort once seen in large numbers across the country and exemplified by the windmill at Thearne in the East Riding of Yorkshire, must have left very conflicting kinds of impression in the minds of those whose job it was to record all windmills (Plate 11). The fundamental differences between mills such as those at Sullington and Thearne should certainly have brought home to any surveyor that there are 'shades of grey' when it comes to recording a windmill in the landscape. Yet if the message of this chapter has so far possibly been one of raising concerns that might have caused mills to be misrepresented by the Ordnance Survey, then it must be said immediately that the protocols that evolved to depict windmills at each of the different scales probably worked well for the great majority of mills. In the last part of this chapter of introductions, a brief discussion is offered for the use of symbols on maps, and why that use may not be quite as straightforward as one might think it should be.

What map symbols do

It has been suggested that the secondary mill literature available today suffers from a lack of critical appreciation of maps as sources of evidence. Of course, evidence for windmills has been acquired from maps and incorporated into the literature, but often this has been done uncritically. This proposition highlights a general point about the use of maps – that on publication there is an unthinking presumption of authenticity for them to be accurate accounts of landscapes. Yet the conveniently submerged realities faced by a mapmaking organisation, and the difficulties of recording features such as windmills, must render that presumption fallible to some degree. As is more often said than remembered, any map becomes out of date from the time the surveyor leaves the ground, and in the nineteenth century the Ordnance Survey condoned lengthy delays between its large-scale surveys and use of the data to provide derivative one-inch sheets. The depictions seen for Belper Mill were taken from the larger scales, which make use of an assortment of means for signifying the presence of a windmill (as we shall see), but it is the use of symbols at the one-inch scale that is particularly intriguing, cartographically speaking. These depictive devices that mimic the generally understood form of a windmill can only be construed, if taken at face value by a map user, as indicating the ground presence of a windmill. The mapmaker clearly means this to be quite unambiguous with no hidden caveat to mislead an unwary user. If prompted to describe these windmill symbols, the user would probably declare them to be geometrical figures that are upright and symmetrical about a vertical axis and of consistency in size and general lucidity, especially if it is a later nineteenth-century map that is being discussed. Further prompting might provoke the comment that this consistency may or may not be mirrored by a corresponding consistency in the form of windmill structures found on the ground. But beyond this recognition that the language of signs and symbols introduces an element of artificiality to a map, nothing more is normally ascribed to this facet of maps and the language of their signs.

The attribute of maps whereby this sort of language colludes both to reveal the details of a landscape to the user but at the same time also to conceal the richness of that detail or to supplant the message that ideally should be on offer has, though, attracted the attention of cartographers. The appeal of map signs and symbols to such people has many perspectives. At a somewhat rudimentary level, much is said about map design where symbols are subject to considerations of colour, proportion and prominence, so that assessments can be made of a symbol's ability to be 'read' from different psychological perspectives. At another level of inquiry, a strand of philosophical thinking that advocates meta-interpretation of written texts to find meaning – the so-called discipline of hermeneutics – is useful here. Taking the wider definition of a text as an assemblage of linguistic characters that may not necessarily closely correspond with what are 'seen' as letters from any alphabet, but that may be 'seen' as more pictographic in their nature, then maps would seem an entirely appropriate subject for such treatment. The face of a map contains many discrete bits of information, whether they are in the medium of lettered text, point, line or areal symbology, or perhaps just lines indicating boundaries and features. All these in the language of systems theorists are the many parts that go to make up a map, which is then deemed to be a whole. One extension to this line of thinking makes use of the paradox that while the individual discrete bits of knowledge on a map are responsible when aggregated for the message of the map, that in turn the message of the map in its entirety is the set of conditions that provides meaning to any one individual element within it. Putting this another way, a completed map, while undeniably the outcome of its ingredient parts being assembled, is also the preconceived model or template that brings *a priori* the 'correct' parts together; this paradox going by the tag of 'the hermeneutic circle of understanding'.

One implication of this is that the collective contribution of all the windmill symbols on a map to the overall message of that map is a function of more than simply their number; there is a subjective element at work here circumscribing the efficacy and worth of the map. Consequently, bias is an integral part of any map, and this bias overlaps and competes with the notion of a map user's psychological ability to 'read' it, so that any 'reading' of a map is not a passive activity but an interactive one. Those who theorise about the compilation of maps with a view to their becoming more effective often do so by applying the expression 'sign-vehicles' to describe the markings on a map, which they then see as so many encoded fragments of information. While this theorising often then gets extended, making use of a language that we can safely bypass, it is at least interesting and, with a moment's thought, understandable that the ground feature cited by many map theorists as being the one that is accorded the most highly iconic, mimetic or pictorial sign-vehicle is the windmill.[43] This contrasts sharply with sign-vehicles at the other end of a continuum of iconicity where a greater degree of diagrammicity occurs. An example of the intricate reasoning that can prevail when trying to theorise about sign-vehicles is the status of churches. Here, the idea of a solid box or circle surmounted by a cross, as later used by the Survey as a symbol for signifying a church with a steeple, is considered to be associative rather than pictorial, as not all churches necessarily have a tower or a spire, or even a cross. Furthermore, in the case of churches, another way to signify one is by using a cross without any other adornment – as seen on earlier OS maps – since, culturally, the cross is symbolic of religion (in a Christian

[43] Some discussion for this is provided, for example, in J.S. Keates, *Understanding maps,* second edition, Harlow: Addison Wesley, 1996.

context at least).[44] Hence, the conventional sign used for churches, we are told, needs to be thought of as emblematic rather than anything else. At least in a study relating to mills with their map symbols strong on pictorialism and mimicry, there are some difficulties we have largely been spared.[45]

This short précis for the nature of map symbols can itself be summed up in three points. First, maps are systemic in nature; second, they inevitably embrace an ethos since the nature of their compilation is always subject to the chicken-and-egg tension of parts and wholes; and third, the sign-vehicles (map symbols in common parlance) used for windmills usually demonstrate an acute degree of mimicry. Using a visually empathetic or isomorphic device to identify each windmill on a map leads to the collective contribution of windmills to the ethos of that sheet being disproportionate. There may, on the part of both the mapmaker and the map user, thus be a greater presupposition of comprehensiveness when locating windmills on those sheets where mills are plentiful in the landscape than for those sheets and landscapes where they are not so much in evidence. Put simply, we might suspect that sheets with a high density of windmills are more accurate in their mill representation than those with a low density. Indeed, much of what has been uncovered in the preparation of this study appears to justify that suspicion. Hence, it can be conjectured that a hard line would have been taken over the appearance of a mill, and whether or not it constituted a notifiable structure, under circumstances of a fading windmill presence in the landscape, more so than had it been sited in an area where working windmills still outnumbered those that had fallen into disuse. The issue of what constitutes a notifiable windmill structure is potentially very problematic and is discussed in passing throughout the study, but more particularly when the mill representation for one county – Essex – is reviewed (see Chapter Five). Before that, detailed considerations of the one-inch Old Series and the New Series occupy Chapters Three and Four. But before any of that, the 'standard' modes of depiction that were thought sufficient for the majority of windmills are discussed next, in Chapter Two. At the end of the study, an epilogue brings the story of OS windmill representation fully into the twentieth century (see Chapter Six). These six chapters make up the first part of this study; a second part looks at a geographical sample of eighteen New Series sheets and, for the landscapes covered by each, offers an analysis of the depictions of windmills provided by the major nineteenth- and twentieth-century map series.

---

44 For a more detailed discussion on how the Survey dealt with the intricacies of depicting churches at this time see Richard Oliver, 'Steeples and spires: the use of church symbols on Ordnance Survey one-inch maps', *Sheetlines* 28 (1990), 24-31; Richard Oliver, 'The one-inch revision instructions of 1896', *Sheetlines* 66 (2003), 11-25; Hellyer and Oliver, *One-inch engraved maps of the Ordnance Survey from 1847,* 84-5.

45 For a comprehensive guide to the theory relating to cartographic design the reader is directed to Alan M. MacEachren, *How maps work: representation, visualization and design,* New York: The Guilford Press, 1995.

2

Styles of depiction

The reader can now appreciate that Ordnance surveyors were confronted with something of a visual diversity when it came to recording the windmill population, and so quite how the ground presence of any mill translated into a depiction on a map is unlikely to be clear-cut. Of necessity, the number of ways a mill could be portrayed on a map was markedly less than the many combinations of visual and functional attributes that the windmill population could offer. This had to be the case, of course, as the prime role of any map is to convert the messiness and infinite variety of an actual landscape into an artificially simplified model. This is then 'read' using a language mutually acceptable to mapmaker and map user. So, the cleverness in mapmaking can be thought of as finding the minimum number of discrete portrayals that adequately captures the essence of the largest proportion of any feature – in our case windmills. At the one-inch scale this was always going to be a problem since the capacity for using a range of depictions for just one landscape feature is extremely limited. Nonetheless, at this scale, and certainly at the larger ones, it is suggested that the Survey did learn to get this balance just about right. By the end of the nineteenth century, it was doing a good job in combining the need for a not inaccurate representation for the great majority of mills with the need for unfussiness on each sheet.

There will, of course, always be exceptions that break a general rule, and to highlight some of the small minority of windmills whose map depictions were conspicuously outside the norms of 'correct' map depiction is not so much being churlish as being rigorous in our inquiry. The reputation of the Ordnance Survey can withstand this onslaught since, on this issue of diversity of depiction, it is very much a case of being able to 'praise the Survey with faint damns' rather than the other way around. That there are anomalous depictions of mills on individual maps if one looks hard enough should not come as too big a surprise. Neither will it be surprising to learn that, among a windmill population still in excess of a thousand at the end of the nineteenth century, there were anomalous mill scenarios that would have defied most attempts at any categorisation. It would be convenient indeed if the minority of non-standard windmill situations were to match the minority of anomalous map depictions that relate to windmills. But, life being what it is, this does not happen nearly as often as one might think it should. Interspersed, therefore, with the following narrative for the standard norms of windmill depiction are examples where mills and map depictions are idiosyncratic, each one further illuminating the rich but sometimes quixotic practice of windmill depiction rather than seeking to condemn it.

Map depictions for diverse scales and dissimilar mills

At its inception, the approach that the Survey adopted for acknowledging the presence of a windmill on one of its one-inch maps hardly represented a break with tradition. In the years before the Survey came into being, and for years afterwards, private cartographers preparing county maps at the same scale had used ideographs, symbolic devices behaving like Chinese characters in portraying the idea of something without having to express the name of it. In preparing a map this would obviously be a time-saving and space-saving measure, the more important at the one-inch scale where space could be at a premium. If aesthetics were also a consideration for the mapmaker, then the reduction in wordy annotations could also make the use of symbols a welcome proposition. The decision to use symbols to depict windmills in particular would have been very understandable. Visually distinctive and, to an untutored eye, very consistent in their form, the apparent uniformity offered by windmills in a late eighteenth-century landscape could happily find its correspondence in the landscape of the map. Whatever misgivings that may have been expressed about rationalising the landscape to make it fit the model of the map, they could have been overlooked more so in the case of the windmill than possibly any other point feature. Easy to mimic, the mutation of common windmills into mill symbols was such that even without a key, legend or characteristic sheet, any map user would have understood this association between map and landscape.

For much of its history the Old Series made use of symbols to signify windmills. These symbols would sometimes be annotated, but mostly they were not. Variants of the symbol meant that between sheets, and even on a single sheet, marginally different symbols could be found, leading to a suspicion that more information was possibly being offered to the user. Certainly, styles of symbol clearly intended to differentiate between post and tower mills came into fashion and, with further study, the patterns of these differing styles could be assessed for congruence with what is known to have been the case on the ground. But, undermining the usefulness of any such exercise, the overriding impression gained from a perusal of successive states of various sheets is simply that the Old Series was not updated to reflect the changing presence of windmills, or much else for that matter, in a consistent way. Sheets published in the early years of the nineteenth century mostly carried their initial state of windmill representation through to their later states. This did not give rise to any great percentage of error until the windmill population went into decline, but by 1880, at a time when the Old Series was still the only one-inch mapping available for much of the country, to be signifying, say for Lincolnshire, the same pattern of windmills as it had done in 1824, more than half a century before, was clearly a bad case of misrepresentation.

There were exceptions to this general observation. At Wisbech in Cambridgeshire, the eight-sailed tower mill there was obviously sufficient of a landmark or rarity when it either replaced an earlier post mill or was re-equipped with a large number of sails that the Survey felt moved to update its map, recording it with a symbol that actually denoted the correct number of sails, albeit somewhat crudely. The initial state of Old Series sheet 65 dated 1824 had shown two windmills to the north-east of the town, both depicted using the more usual convention of that time, a symbol visually akin to a post mill (Fig. 2.1.1). Much later, with an electrotyping date of 1867, Margary state 12 (of 14) for this sheet shows the situation to have changed.[1] The symbol akin to a post mill has been retained – or rather, replaced by a

[1] See, by way of example in this first case of wanting to identify a Margary state, the diagnostic information given

very slightly different post mill symbol – to which the extra sails have then been added and, lest the map user be unaware of this subtle addition, an annotation to suit the occasion has also been added (Fig. 2.1.2). Strangely, the opportunity had not been taken at the outset, or in the intervening years, to make use of the tower mill symbol that had, after all, started to appear on Old Series sheets when the initial state of sheet 65 was published – there are two such symbols shown, for example, on this state within close reach of the town.[2] The Survey may have been aware of this incongruity on its later states at a time when this mill is safely known to have long since been a tower mill.[3] But either it elected not to highlight an earlier error in providing a post mill depiction for what probably was a tower mill *circa* 1820 or, as is the more likely scenario, it failed to give sufficiently accurate information to the engraver when the later states were being prepared.

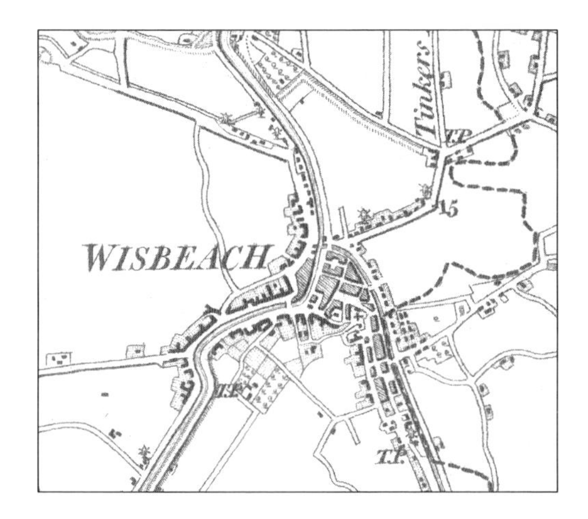

Figure 2.1.1. Windmills at Wisbech, Cambridgeshire: one-inch Old Series sheet 65 (circa 1825).

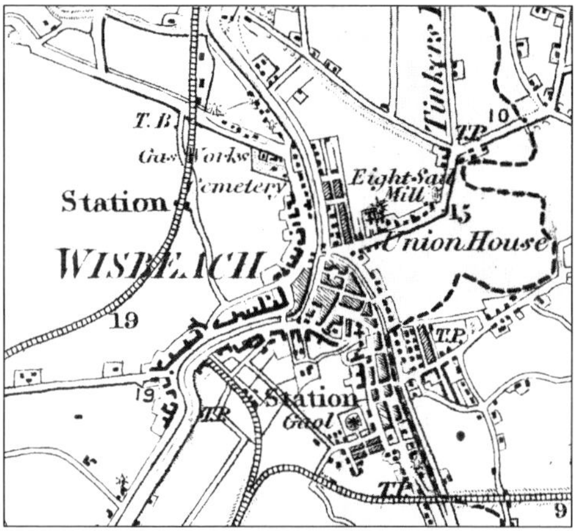

Figure 2.1.2. Windmills at Wisbech, Cambridgeshire: one-inch Old Series sheet 65 (circa 1867).

The representation of this windmill is unique in that none of the other windmills in the country suspected of having eight sails at that time was recorded as such by the Old Series.[4]

in Harry Margary, *The Old Series Ordnance Survey maps of England and Wales,* volume V, *Lincolnshire, Rutland and East Anglia,* Lympne Castle: Harry Margary, 1987, with *Introductory Essay* by J.B. Harley, and *Cartobibliography* by J.B. and B.A.D. Manterfield, pp xxxviii ff.

[2] Making this change to the windmill depiction was part of a wider package of revision for the Wisbech area that had been added to the map by 1853: it was no doubt linked to the arrival of the railway there in 1848.

[3] The earliest known picture of an eight-sailed tower mill occupying this site dates from 1840. One source dates the construction of this mill as early as 1778. See Arthur C. Smith, *Windmills in Cambridgeshire: a contemporary survey,* Stevenage: Stevenage Museum Publications, 1975, 5.

[4] Altogether, seven other sites had windmills known to have been equipped with eight sails: these were at Diss in Norfolk, at Eye in Northamptonshire and at Holbeach and Market Rasen as well as Boston (Tuxford's Mill) all in Lincolnshire; the late building of a tower mill at Much Hadham in Hertfordshire and the re-equipping of a tower at Heckington in Lincolnshire in 1892 with the machinery from Tuxford's Mill both postdate the Old Series last states. See Wailes, *The English windmill,* 100-1. See also Hanson and Waterfield, *Boston windmills, passim.* Of all these mills only the one at Market Rasen was similar to Wisbech in being close to new railway construction. At the time of the railway coming to Market Rasen (again 1848) this windmill had yet to be equipped with its complement of eight sails (*circa* 1868). In any event, there was precious little room left for adding a Wisbech-style annotation since by 1853 the word 'Gashouse' had been added to the map, and this took up the free space next to the mill symbol.

Furthermore, other multi-sailed windmills, of which there were more than a few, either five-sailed ones or six-sailed ones, were also only selectively acknowledged using annotations similar to that used at Wisbech. So, whether simply to conclude that a few cases of unusual mills were recognised in an inconsistent manner, or whether this situation forms a basis for arguing that systemic errors exist across the sheets of the Old Series, is a matter for debate. Assuming that the Survey had an explicit wish to record all windmills coherently, then any sort of inconsistency or anomaly can either be viewed benignly as idiosyncratic and perhaps even adding to the character of the map or, from a more purist viewpoint, it may be seen as an aberration which, alongside others, may seriously be undermining the integrity of the sheet. This is perhaps a pedantic judgement to have to make and, in any case, certainly one that can be deferred until after other issues of misrepresentation have been discussed.

In its later years then, the Old Series became seriously wanting in its representation of windmills. The principal reason for this, it would seem, was that on post-initial states of any sheet, the only reason a mill symbol appears to have been changed was if the construction of a railway or a related building demanded the destruction of the mill. When re-engraving the map, the symbols for such mills would be removed though this did not always happen and certainly it did not happen for windmills that had succumbed for other reasons. All this was happening, of course, amid the protracted process of producing one-inch mapping not just for England and Wales, but also for Ireland and Scotland. In addition, the Ordnance Survey was having to cope with the upheaval from having altered its survey regime to one whose focus was now geared to the larger scales. Moreover, the impending renewal of the one-inch through the New Series was diverting attention from the Old Series. Yet despite these continuing aims at improvement, it was only eventually that the representation of the windmill on the New Series came to be better than on later states of the Old Series. This would happen when, for various reasons, use of the windmill symbol was revived *circa* 1888 after a period of neglect and it was then uniformly applied to later sheets of the New Series. Though the capacity to differentiate between, say, post mills and tower mills was lost at this point, this move did bestow on the New Series some homogeneity of representation for windmills based on updated information.

Before 1914 the New Series underwent two complete revisions. The first of these led to the Revised New Series sheets that were surveyed and published during the 1890s, and the second led to the Third Edition sheets that we broadly associate with the following decade.[5] The level of refinement in how windmills were shown was enhanced on both these editions. The standard New Series symbol was strengthened in appearance and the use of standard annotations was introduced to try and convey to the user some impression of a windmill's condition. On the Revised New Series, symbols depicting windmills were annotated with *Windmill* presumably if the mill was capable of work (or nearly so) or, if derelict, with *Old Windmill* although use is made of the word *Disused* on more than one occasion. Several symbols are left unannotated due either to a lack of space on the map or, clearly, from the evidence of the sheets, to signify drainage or marsh mills that would have looked much like cereal-grinding mills. The lack of annotation for a windmill symbol could also be signifying a site that was equipped with the new style of metal windpump. No misunderstanding is

[5] The Survey was also having to cope with providing a revision for Scottish one-inch sheets at the same time. Revision for the equivalent Irish mapping was not started until 1898.

attributed here to the Survey for what it considered, at this time, to constitute a notifiable windmill. In using symbols, the Survey simply did not differentiate between mills in the truest sense of the word – structures used for milling cereal or other produce – and all the wind-driven drainage 'mills', once to be found in their hundreds in the eastern counties and culturally thought of as mills, but whose *raison d'être* was to pump water rather than grind anything.[6] If it can be accepted for now that the Survey had no motive for differentiating between grinding mills and pumping 'mills' and that, indeed, appearances were everything, then this is an issue that need not detain us. The more so as the embodiment of 'millness' exhibited by, say, the drainage mill that still stands at Thurne Dyke on the Norfolk Broads encourages us to adopt just such a position (Plate 12). Of all the drainage mills that could have been chosen as an example at this point in the story, this is as typical as any. Today it is something of a landmark, but when first seen by the surveyor only its slender proportions and obvious lack of space in the cap for the requisite machinery of a grinding mill would have marked it out as a drainage mill rather than the more generic form of mill. Even the existence of a scoop wheel or some other arrangement for moving water away from where it was unwanted to where it needed to go, is not always obvious, as Plate 12 demonstrates. All in all, the sheets of the Revised New Series are clearer than their predecessor sheets and do provide *some* inkling of the many distinctions that can be made between types of mill structure. They also go to some length to ensure that windmills in very congested built-up areas get recorded, though by no means are all such mills shown.

When revised again and issued as Third Edition sheets, distinguishing between these and sheets of the previous edition would have been reminiscent of trying to tell apart the states of earlier Old Series sheets, so similar is their style. Inevitably, on the Third Edition sheets, the windmill symbol count is generally reduced given the by then familiar pattern of mills falling into disuse, but the style of annotation from the earlier edition is retained. Obligingly, the symbol for a windpump was introduced to cater mainly for the annular-sailed skeletal metal farm pumps then appearing, but only in a small number of cases was it retrospectively applied to the older conventional pumping mills such as Thurne Dyke Mill. The all-metal structures that were newly-introduced at this time could vary from being mechanically very simple and not so large, to being impressive both in size and their use of windmill-related technologies such as the fantail (Plate 13). The annotative strategy for recording a windmill's condition on both the Revised New Series and the Third Edition appears to have been mindful of the mill's then newly-developed status as a popular architectural icon. Indeed, from this time onwards, the use of symbols by the Survey for its one-inch scale in respect of windmills, annotated or otherwise, will be shown to have been guided, partially at least, by the need to provide evidence of a windmill's status in the landscape seen from this new

[6] On the other hand, it might be thought that the Survey *was* making a distinction between these two sorts of mill by virtue of withholding annotation for pumping 'mills', except this was not done consistently. The failure always to draw a cartographic distinction between grinding mills – colloquially known as corn mills – and pumping mills at the one-inch scale may be rather vexing to purists who would prefer windpumps not to be thought of as a type of windmill, but the truth is that within the secondary mill literature this linguistic nicety is not always made clear. Not wishing to re-open a battle of semantics that was decided years ago, the remainder of this study is content to fall in line with the literature by letting the context in which a windmill is being described inform the reader as to which type of mill is being referred to. (This is a slightly altered position from that adopted by the writer in Bill Bignell, 'Conventional signs and the Ordnance Survey: the case of mills and the New Series', *Sheetlines* 35 (1993), 10-13.)

perspective.

It will be appreciated easily enough that if the one-inch New Series edition, the Revised New Series edition and the Third Edition were all of equal probity in the way they recorded windmills, then comparing each of the three depictions for any individual mill could provide a simple account of that mill's history. Moreover, all such accounts for all mills could then be collated to generate regionally contrasting accounts of windmill presence in the landscape over a period of some thirty years. Given, however, that the New Series was not diligent in its recognition of windmills in its first few years, this means that each individual mill account has to be carefully assessed for what it can tell us. Yet, for the great majority of windmills, this is still a worthwhile exercise. In the process, different scenarios of windmill presence become apparent. Two not uncommon ones can be pointed out straightaway. First, when a windmill site is identified on an early New Series sheet, either by the word *Mill* and a setting of an exposed, maybe hilltop, position, or indeed by the word *Windmill*, this might then be confirmed later by a combined symbol and *Windmill* annotation on the Revised New Series. Later again, the mill might have altered in status or even vanished from the Third Edition. Second, a situation may occur where a symbol on a Revised New Series sheet, perhaps even annotated as *Windmill*, was being used to note a windpump and this later becomes obvious when the structure is more accurately declared on the Third Edition. Both of these putative situations occur together at Owslebury in Hampshire. The New Series sheet carries survey dates of 1869-70 and records a black common-sailed tower mill close to the village using the word *Windmill* (Fig. 2.2.1). In 1870 a brick-towered windpump, looking for all the world like a conventional windmill, apart from having an annular sail, was built next to the existing mill to raise water for the village's needs. The Revised New Series sheet of 1893 unsurprisingly shows two symbols annotated with *Windmills* (Fig. 2.2.2). Only with the publication of the Third Edition in 1904, which was now able to make use of the new symbol for recognising a windpump, together with the wording *Old Windmill* for the earlier mill, is the situation adequately outlined (Fig. 2.2.3). One can maybe speculate, though, on whether or not the engraver would have been as assiduous in declaring the non-working state of the older mill had there not been a clear space in which to insert the word *Old*. Indeed, one might even suspect that, since the root word *Windmill* shows no indication of having been repositioned on the Revised New Series when the pump had to be declared, the spatial accuracy of the symbol used to depict this mill may have been compromised. Nonetheless, for Owslebury the one-inch sheets presented a fair rendering of ground-truth accuracy for this scene of milling and pumping activity, notwithstanding the disregard for any functional differences between windmills and windpumps on the Revised New Series. Acknowledging the level of effectiveness in defining milling activity for Owslebury to be only marginally flawed – and even that depends on the remit for windmill recognition that the Survey gave itself – this typifies a standard of recognition that ideally should have been repeated across the country. But, as we have begun to see, there were mills whose circumstances resisted straightforward categorisation, and consequently whose depictions on the three earliest New Series editions did not match up to the model of 'correctness' shown at Owslebury. In many cases, the difficulty becomes more apparent at the margins of windmill presence – situations where a mill's continuing survival is perhaps under threat, or the social context in which it had been thriving is rapidly altering.

Caught between the many examples where annotated map symbols were sensibly used on the early editions of the New Series, and a few sites that had the potential to make life awkward for a surveyor trying to classify what he was seeing, there were more than a few cases where mild dissension rather than outright discrepancy occurs between depictions on the New Series, the Revised New Series and the Third Edition. The small town of Brigg in the north of Lincolnshire supported a number of windmills in the nineteenth century, some of which provide for one such instance of gentle ambiguity. The fluctuations in the mill population on the western edge of town, and the associated restyling of mill symbols and

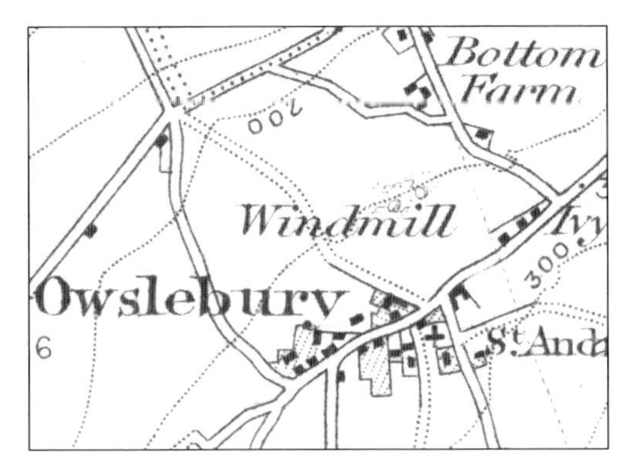

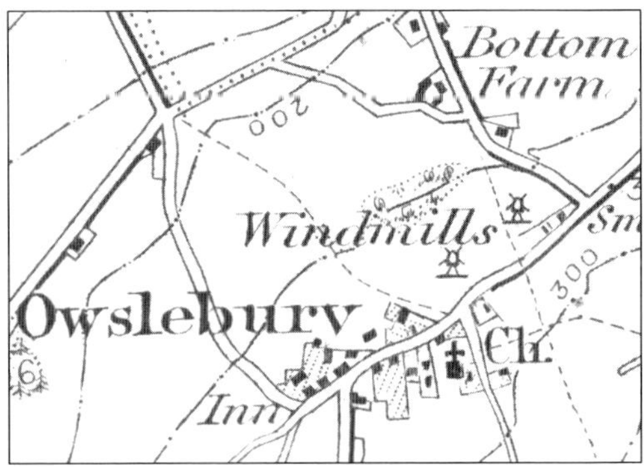

Figure 2.2.1. (above left) Windmill at Owslebury, Hampshire: one-inch New Series sheet 299 (1877).

Figure 2.2.2. (above right) Windmills at Owslebury, Hampshire: one-inch Revised New Series sheet 299 (1895).

Figure 2.2.3. (left) Windmills at Owslebury, Hampshire: one-inch Third Edition sheet 299 (1903).

annotations across these three editions no doubt made perfect sense at the time. A century on, an element of shrewd calculation is clearly necessary to unravel the individual stories for at least two of the five windmills shown by a symbol on one or more of the editions. One mill at least – Bell's Mill, the most southerly of the group – continued to work well into the twentieth century despite its forlorn treatment on the Revised New Series in the mid-1890s. Another, at Mill Place, missing from the New Series is later described as *Old*. Turning to the larger scales does not really clarify matters. This second mill, now recorded on the Revised New Series with a symbol and a share in an *Old Windmills* annotation, had existed at the time of 1:2500 survey but was embedded within a complex of buildings that may have curtailed its ability to operate by wind. With these buildings then apparently gone from the Revised New Series, this could have allowed a windmill symbol to be inserted more easily but this scenario of the surrounding buildings having disappeared is not confirmed by the second edition 1:2500. Moreover, there is no clear reason for not having provided Bell's Mill with its own *Windmill* annotation on the Revised New Series, as was done on the Third Edition. The annotation on the Revised New Series is awkward and inaccurate, so giving us no other option than to observe that the circumstances of milling by wind in this part of Brigg were inexactly, if not wrongly, declared on these three editions (Figs. 2.3.1, 2.3.2, 2.3.3).[7] Moving further into the realm of windmills that clearly threaten our capacity for being able to infer accounts for them from the evidence of just the three editions, Farndon near Newark offers a peculiar set of depictions. To the east of this village, the New Series sheet offers a rather misshapen mill symbol that is later partly obliterated on the Revised New Series by an *Old W.Mill* annotation that relates to a new mill symbol. A second mill symbol makes its first appearance on this edition near the others and is given a *Windmill* annotation. This third mill and the first mill are then not shown on the Third Edition, while the second mill seems to have been upgraded by this edition since it is now accorded a *W.Mill* annotation. Chancing across this sort of contradictory information on the three editions happens only rarely, and when it does occur, frequently it is because of a major shift in windmill depiction for New Series sheets attributable to a change of OS policy that occurred in 1888.

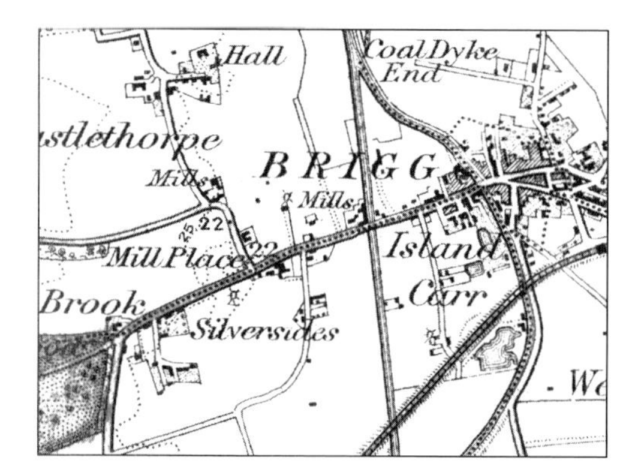

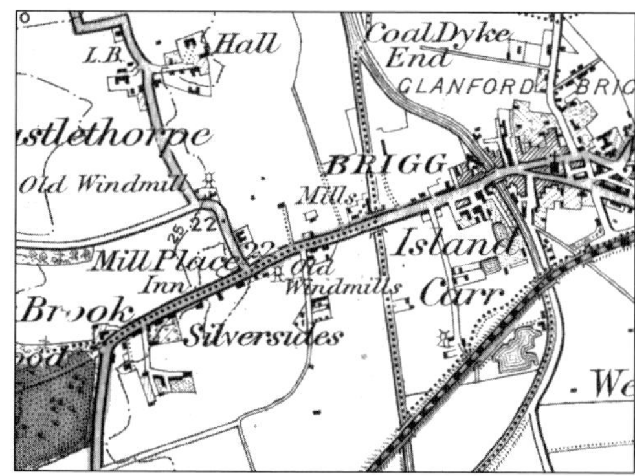

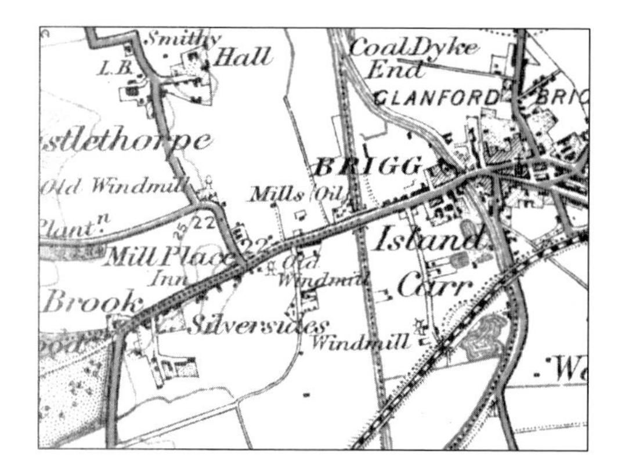

Figure 2.3.1. (above left) Windmills at Brigg, Lincolnshire: one-inch New Series sheet 89 (1890).

Figure 2.3.2. (above right) Windmills at Brigg, Lincolnshire: one-inch Revised New Series sheet 89 (1898).

Figure 2.3.3. (left) Windmills at Brigg, Lincolnshire: one-inch Third Edition sheet 89 (1907).

---

[7] Keen-eyed readers will have noticed a change to the 'look' of the Third Edition at Owslebury (Fig. 2.2.3) from the earlier New Series and Revised New Series. This is due to the Third Edition having been the one to embrace the changeover between sheets being engraved – where each sheet was 'pulled' from a master – and being lithographically printed, which was much more convenient for the mass production of maps. The family of Third Edition sheets was not the first to make use of printed colour but when, after the issue of its regular sheets – which were engraved – printed, coloured variants were issued, the results could still be rather crude. If this was not quite the case for the Third Edition colour-printed sheet used here to show the windmills at Owslebury, it does explain the Third Edition image of Brigg, for which a colour-printed sheet has also been used and on which it might be thought that the colour registration with the feature outlines leaves something to be desired (Fig. 2.3.3).

It has been suggested that using symbols on a map stems in part from a need to be prudent in the use of space on scales such as the one-inch. It has also been suggested that since the appearance of a windmill lends itself to an unambiguous design of symbol, it is a sensible candidate for the use of this approach to map construction. For the time being, let it be presumed that windmills warranted their recognition by nineteenth-century one-inch maps, if not for any intrinsic worth, then possibly because of their value as noteworthy features that allowed map users to orientate themselves. The alternative to using a symbol for depicting a windmill would have been to treat this type of structure in just the same way as any other type of building, and record its ground plan. At a stroke this would mean that windmills would have much less visibility on a map. But if they were needed, for any reason, to remain as a notifiable class of structure, then their ground plans could be annotated and so a very different type of map to one using symbols would result. The Survey was to make use of both methods for depicting windmills at the one-inch scale.

The representation of windmills on later states of Old Series sheets has been noted as flawed due to a general lack of updated information finding its way onto these sheets, but there is a broader reason for reproaching the Old Series towards the end of its life. From just before 1850, after a lull in publication, the surveys that were undertaken or about to be undertaken north of the Preston-Hull line for the larger scales were available for assisting in the preparation of the remaining one-inch sheets. However, on the last sheets to be issued, the depiction of windmills using symbols all but vanishes. Instead, the alternative method of recognising a mill is followed – treating it just like any other sort of building, and recording a ground plan for it. This means that, at the one-inch, mills appear simply as dots. Since the facility to augment this style of depiction with an annotation was seldom used, any capability of assessing the presence of windmills virtually disappears. This could give the illusion of a landscape quite devoid of windmills when this was not necessarily the case.[8] This restrained way of recording windmills was carried over to the early sheets of the New Series, which began to appear in the mid-1870s, supposedly as by-products of surveys already carried out for the larger scales. The frequency with which annotations were provided, alerting the user of an New Series sheet to the presence of a windmill, had however improved markedly by the mid-1880s. Then in 1888 the New Series reverted to the use of unannotated symbols for showing mills.[9] There had been a period of time, therefore, from before 1850 until after

---

[8] On sheet 91SW, nominally published on the last day of 1847 and showing an area of the Lancashire Fylde, the only indication of a windmill is that at Hambleton though at least another half-dozen were working in this area at this time. Looking alternatively at sheet 94SW, nominally published in 1858 and covering part of the East Riding of Yorkshire, some windmills are shown and those on Beverley Westwood, for example, are shown quite well but many others are missing.

[9] The transition seems to have been fairly smooth and took place within the space of months rather than years. Sheets published in 1887 are consistent in their use of annotated ground plans and seemingly proficient in the numbers of mills being recorded. Equally, some sheets published in 1889, particularly those issued in the last few months of that year, seem thorough in their recording of mills with the use of just unannotated symbols, but a few sheets, for example sheet 176 covering the north-eastern part of Suffolk around Lowestoft (issued May 1889), make very limited use of symbols as their only means of recognising windmills. One is prompted to think either that the reintroduced use of the windmill symbol was sparingly applied for a few months or that the time interval between, possibly a pre-1888 draughtsman's master copy, and publication of the sheet, say eighteen months later, caused this inadequacy. At least one sheet is caught in the middle having few symbols and with many windmills left poorly recorded as ground plans – for example, sheet 113 issued May 1888. Notably, sheet 126, which covers Farndon near Newark, was also published in 1889.

1880 when windmills had been all but invisible on the one-inch, and this coincided with the period of the windmill going into serious decline and before it became re-established in the public mind.

Recording a windmill by reference to a ground plan as the alternative to using a symbol was the preferred option, indeed a defining requirement, for the 1:2500 scale. A narrative for the events that led to the county-by-county rolling programme for the preparation of maps at this scale was offered in the last chapter, and this included the associated story of the derivative six-inch. The use of ground plans to record mills, accompanied possibly by annotations, and the implications for the effectiveness of windmill representation will be a recurring theme of this study. So, an introduction to this area will be useful, to complement the remarks just made about the normal and adequate recognition of most windmills when using symbols at the one-inch scale.

By 1860 the OS large-scale map provision for England consisted of six-inch coverages of Lancashire and Yorkshire and the beginnings of the 1:2500 county coverages. These had begun with the four northermost counties before the survey teams switched their attention to the southern coastal counties to re-survey these, now at 1:2500. On the six-inch sheets of Lancashire and Yorkshire, one noteworthy convention was holding sway in the depiction of windmills. This took the form of a circle with a flattened cross abutting it.[10] This is arguably what one would see in plan looking down at a tower mill that still had its full complement of sails, remembering that the windshaft in a windmill – the axle that supports the sails – is slightly inclined to the horizontal to improve the balance of forces on it. Instead of a circle, a square or rectangle sometimes appears instead, and the sensible speculation is that these were used to depict post mills. For the windmill-laden countrysides of both Lancashire and Yorkshire, there is evidence that the use of the two styles of depiction correlated to the two types of windmill as they were found on the ground – smock mills being unknown in these two counties. Since, on a six-inch sheet, a windmill depiction had to be close to, or at, true scale in size, these depictions are very small and need annotating in order to be easily seen with the naked eye. Due to this, and despite the high quality of the engraving found on these early six-inch sheets, examining them for signs of windmill presence is not particularly easy. The manner in which a couple of post mills at Flatfield, east of Howden in Yorkshire, are shown illustrates the diminutive size of the style of mill depiction being used at this new scale (Fig. 2.4.1). Occasionally, if the lofty locations where many windmills were placed were also adopted as sites for trigonometrical stations, and the mill's role as a sighting target for surveyors engaged in secondary triangulation work also had to be recognised, then a mill's depiction could become even more intricate. The tower mill immediately south of the two post mills at Flatfield reveals the sort of depiction one comes across in this type of situation. Here, the mill's ground plan has been overlain by the kind of triangular symbol that was in general use for indicating a control station. Occasionally post mills were requisitioned for this work too. At Rawcliffe, also close to Howden, the combined normal depiction of a post mill with that of a triangulation station is certainly elaborate, probably to the point of obfuscation (Fig. 2.4.2). But, despite the level of scrutiny required to examine these sheets,

[10] As a means of depicting a windmill, this was a convention new to England. It had been imported from the six-inch work in Ireland where it had earlier been extensively, but not exclusively, used. See, for example, the six-inch sheets of the coastal areas of County Wexford.

they provide some of the most valuable evidence for windmills at county level that we have due to the all-inclusive nature of their mill representation through use of a large scale – for the first time on a wide basis – and because they reach further back in time than any other Ordnance Survey large-scale maps. That all-inclusiveness extended to the recording in some cases of mill mounds where the windmill itself had disappeared. For a mound to warrant recognition after its mill had disappeared, and moreover to be annotated as a redundant mill mound (as was done) demonstrates an assiduousness to the task of recording the landscape that is certainly worthy of praise. An example can be seen on Yorkshire six-inch sheet 210, published in 1855, adjacent to the five windmills to be found on Beverley Westwood. This practice of acknowledging mounds at the larger scales was to continue, but increasingly mill mounds became a feature that the maps felt they could overlook. If a mound was especially prominent, or maybe of the type into which the lower floor of a smock mill was built such as one of the pair of windmills at Woodchurch in Kent, it could survive cartographically along with its mill until maybe both were cleared away.[11] But less frequently encountered is cartographic evidence from later in the nineteenth century of mill mounds that had lost their mills, or of mounds that were not so prominent as the one at Woodchurch. Typical of the latter category was the slightest of mounds supporting the charismatic old post mill that continued to work at Thorne in Yorkshire until at least 1908 (Plate 14). This was without doubt a mound that could not be considered anything other than the most insignificant of hummocks once its mill had gone. Yet this mound was recognised not only by the first two editions of the 1:2500 when the mill was still in place, but also by the third edition once the mill had disappeared (Figs. 2.4.3 and 2.4.4). So it was that a connection with a milling past was tenuously kept alive for this one-time mill well into the twentieth century. But talk of early twentieth century revised 1:2500 mapping, let alone a third edition, is moving too far ahead.

The earliest of the 1:2500 county surveys to have got underway before 1860 had been that of Durham, started in 1854. For *cognoscenti* of windmills, Durham does not disappoint, but the circle-and-flattened-cross style of windmill depiction used for the Lancashire and Yorkshire six-inch sheets a decade earlier was only used for a few of the mills recorded here on the earliest 1:2500 sheets. One that was given this style of depiction was the tower mill at Whickham near Gateshead (Fig. 2.5.1). But if the convention for large-scale maps was being followed here, where ground plans are employed instead of symbols, then these depictions,

[11] It should be explained that the supporting trestle of an early medieval post mill could be encased in an earthen mound for extra stability. One result of doing this was that the timbers would often rot – a situation not designed to enhance any windmill's long-term survival prospects! But it was quickly realised that the simple expedient of constructing the mound first gave a mill placed on it more height, and so more chance of catching the wind. By the nineteenth century, however, windmills – especially of course the new tower mills – were of such a design and weight that a better option for raising them, if this was felt necessary after their initial construction, was to jack up the cap at the curb and insert extra courses of brickwork. This would lead to these mills having their upper stories rise vertically rather than continuing the slope of the lower, original, stories. The roundhouses of post mills, which had integral piers that supported the trestle, could also be raised by inserting extra brick courses so that they might eventually became equal in height to the body of the mill. All in all, windmill mounds were a legacy from earlier centuries but many older mills of course continued to have them. Just to complicate the picture, mill mounds also provided easy access for reefing the simpler types of sail, such as cloth or spring sails. This use of a mound might be found at mills that in all other respects were relatively sophisticated, though whether such mills were making use of old post mill mounds, or whether their mounds were raised at the same time as when the mill was built is a moot point.

where the circular plan is drawn to scale, could be suggesting that all windmills so recorded are those that still had their sails. This hypothesis breaks down as the few mills depicted this way do not correlate with those Durham mills suspected of still having been at work by sail at this time.[12] The other option is to consider this circle-and-flattened-cross depiction as a symbol. Either way, this early situation represents an interesting case where the concept of a map symbol and that of a ground plan became merged. This style of depiction, which had been in widespread use on the six-inch, but then only partially brought to the larger 1:2500

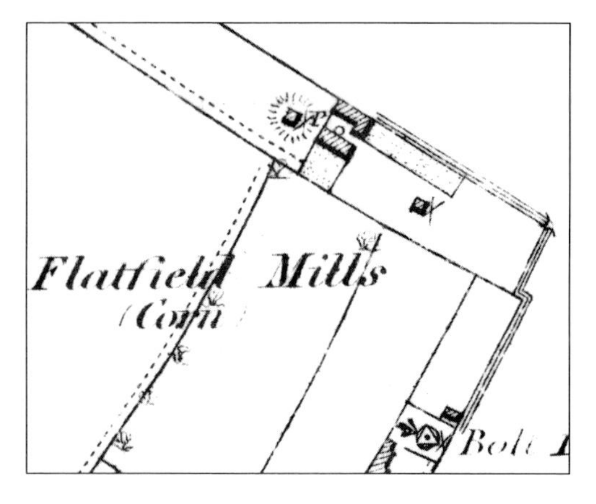

Figure 2.4.1. Windmills at Flatfield, Yorkshire: 1:10,560 (six-inch) sheet 238 (1855).

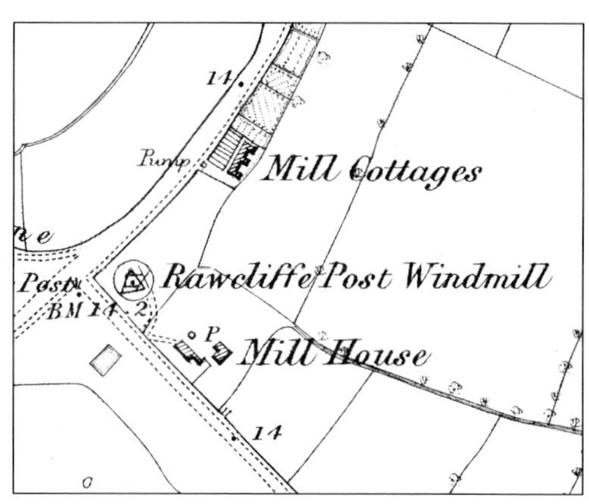

Figure 2.4.2. Post mill at Rawcliffe, Yorkshire: 1:10,560 (six-inch) sheet 252 (1853).

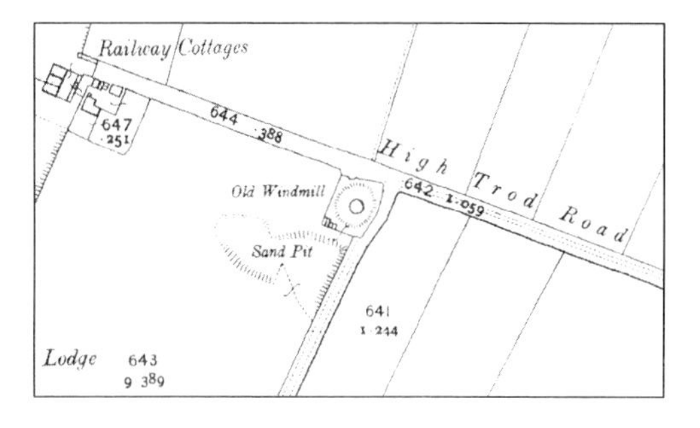

Figure 2.4.3. Post mill at Thorne, Yorkshire: 1:2500 sheet 266.5 (1906).

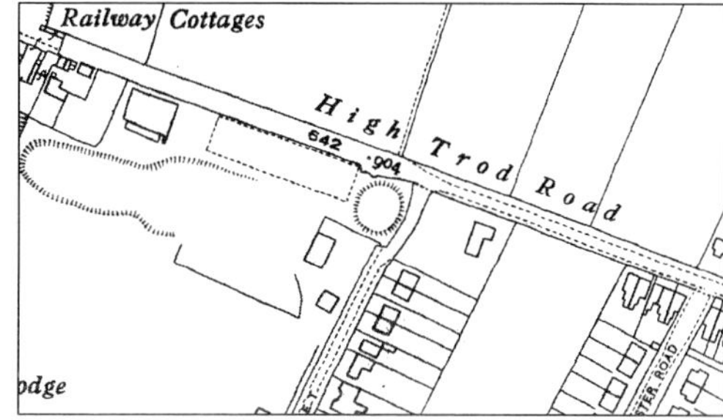

Figure 2.4.4. Post mill at Thorne, Yorkshire: 1:2500 sheet 266.5 (1932).

scale, also found its way onto some of the Old Series sheets of Yorkshire. This occurs, for example, at Sheriff Hutton and at Dunnington, both in the East Riding, with the depiction

[12] As well as this mill at Whickham, one of three windmills at Easington and a mill near Washington were depicted in this soon-to-be-discontinued manner. Quite why these three particular windmills were selected for this treatment is something of a mystery. Geographically, they were well separated from one another; each was reasonably close to other windmills, and they had no known common characteristic that was not shared with other mills. Any perception that they were somehow working in a different way to other mills seems very unlikely given that other windmills in Durham are known to have been working by sail, or by other means, in the mid-1850s. The only pointer to a possible explanation is that the new 1:2500 sheets were being prepared by those earlier responsible for the six-inch scale mapping and it may have taken a short while to realise that the six-inch convention for windmills could, in light of the typical 'footprint' of a windmill at the 1:2500 scale, sensibly give way to an annotated ground plan at the new larger scale. Added to this is evidence that the order in which the 1:2500 Durham sheets were prepared did not follow the sheet numbering sequence.

at Sheriff Hutton avoiding erasure for decades, and eventually making it onto the Revised New Series. Odder still is the isolated case of a circle-and-flattened-cross style of depiction on New Series sheet 273 for Borstal Hill Mill at Whitstable in Kent (Fig. 2.5.2). This was one of the earliest New Series sheets to be published (1878) and the depiction may have been an experimental one, especially as the 'circle' was filled in. This was during the period of scant windmill disclosure on early New Series sheets that had a further decade to run, yet it was long after this style of depiction can be presumed to have otherwise disappeared.[13]

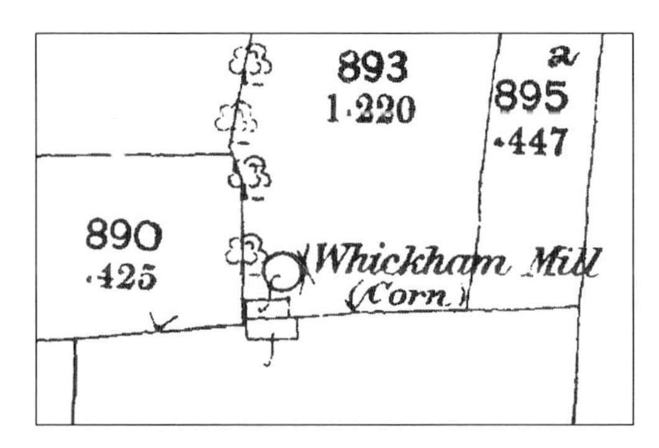

Figure. 2.5.1. Windmill at Whickham, Durham: 1:2500 sheet 6.2 (circa 1857).

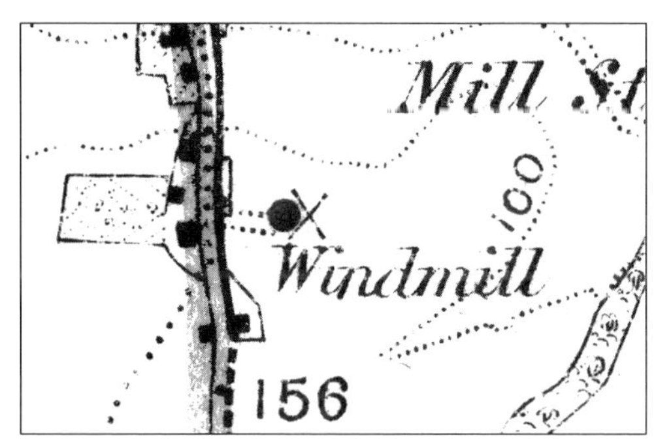

Figure. 2.5.2. Windmill at Borstal Hill, Kent: one-inch New Series sheet 273 (1878).

With so little to go on, searching for any meaning behind the isolated use of this type of depiction from mid-century onwards, whatever the map scale, is going to elude us unless we are somehow able to identify characteristics of the windmill population matched by the few survivors accorded this type of recognition. In reality, though, this course of action would probably turn out to be fruitless and, from the evidence we have, it seems much more likely to be a case of a once-standard means of portrayal stubbornly refusing to disappear totally until long after its 'natural' time. It is maybe inevitable that some instances of this protocol, which underpinned hundreds of mill depictions over the best part of a decade, survived the notice calling for its discontinuation and the subsequent agenda for its erasure.

As difficult as it may be to fathom this conundrum from the early days of OS large-scale mapping, there are luckily no others to have to consider as the 1:2500 quickly settled into its conventional arrangement of showing true ground plans. This meant that a scaled circle was used for tower mills and for post mills that had roundhouses. Smock mills were shown by a polygon-shaped plan that frequently could appear slightly distorted rather than be perfectly symmetrical. Moreover, in the majority of cases for all three types of mill an annotation is provided to leave the map user in no doubt as to the presence of a windmill, and possibly its use. The depiction of a windmill at Sprowston near Norwich on the initial edition of the 1:2500 is simple and clear and affords an idea of the sort of depiction that could normally be expected, and which indeed was provided for the majority of mills (Plate 15). The hand-colouring applied to this mill's depiction was provided as an 'extra' that had to be paid for,

[13] One answer to this incongruity may lie in the continuing use of this type of depiction on continental maps. In particular, the French 1:80,000 map series was making use of it at this time and it is a symbol that is found on modern continental map series. Ironically, on neighbouring New Series sheet 274, the single surviving mill of a pair of windmills that once stood near Minster is shown using a very archaic-looking symbol reminiscent of the Old Series.

but its use can offer clues to a mill's circumstances. For most mills, this addition of colour to a map is helpful, but on occasions it can serve to raise more questions than it answers.[14] Less than a mile from the windmill at Sprowston and shown on an adjacent sheet are two apparently non-working mills (Plate 16). On the hand-coloured version of this sheet, the colouration of one site with contrasting carmine (indicating brick or stone-work) and grey (indicating woodwork or iron) is seemingly helpful, but a possible oversight in not recording the material used at the other site is not. In any case, one is left to ponder the difference in status suggested by the two different annotations of *Old Windmill* and *Windmill (Disused)*. To redress any suggestion here that the great majority of 1:2500 sheets did not do a good job in explaining themselves, another example of two windmills that stood in close proximity to one another can be offered. Well away from Norwich, a pair of urban mills at Stepney, not far from the centre of Newcastle upon Tyne, appear on a Northumberland sheet (Plate 17). The two depictions are coloured and annotated so as to leave much less to the imagination. Both tower mills were stone-built and both are thought from other pictorial evidence to justify the depictions they were given at the time of survey (1855-8). This more clear-cut type of representation was applied to the great majority of windmills once the 1:2500 had got into its stride. With the standard means now established for depicting windmills at the 1:2500, we are left to consider the relatively small number of mills, regardless of type, that were recorded at this scale using more idiosyncratic forms of depiction.

Even then, some of the more unusual depictions are not necessarily hard to decipher. Returning to the Norwich area, on the same sheet that shows Sprowston Mill another mill is depicted by an annotation of *Old Windmill* against four diminutive black squares. With just a little thought, it will be realised that these four square depictions are the ground plans for four brick piers of an open-trestle post mill similar to the one at Mumby. The depiction of Mumby Mill itself at this scale does not challenge our knowledge of it as an open-trestle post mill: indeed, the rectangular-shaped plan with four additions implies exactly that (Fig. 2.6.1). But the conclusions to be drawn from some other mill depictions are not always quite so straightforward. The circular plan for a windmill at Wetwang in the East Riding of Yorkshire has an added feature that could well be the steps for a post mill shown in plan, suggesting this to be a post mill with a roundhouse. This inference is not wrong, except that we may further suppose that recording the steps suggests the mill to be fixed in orientation and, if not able to turn to face the wind, presumably to be no longer working. There again, this mill had the unusual feature of a platform that extended back over the steps for better access to the sail-reefing mechanism, but it seems extremely unlikely that any OS surveyor would notice such a trivial element of a windmill's functional appearance and make it the reason for an abnormal depiction (Plate 18). Furthermore, this 1:2500 depiction of the mill at Wetwang is not an early one but a second edition one of 1910, by which time, one might think, such idiosyncrasies would have been smoothed away (Fig. 2.6.2). Given that this mill was still at work and looking very pristine in the early years of the twentieth century, it is

[14] The hand-colouring, when provided on this edition, signifies the types of material used in the construction of a building. This optional colouring for the 1:2500 was done by the Survey itself. But other hand-colouring found on small-scale sheets of this period is post-production ornamentation added by map-selling agencies. Primarily, this involves each county having a colour wash applied across it with the county boundary being emphasised using the same colour. By the time of the New Series, the agencies seem to have come to an agreement about which colour (from a selection of just five) to apply to each county.

much more difficult to appreciate what prompted the cartographic treatment of Wetwang Mill than to understand, for example, the treatment accorded Mumby Mill.

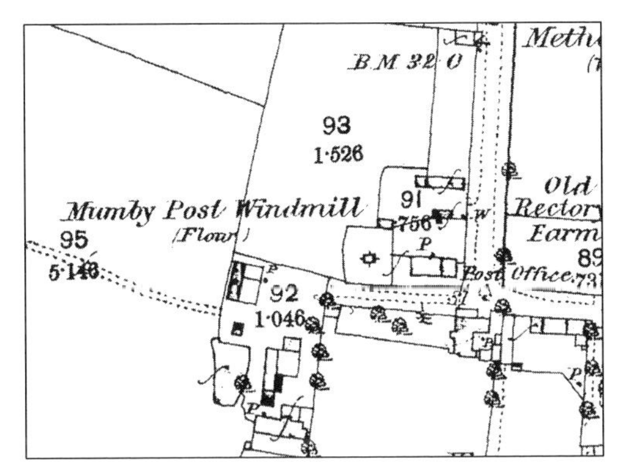

Figure 2.6.1. Post mill at Mumby, Lincolnshire: 1:2500 sheet 67.13 (1889).

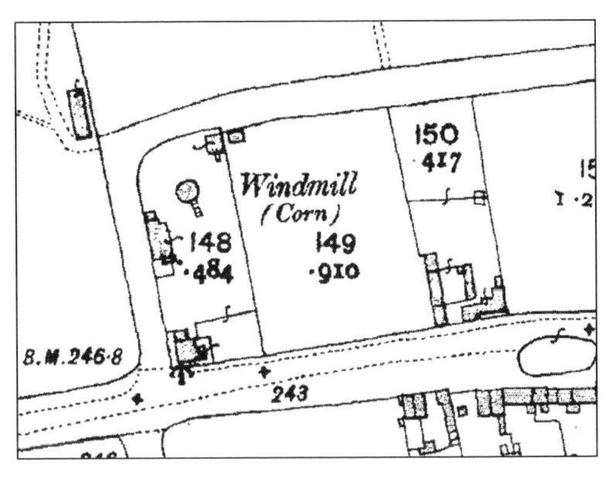

Figure 2.6.2. Post mill at Wetwang, Yorkshire: 1:2500 sheet 160.8 (1910).

Other depictions that appear atypical can be the basis for shrewd guesswork. What, for example, prompted the precise shape of the grey-coloured ground plan of a mill at Hucknall under Huthwaite in Nottinghamshire? (See Plate 19.) In reply, familiarity with other post mills of the area through their photographic record (including the surviving 'Cat and Fiddle Mill' at Dale Abbey) provides a possible clue. The custom here was for the roundhouses of post mills to be of a smaller diameter than usual, so that the crosstrees and their supporting piers protruded through the walls to be clearly visible to an outside observer. This feature could be seen, for example, at Club Mill near Hucknall Torkard before it and the rest of the mill disappeared in 1930 (Plate 20). But, if Hucknall under Huthwaite Mill appears to be cartographically ambiguous, the depiction of Wighay Mill only a mile from Torkard offers quite unambiguous evidence for some unusual structural feature. Possibly this feature was the roundhouse-related one exhibited by Club Mill, or perhaps Wighay Mill was an open-trestle mill (Fig. 2.7). The similarity between the depictions of Mumby Mill and Wighay Mill might suggest the latter to be more likely: otherwise it would be the case that the Survey was making no distinction between open-trestle mills like Mumby Mill and some Midlands-style post mills like Club Mill, in which case the extra ornamentation being added to a 'standard' depiction makes it disingenuous and serves no purpose. But, confusingly, many windmills known to have been like Club Mill are recorded quite conventionally with a purely circular ground plan, and the style of ground plan seen at Hucknall under Huthwaite is quite rare, although the 'Cat and Fiddle' Mill at Dale Abbey is recorded in this way. For the depiction of the windmill at Hucknall under Huthwaite, and for a fair number of similarly enigmatic depictions found elsewhere, it can be hard at this remove in time to infer the complete truth about them unless other forms of evidence for the mills concerned become available.

Our account of ground plans, whose precise meanings in their recordings of windmills are unclear, could easily be developed using a modest fund of examples to form the basis of a chapter all of its own, and very engaging this could be. It would also support our growing perception that the whole subject of representation of windmills by the Survey is a rich and multi-faceted one. Yet, for the great majority of depictions given to windmills, it is part of their nature that they *are* informative, so any lengthy development of this current theme is probably unnecessary. Furthermore, if wanting to understand the 'big picture' of windmill representation, it could easily distract the reader from appreciating that insights into the broader themes of mill representation are more likely to accrue from examining groups of windmills rather than individual ones. Why should this be the case? Well, the statisticians' argument would demand that the sample size of a windmill population be above a certain threshold for any finding to be valid. Under this orthodoxy, any individual aberration would not so much nullify the broader trend as simply be seen as an error on the part of, say, an individual surveyor, draughtsman or engraver, and so safely able to be disregarded. Another argument would have it that the contribution to a map by its representation of windmills is a function, in more than one sense, of the numbers of mills involved, as was strongly hinted at in the last part of Chapter One. Mindful then that the study we are engaged upon should eventually veer towards examining groups of windmills rather than individual ones, further inspection of single mill representation will be limited to just three last examples of irregular depiction. Any subsequent consideration of an individual mill should only be in the light of its contribution to the mapping of a group of mills, or where it can illuminate a particular difficulty of windmill placement that the Survey had to deal with, or where it can clarify the intricacies of annotation applied to ground plans. The topic of annotations deserves lengthy consideration and this will be tackled later, but the current overview of irregular depictions for individual mills at the 1:2500 deserves its last three examples and, lest it be thought that the northern and eastern counties seem to have held something of a prerogative for these, the first one comes from Devon.

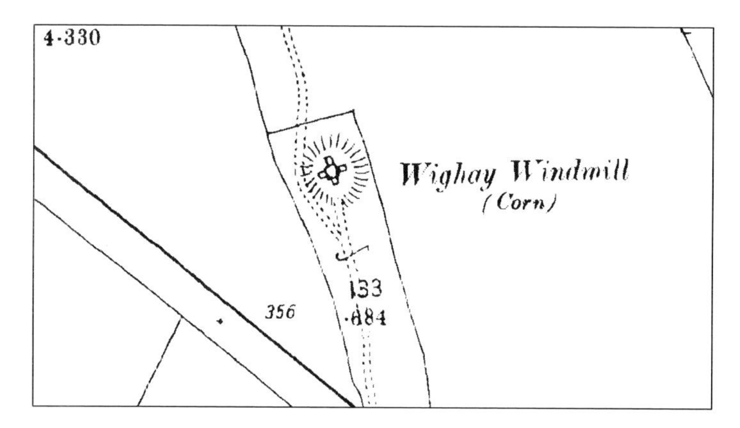

Figure 2.7. Post mill at Wighay, Nottinghamshire: 1:2500 sheet 32.12 (circa 1880).

Scant remains from the shell of a windmill survived until recently at Northam, north of Bideford, and the depiction of this mill on the first edition 1:2500 map of 1888 is certainly enough to catch the attention (Fig. 2.8.1). It is known that the machinery from this mill was dismantled and removed at some stage and the empty tower left to become derelict. What remained of this tower in 1919 collapsed in a storm that year: photographs taken of the old tower before then show it as truly derelict with one half having fallen away. This, of course, would match the 1:2500 ground plan so, given that the survey for the 1:2500 came late to this part of Devon, that the depiction is repeated for the second edition of 1904, and that the Revised New Series map of 1897 does not record this mill, it suggests that the Survey is recording here a very severely derelict mill tower that cannot have had any visual vestiges of once being a working windmill. Yet, this partial ground plan is accompanied by a full-blown annotation. It is worth noting at this stage, therefore, that in a part of the country that only had a small windmill population, the Survey went to a lot of effort to be seen as recording a one-time windmill where the ground remains were nondescript to say the least.

For the next example, we move to a windmill at Herne Bay in Kent whose ground plan and accompanying annotation are far from explicit (Fig. 2.8.2). The site of this mill was only a few yards from the sea and this could easily have contributed to the likelihood of a map user failing to realise that it was a windmill that was being recorded (Plate 21). Incidentally, this windmill is a good example of one that continued to work until it disappeared, almost overnight, when it was sold in 1878 to be demolished and for its site to be developed. Any impression that might have been formed by the reader that windmills were liable to decay gradually after they had ceased working, and then to linger on the edge of existence for years, if not decades, is shown to be false by the story of this windmill. The last choice of a 1:2500 windmill depiction is, if nothing else, ending with an impressive mill. Of the pair of large tower mills that stood in the western suburbs of Great Yarmouth, one – Press's High Mill – is widely considered to have been the tallest windmill in the country. This was truly a mill whose size and capacity can only be described using superlatives. Clearly immune to the judgement of history, though, and the significance and grandeur of what it was facing, the Survey contrived to show Press's Mill and its neighbour, Green Cap Mill, in something less than eye-catching fashion, using depictions that are much less definitive than many used for mills a fraction of the size of these two (Fig. 2.8.3). To summarise for this last section: the precise thinking that led to the depictions of the windmills discussed being as they were will

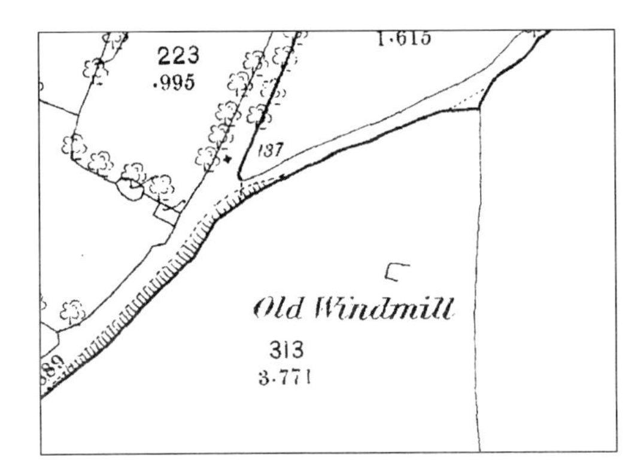

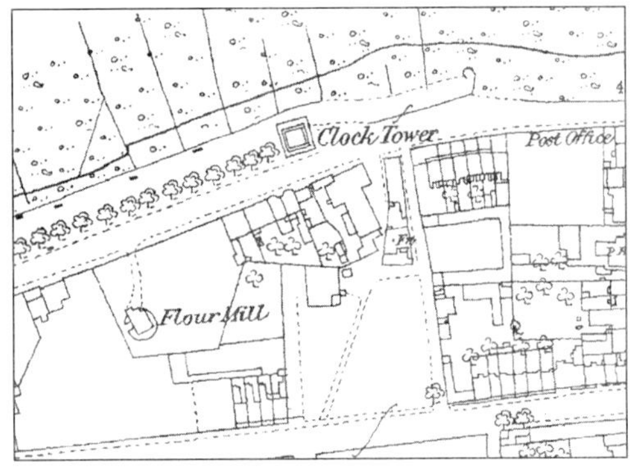

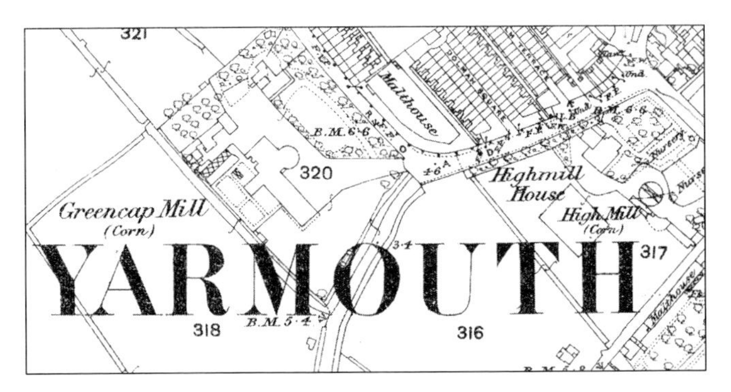

Figure 2.8.1. (above left) Tower mill at Northam, Devon: 1:2500 sheet 12.14 (1888).

Figure 2.8.2. (above right) Smock mill at Herne Bay, Kent: 1:2500 sheet 23.12 (circa 1874).

Figure 2.8.3. (left) Tower mills at Southtown, Great Yarmouth, Suffolk: 1:2500 sheet 2.3 (1887).

probably remain closed to us. It is not even as if these mills were largely from counties that were among the earliest to be surveyed or from counties that had a small, marginal windmill population. Any excuse that OS staff may have been experiencing teething problems, for example, in the recording of these windmills and many others that have not been discussed would not seem terribly plausible.

To complicate matters, the circular or polygonal ground plan for the brick base of a mill (usually a tower mill or a smock mill) is frequently found embedded within, or adjoining the outlines of other buildings. In this situation, the depiction is a composite one and so not a windmill depiction in the sense already discussed. This additional facet of a windmill's map depiction became necessary if a surveyor found outbuildings attached to a mill, since they too had to be recorded. The type of situation occurring at Greatham Mill, where a simple corrugated-iron lean-to has been attached to the mill, is certainly not rare. Not too far from Greatham, a variation on this theme was demonstrated by a handful of simple tower mills, each of which was built on top of a substantial circular stone base that extended far enough out to provide a sail-reefing gallery at second floor level. Fulwell Mill near Sunderland still shows this feature. This probably explains the early 1:2500 depiction that was used for a mill not far from Fulwell at Houghton-le-Spring (Fig. 2.9.1). But the positioning of sizeable and possibly quite misshapen extensions around windmills can be found across the country. Consequently, variations from the 'standard' 1:2500 windmill depiction, where each is trying to accommodate just such an addition to a mill, and so having to show an altered footprint or ground plan for a building, are found across the country. The smock mill at Woodham Mortimer in Essex, for example, is seen to have a somewhat irregularly-shaped brick base, both in a photograph taken in 1891 (Plate 22) and on the first edition 1:2500 (Fig. 2.9.2). In our modern age when the windmills that remain have undergone preservation, or have at the very least been tidied-up, many of these trappings have been seen as irrelevant clutter and removed. One exception to this is the post mill at Herstmonceux in Sussex, where the semicircular addition to half of its roundhouse, explicitly recorded in 1874, was included within the windmill's restoration that saw new sails fitted at the end of 2005.

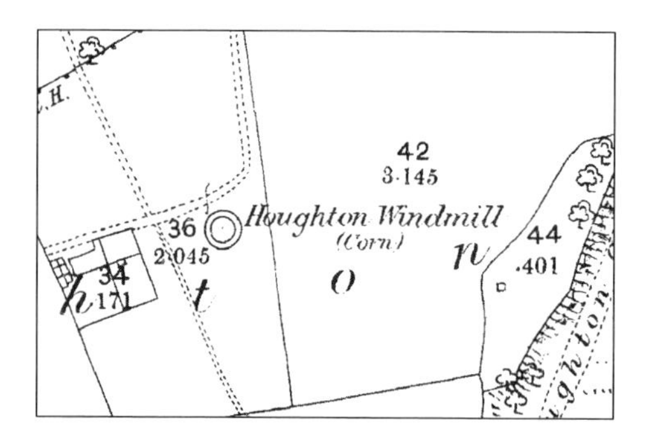

Figure 2.9.1. Windmill at Houghton-le-Spring, Durham 1:2500 sheet 13.16 (1857).

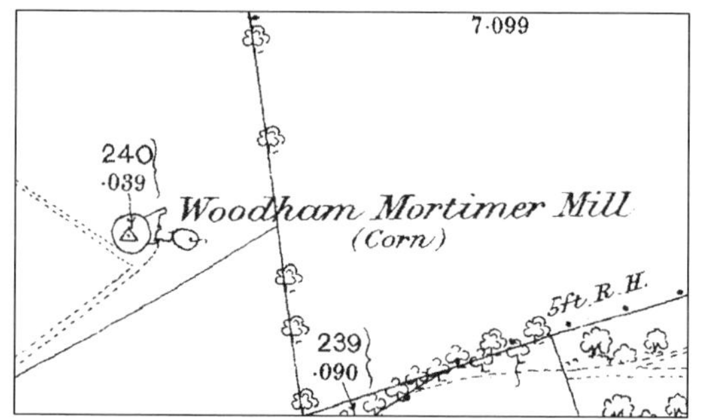

Figure 2.9.2. Smock mill at Woodham Mortimer, Essex: 1:2500 sheet 54.9 (circa 1875).

So the depiction of windmills on the large 1:2500 scale, it would seem, can be less than clear on at least two accounts. Any adjacent buildings that were very close or even attached to the mill could impair an obvious depiction, and there was also always the chance that the surveyor or the engraver would exercise some prerogative for individualism. The uniformity of mill depictions does not seem to have been of outstanding concern, so the danger of the latter happening was never far away. Nor was there any guarantee of a lucid mill depiction because of a weight in numbers. Windmills from counties that supported several mills often got short-changed. The two mills together known as Draper's Mills, for example, that stood close to each other near Margate in Kent, are less than explicitly depicted on the 1:2500, but on this occasion this is due to an inadequacy of annotation (Plate 23). It is to the different styles of annotation that we can now turn. If one is prepared to look hard enough, a variety of annotations used for denoting the presence of a windmill can be found on the 1:2500, and this topic probably deserves more of an extended study than we can offer it here. The subject becomes the more fascinating if the vagaries of the supposedly derivative six-inch scale are also included, since the key potential for the two large scales to be in disagreement with one another comes down to the availability of map space. A suggestion that there may sometimes have been pressure to curtail an expansive 1:2500 annotation to suit the smaller six-inch should therefore be taken seriously.

We can open this discussion by observing that for both scales the descriptive root word *Windmill* is nearly always augmented by a bracketed description relating to the function of the mill. Hence *(Flour)* and *(Corn)* are both commonly found as descriptions and appear to have been interchangeable in the way that they were applied, simply being the input and output commodity for a majority of grinding mills.[15] For those grinding windmills still at work when visited by the OS surveyor, one or other of these descriptions is used for the great majority of them, but in addition many windmills were used for industrial grinding purposes. Visually, these mills were little if at all different from their corn-grinding brethren, and often a mill could be adapted from one role to the other, and then quite possibly back again. Little wonder that a surveyor was often unaware of what exactly the mill he was looking at was actually doing, and so *(Corn)* is applied rather more often than technically it should have been. But there are exceptions: two mills – at Little Milton and Wheatley in Oxfordshire – are each described by *(Ochre)*, implying that they were contributing to the paint industry. A windmill at Lyth Hill near Shrewsbury is described by *(Flax)*, suggesting its usefulness to the nearby rope works, and in the same county Hawkstone Mill is described by *(Corn & Bone)* though, if this is to be believed, presumably the miller at this most flexible of windmills (earlier it had been used to produce linseed oil) was careful to keep his types of produce apart! Together with a few windmills, such as the one that still stands on the north foreshore of the River Humber at Hessle, described by *(Whiting)* and other mills described by *(Sawing)* – an example can be seen on the 1:2500 sheet covering the northern suburbs of Norwich – sample descriptions such as these do not begin to embrace the broad range of industrial uses to which windmills were put during the nineteenth century.[16] The use of a variety of bracketed descriptions was an attempt to capture something of that breadth of

---

[15] It may be prudent to point out that the use of *Corn* here is as a generic term whose true meaning can be to denote the leading crop of a district. Hence in England this is usually taken to be wheat; in Scotland it is more likely to mean oats, and elsewhere it can be taken as meaning maize. For England, its use legitimately includes the grinding of barley or the production of provender for animal feed.

[16] For further discussion of the industrial uses to which windmills were put in nineteenth-century England, the reader is directed to Roy Gregory, *The industrial windmill in Britain,* Chichester: Phillimore, 2005. This book has been a welcome addition to the secondary windmill literature, rescuing as it does a significant part of the one-time windmill population from historical oblivion.

usage, and is often made all the more complex by linking the description to the name of the mill. The theme of industrial windmills and how the Survey dealt with them will be touched upon again.

Looking at the general treatment of windmills: if beyond use or even derelict, then *Old Windmill* was often used as a complete description, but the use of either *(Ruin)* or *(In ruins)* attached to *Windmill* is also found. Appending *(Disused)* to the root word *Windmill* was not abnormal for a mill that was no longer working, but less use was made of *(Derelict)*. Rarely found within the same set of brackets is a combination of *Disused* and the former function of the mill such as *Corn* or *Flour* – Belper Mill was one exception. Complicating matters further is the fact that the word *Windmill* does not always appear as part of the annotation for sites known to have supported a windmill. Instead, the expression *Corn Mill* is seen (or *Cornmill* though this was less popular) so, because these labels were also used for obscure sites about which we may know little except that the lack of a watercourse and the general topography suggests the presence of windmill remains, this can easily lead to confusion. For some mills so described, there is a promising explanation. Conceivably, if a windmill-based business outgrows the capacity that the windmill offers, or reliance on the wind is deemed precarious and the business has therefore turned to the use of steam, oil or gas as its power source, then the essence of that site has become disconnected from one of 'windmilling'.[17] This despite the fact that the windmill body stays put with its stones still very much in use, but probably having had its sails dismantled. The new primary power source would then be housed close by. This would be a different type of situation to that of a windmill that had stopped trading and been left to become ruinous, thereby eventually also losing its sails. Under these conditions, this would not necessarily diminish its essence of being a windmill in a landmark sense, so explaining numerous cases where *Windmill* is retained for the large-scale annotation for a disused mill. These circumstances where a windmill was adapted to use more reliable sources of power provides one explanation for where the annotative word *Windmill* has been replaced by *Corn Mill*, but by no means does it explain all such cases.[18] An example that endorses something of what has been said is not so hard to find. The smock mill that still stands at Upminster in Essex carries a full complement of sails and its history of having been – and its current status as – a traditional windmill is not in dispute. Yet, as with many windmills in the nineteenth century, its owners sought to enhance its output by adding a complementary power source, and this had been done here relatively early in the century. Picking up on this, the Survey felt moved to provide a 1:2500 annotation here (one that, in this instance, was replicated on the six-inch) of *Corn Mill (Steam & Wind)* so firmly subjugating the capacity of this mill to work by wind to one, cartographically at least, of

---

[17] The word 'essence' is used in this text only occasionally, and then usually to emphasise the worth of trying to establish something profoundly 'true' of a windmill or its location. There is no expectation that the Survey should have got involved with such metaphysical ruminations: quite sensibly, all that would have concerned a surveyor would have been a mill's appearance. Besides, by early twentieth-century thinking, how something 'appeared' was not to be thought of as a lesser, or even a deceptive, form of reality behind which its 'essence' could be sought, so there is no intellectual argument potentially going unexamined here – and an explanation of what *can* be meant by the word 'essence' easily constitutes another Pandora's box of ideas, which we can again, justifiably and thankfully, leave unopened.

[18] Yet again, a windmill still working by sail as part of a large milling complex and instantly recognisable by its circular ground plan could find itself annotatively amalgamated with a steam mill. Together they would simply be annotated *Flour Mills*. This happened for example at Marton, north of Gainsborough in Lincolnshire.

being powered by steam (Fig. 2.10.1). This may have owed something to the survey date of the relevant sheet (1868) being sufficiently early for the incidence of windmills having a secondary power source to still be minimal. Or that this particular mill was unusually well-established in its use of steam power, such that the surveyor considered that what he was seeing was sufficiently progressive to merit such an generous annotation. Whatever the reason, the Survey could not be expected to devote as much map space to every windmill over the following decades that decided to follow Upminster's example, so this remained a very sparingly applied style of annotation.[19] Meanwhile, in neighbouring Hornchurch, the situation was slightly different. On the same sheet of the first edition 1:2500 that depicts Upminster and its windmill, the recognition of Hornchurch Mill is normal and clear and carries the standard annotation of *Windmill (Corn)*. Close by, and obviously part of the same business enterprise, there was another mill that had the annotation *Steam Mill (Corn)*. The information provided at the six-inch scale for the windmill on this occasion is less than a faithful replication of the larger scale – it was depicted using the circle-and-flattened-cross device, which by this time (1872) was a very anachronistic way of doing things – Fig. 2.10.2. But at both scales the two annotations for the mills are quite separate though they and the mills they describe can be seen as conjoined, and we are left to wonder whether or not an amalgamation of these two annotations would have been a more aesthetic option, and one, moreover, that did not need to compromise on the information given. When it afterwards came to the revised editions of the 1:2500 and six-inch scales, this windmill at Hornchurch survives though, in a slight debasement of information on the derived six-inch, it loses its tag of *(Corn)*. The steam mill on the other hand is reduced simply to *Corn Mill* at both scales. Elsewhere however, on this sheet other windmills are treated to normal depictions and the overall verdict for it has to be that ground-truth accuracy has been well served here, even if a little awkwardly in respect of two windmills.

The capacity of the 1:2500 – and to some extent the six-inch, remembering that much of what is being said applies to both scales – to show more annotative detail than the one-inch also meant that windmills that in reality were windpumps could be described in such a way that this difference was made apparent. Usually this was achieved using *Windmill (Pumping)*, or just *Pumping Mill* as an annotation. Where *Pumping Mill* is used alongside a small unfilled circle at either scale, especially on earlier sheets, a wind-driven structure is being indicated. On later sheets, though, where the possibility exists that a power source other than wind is being used, then the simple annotation *Pump* is more likely to be found. But on occasions none of these annotations is used, so distinguishing between a corn mill and a windpump has to be judged from the context of the map depiction. The placement of a mill by a water channel, or perhaps within a brickyard or other type of industrial setting, is usually sufficient to give the game away. There are always exceptions, of course, over which care needs to be exercised. At a brickyard near Lowestoft in Suffolk, a windpump was recognised by a small

[19] Variations of the annotation used for Upminster Mill can be found elsewhere and had earlier been used by the six-inch scale, notably for showing the changing circumstances of the many windmills in the Holderness Road and Dansom Lane area east of the centre of Hull. A couple of windmills were described here by using an annotation of *Wind & Steam Mill (Corn)* that, in a reversal of fortunes from Upminster Mill, is apparently emphasising wind as the more important source of power. But it has to be remembered that this was still only the middle of the century and although the relative importance of different power sources was beginning to alter for urban mills at this time, it would not do so for many rural mills until much later.

(coloured) circle, but with a simple annotation of *Windmill* (Fig. 2.10.3). In this case though, it is the setting of a small industrial works that implies the recognition of a windpump more so than of a windmill. The corollary of this is that there were sites where an annotation of *Windmill*, but without any bracketed description, was used to signify a normal grinding mill; again it is the siting of the mill in the landscape that alerts us to this likelihood. On the Isle of Axholme in Lincolnshire, three windmills standing on isolated hilltop positions are each annotated simply by *Windmill*. Yet just a mile away, all the windmills in Epworth are treated conventionally in having a bracketed description of the use to which they were put, as well as *Windmill*. The only conceivable reason why the three Axholme mills may have been seen as different to the sizeable number of local corn-grinding mills, is that they were used from time to time to grind gypsum instead of cereal, but even if the OS surveyor ever uncovered this fact, it is hard to understand why this would have had such an impact on the annotative recognition of these windmills.

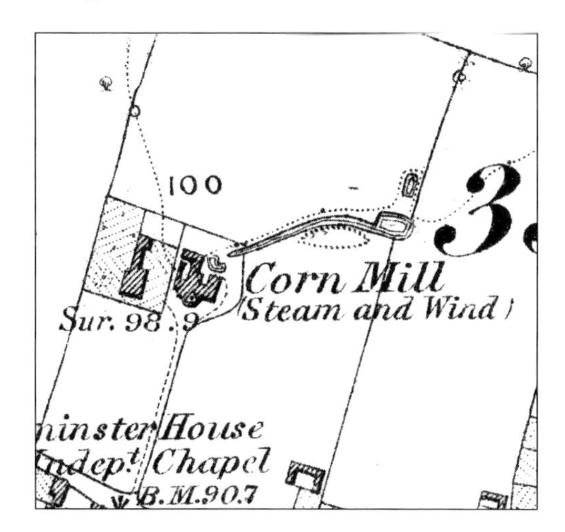

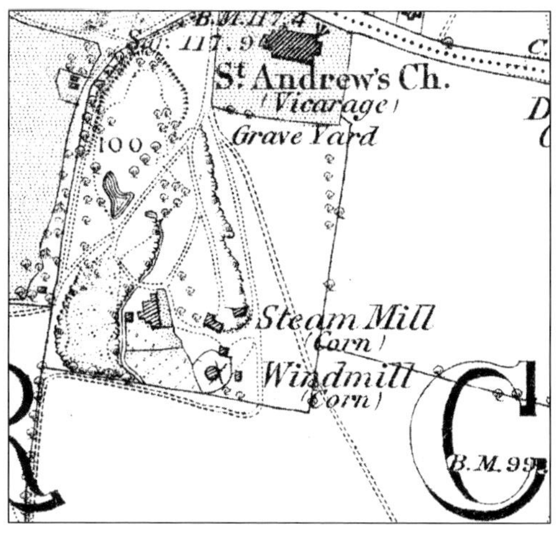

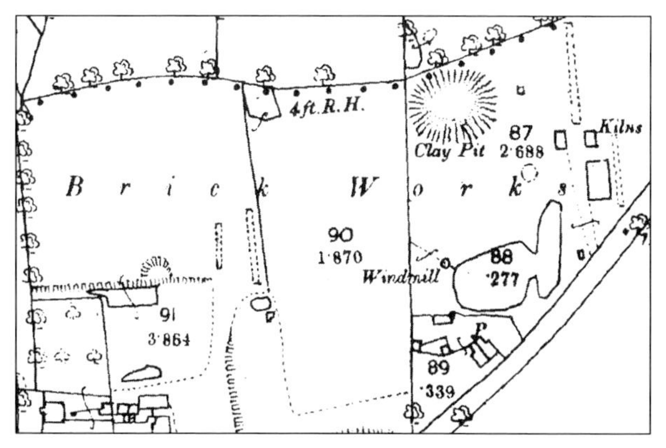

Figure 2.10.1. (above left) Upminster Smock Mill, Essex: 1:10,560 (six-inch) sheet 75 (1872).

Figure 2.10.2. (above right) Post mill at Hornchurch, Essex: 1:10,560 (six-inch) sheet 75 (1872).

Figure 2.10.3. (left) Small pumping mill at Lowestoft, Suffolk: 1:2500 sheet 10.3 (1885).

As well as the probability that an annotation will incorporate a bracketed description of a windmill's function, there is also the possibility that a local name for the mill will appear as well. Indeed, as preparation of the six-inch got underway for Lancashire and Yorkshire in the 1840s, it must have been decided that the naming of mills would become usual practice, since just about every windmill in these counties was accorded a name. So, if a small village or hamlet had just the one windmill, then the name of that settlement is added to the root word *Windmill* to provide a proprietorial and more majestic-looking annotation. Unhelpfully though, adding the name of a settlement often meant that *Windmill* was compressed to *Mill* and in such cases, unless a circle or rectangle with the flattened-cross sign is present, there is room for ambiguity. If a mill was incorporated within another building, or had an irregular ground plan, this problem is exacerbated and since, on a six-inch map, the scaled diameter of a windmill is minuscule, this problem of an abridged annotation becomes very real, and there is great danger of mills being overlooked. The clutch of windmills once surrounding Liverpool, whose early six-inch depictions are imprecise, illustrate this point well, and there are many other windmills whose depictions are equally ill-defined. In such instances it is often only comparison with other scales that can resolve this difficulty but, in the absence of contemporary 1:2500 sheets for Lancashire and Yorkshire, it is paradoxically left to the smaller one-inch mapping of private cartographers such as Bryant and Greenwood to meet this need. More will be said later on this difficulty of the six-inch, when comparing different scales is discussed as a particular stratagem. The use of names led to a problem, of course if, maybe, a large village had more than one windmill, and the Survey had to find a solution to this. One answer was to use names that each mill had acquired locally. These may have been semi-formal and reflected each mill's business origin. Hence we find *Union Mill*, *Subscription Mill* and *Anti-Mill* applied to some of the more urban mills, but there was always the option of using millers' names in the same way that was discussed earlier for the post mills at Brill. The Survey may have had doubts about the wisdom of working with this rather vernacular style as millers were used to moving around between mills and, very wisely, this approach did not find widespread use. The small town of Thorne in South Yorkshire provides one exception; of the six windmills shown on the six-inch in the 1840s, the annotations of five incorporate the names of their millers. The resulting map has rather an antiquarian feel to it that even then the Ordnance Survey may have felt uneasy about this adding of names. Any such unease would have been made worse for the Survey had it known that, within months of being in this area, the windmill it recorded as Bradberry's Mill – the one seen in Plate 14 sitting on the slightest of mounds – was to become known instead as Bellwood's Mill, and would remain so for the next sixty years.

Another solution to the naming of mills is seen at Stainforth, not far from Thorne. Of the two mills here, one was a post mill so, while the other mill was described as *Stainforth Windmill (Corn)*, this post mill luxuriated in the annotation of *Stainforth Post Windmill (Corn)*. For some mills, the name chosen did not even have to be of anywhere that was populated. The name of the field in some cases sufficed; just north of Bubwith in Yorkshire's East Riding we find *Intake Field Mill (Corn)*, which is odd because there was no other windmill in Bubwith from which this mill had to be differentiated. These engraved maps for Lancashire and Yorkshire have been commended already for their attention to fine detail, particularly the showing of mill mounds, with or without the annotation of *Site of Windmill*. (Fig. 2.11.1.) We can only be grateful too that a few mills, which had vanished before the years of survey, still contribute to the map through use of this same annotation even though there was no ground feature at all, such as a mound, to warrant this. Further help for those interested in windmills today – a century and a half further on – was unlikely to have been intentional but, for example, the annotations *Formby New Windmill* and *Formby Old Windmill* used to recognise two mills in this Lancashire village tell us which was the old post mill and which was the newer tower mill without needing to consult any other source. Such is the type of rich evidence that the early six-inch sheets offer for a large number of windmills but, with a new regime for producing larger-scale sheets in place, the custom of providing names for windmills gradually weakened. Looking beyond the often lavish depictions for the windmills of Lancashire and Yorkshire encountered on the six-inch, these maps sometimes show a prescience for the way that mills would eventually be annotated by the 1:2500 and derivative six-inch for all counties south of the Preston-Hull line. Beverley in the East Riding was host to five windmills on the Westwood at the time of the early six-inch survey. Yet, despite all the foibles and richness of depiction seen for other mills in this area, four of the Westwood mills were annotated simply with *Windmill (Corn)*. (See Fig. 2.11.2.)

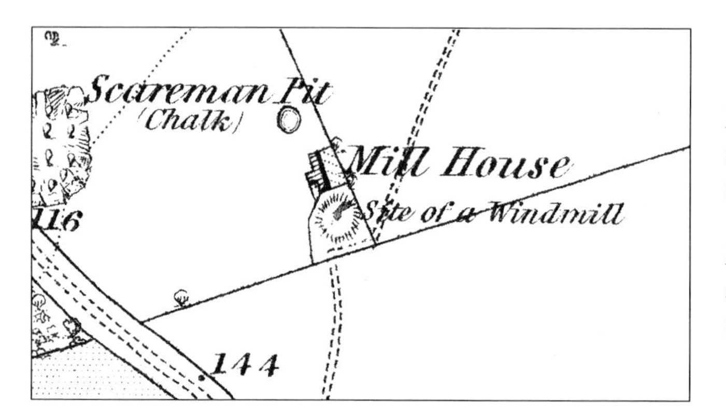

Figure 2.11.1. (left) Mill mound near Beverley, Yorkshire: 1:10,560 (six-inch) sheet 210 (1855).

Figure 2.11.2. (below) Tower mills near Beverley, Yorkshire: 1:10,560 (six-inch) sheet 210 (1855).

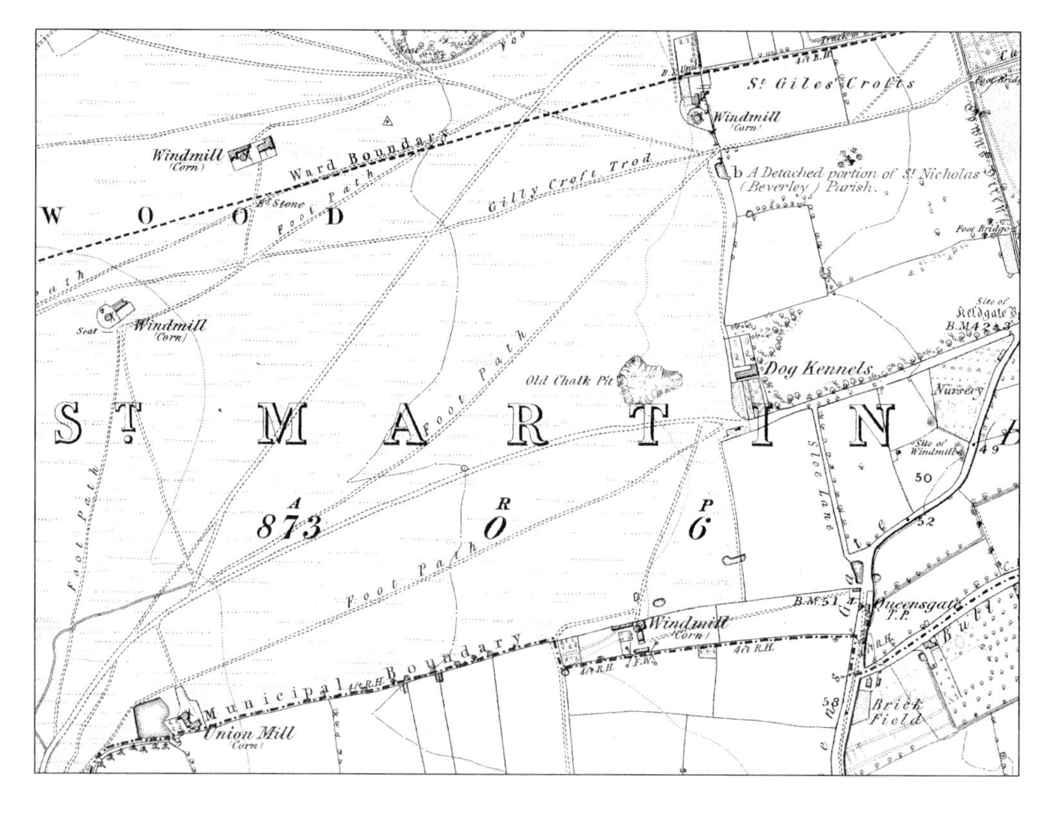

As OS plans for its rolling 1:2500 survey took shape, features that had given life to the Lancashire and Yorkshire six-inch sheets and the very earliest of the 1:2500 sheets, such as those for Durham, became more muted and idiosyncrasies like the flattened-cross sign for windmills were quietly dropped. As far as the depictions of windmills and their supporting annotations were concerned, the effect of this moving ahead was that the recognition of mills below the Preston-Hull line became more streamlined and more consistent. Yet it still appears that, when it came to the large-scale representation of windmills, there were no rigid guidelines to stop the Survey from injecting a selection of enigmatic depictions into it. This clearly happened in the cases of mills at Great Yarmouth, Northam, Wetwang and Wighay. But, after 1860, the crucial factor is that, unlike Lancashire and Yorkshire, whose earlier six-inch editions stood alone, other counties were now to be mapped at two scales. Each scale had to withstand comparison with the other yet be proficient at showing topographic detail commensurate with map space. So, for example, the issue of the first edition 1:2500 sheets for Bedfordshire (1876-82) was immediately followed by issue of the first edition six-inch sheets and *in theory* both sets of maps convey the same information since the six-inch was a direct reduction from the 1:2500.[20] So, while the Lancashire and Yorkshire six-inch sheets, on which the Survey continued to cut its large-scale teeth after its work in Ireland, are very significant for those interested in the Survey's recognition of windmills, the later six-inch sheets for other counties also have much to commend them. This remains the case despite the fact that the presumption that they were just reduced versions of their precursor 1:2500 sheets, and thus the same in every respect, *is* questionable.

Generally speaking, though, the six-inch *does* offer information on windmills that is the equivalent of that found on the 1:2500. That the six-inch sheets are direct reductions of the 1:2500 ones is apparent from the overall impression given by any six-inch sheet issued after 1860. The size of much of the writing and the amount of detail shown mean that legibility for normal use is compromised. A partial solution appears to have been the rewriting of names and annotations for the six-inch so that they become easier to read but, inevitably, the congestion on a 1:2500 urban sheet means that some of the lettering on it disappears altogether. Normally though, if a windmill is provided with a name at the larger scale, then this is repeated at the smaller one; if a mill has no name at the larger scale, then it will not have one at the smaller one; whatever bracketed description is given at the larger scale, it will be repeated at the smaller one unless, occasionally, it is removed altogether.

When space on a six-inch sheet was really at a premium, the worst that could happen was that a windmill annotation on the 1:2500 simply did not make it onto the six-inch sheet. Although a scaled ground plan – effectively just a dot – will be in place and can probably be construed as being the mill if it is known exactly where to look, this is hardly the point. The loss of annotation renders the mill invisible and so the six-inch sheet is immediately of less value than its larger-scale counterpart. It cannot be said that this occurs often, but neither does it happen in just isolated cases. Mills as far apart as Hastings in Sussex and Rochester in Kent; Soham and Wisbech, both in Cambridgeshire; Alford in Lincolnshire and Sneinton in Nottinghamshire; Newcastle upon Tyne and Bowness from northern England, all suffer

[20] It has to be remembered though that this rule breaks down for Lancashire and Yorkshire. Anyone hoping for a direct comparison using the first editions of each scale for, say, Yorkshire is in for a surprise, since the sets of maps here would be a six-inch set dated 1844-53 and a 1:2500 set dated 1887-93 (although the second six-inch then becomes available for a direct comparison).

from this trait.[21] Lesser differences between the two scales brought about by lack of space on the six-inch include cases where a pair of mills are close to each other, but still separately annotated on the 1:2500. However, they are then given a joint annotation on the smaller six-inch – frequently *Windmills (Corn)*. But, like most things, this seems to have been applied using a local level of discretion. At Outwood in Surrey, it was not done for a post mill and adjacent smock mill, but was done for a pair of post mills at Cross-in-Hand in neighbouring Sussex, though some might well argue that there were differences in operating conditions between these two pairs of mills that could account for this. Such differences between the scales due to lack of space are usually minimal and, as with joint annotations, the result of somebody applying common sense. There are, however, times when the differences are not due to restricted map space. At Stanton in Suffolk, of the four mills described as *Windmill (Corn)* on the 1:2500, three are simply described as *Corn Mill* on the six-inch. This was not done for the fourth mill or for any neighbouring windmills. Indeed, in a large sample of 204 first edition six-inch quarter-sheets for Suffolk, only five *Corn Mill* annotations are used in the recognition of windmills.[22] These were the three mills at Stanton and two of the three post mills that stood in close proximity at Swefling – the third mill here, likely to have been more prosperous than the others, is annotated conventionally. This is in sharp contrast to the use of *Windmill(s) Corn* on 226 occasions, with or without a place name. This is clearly an odd finding; the villages of Stanton and Swefling are anything but close to one another and, once again, there would appear to be no easy answers to this. At least when it came to the first revisions at each scale, these mills lose their exclusivity. Hence, other than in something rather more than just a few cases, the potential for the 1:2500 depiction of every windmill to be fully cascaded down to the six-inch was very largely realised – the only remaining issue being the need for the map user to have good eyesight! The two windmills that stood close to each other at Shrewley in Warwickshire well into the twentieth century provide just one example where, as was the case with Belper Mill, their six-inch depictions are faithful copies of the depictions found for them on the 1:2500 (Fig. 2.12). One of these mills is supposedly working, although the circle indicating it is all but lost within a triangulation-station symbol – a common enough occurrence at both large scales – and the other is a non-working open-trestle post mill similar to Mumby Mill, whose four individual supporting brick piers are diminutively shown in ground plan. Despite the limitations of 'reading' the detail on the six-inch sheet compared to the 1:2500 sheet, the smaller scale still seemingly offers a coherent picture of the milling activity happening on this joint site.[23]

---

[21] In an unusual turn of events, Belloc's Mill at Shipley in Sussex is shown by the second edition of the six-inch, but not by the 1:2500. This is nothing to do with a space restriction on the sheet – hardly likely at this scale anyway – but probably stems from the Survey having been alerted to an oversight in failing to record this windmill on the larger scale, and then quickly having to insert it on the smaller scale in the short interim between publication of scales. This mill, it will be remembered, was not built until 1879, long after the initial large-scale survey for Sussex had occurred (1869-75).

[22] These quarter-sheets cover East Suffolk and much of West Suffolk. Unless a particular quarter for a full sheet is noted, all quarters of a sheet are used. Sheets included are 1SE, 2, 3NE, 3SE, 4, 8SE, 9NE, 9SW, 9SE, 10, 16SE, 17-19, 23-28, 29NW, 29NE, 29SW, 34-39, 40NW, 40SW, 44-50, 51NW, 51SW, 54-60, 60ANW, 63-9, 75-7, 78NW, 78NE, 78SW, 82-4, 88NW, 88NE, 89NW, 89NE and 90NW.

[23] Indeed, it can sometimes happen that a windmill is unambiguously recorded by the six-inch with the use of a *Windmill* annotation, but all that is shown by the 1:2500 is an annotation of *Corn Mill* with a less than distinct ground plan – an example of this can be seen at Hibaldstow in Lincolnshire.

After 1860, no doubt partly prompted by the problem of fitting annotative detail into the available space on a six-inch sheet, the fashion for having the name of a windmill as part of an annotation went into decline. But it did not disappear altogether. The large sample of 204 six-inch quarter-sheets for Suffolk, for example, offers 30 cases where a name is part of an annotation, but this is still less than one in seven of all annotations. We can only guess at why the penchant for place-naming a windmill might have still had its adherents. It may have been pure serendipity on the part of the surveyor, or possibly of the draughtsman or engraver wanting to occupy an otherwise featureless part of his map, and improve the 'look' of it. Or it could have been that mills located at crossroads, rather than those tucked away in the landscape, were potential navigational aids, and for that reason given more prominence on the map.[24] In the windmill-rich landscape of East Suffolk, the minority of mills recorded not just by *Windmill (Corn)* but by a name as well includes Saxtead Green Mill, still sited at something of an important road junction. Others of the 30 named mills from our Suffolk sample can be considered to support this proposition reasonably well. There again, there is evidence that named mills were often close to a parish boundary, so maybe a draughtsman was ensuring that the user was fully aware of a windmill's allegiance. As well as place names assuming less importance as the century wore on, other kinds of flamboyant annotation used earlier in the century are seen much less frequently towards the end of it. But there were, however, late examples: *Town Mill (Windmill) (Flour)* is used to describe a large smock mill adjacent to the walls of the prison at Lewes in Sussex. In another overblown example, *Clayworth Mill (Windmill Flour)* can be seen on sheet 7SW from the first edition six-inch of Nottinghamshire, and the annotation *Tower Windmill* is occasionally seen, but its use is not limited to any one part of the country. Much rarer than it was before 1860 – and even then it was rare – was the inclusion of the miller's name into an annotation, as was done *en masse* at Thorne. An example can be seen at Long Crendon in Buckinghamshire where one of the two post mills standing above the village is annotated as *Smart's Windmill (Corn)*. One saving grace for all of these survivals of florid yet rather endearing six-inch annotations is that they never stint on using the word *Windmill*, so that the map user does not have any ambiguity to unravel on that account.

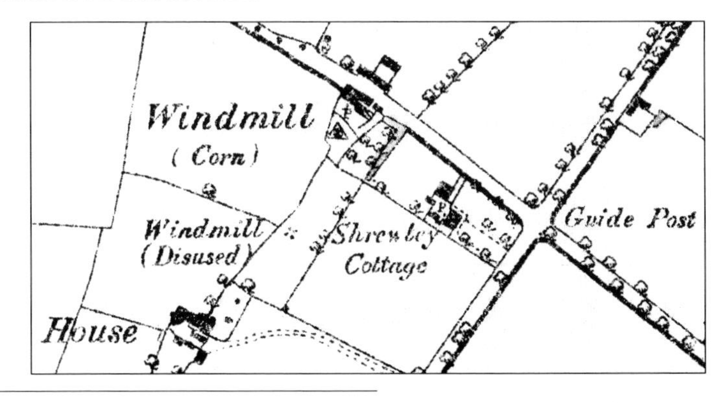

Figure 2.12. Windmills at Shrewley, Warwickshire: 1:10,560 (six-inch) sheet 32NE (1891).

[24] Passing mention was made in the last chapter of the plight of the OS historian often being a bleak one. A key reason for this is that many OS internal records relating to nineteenth-century working practices were lost when the Southampton headquarters was severely bombed in 1940. Among the material destroyed was the set of Object Name Books for all counties other than the first few to be surveyed. These may well have shed light on the uncertainty under discussion. Incidentally, the same air raid caused the writer's mother to be effectively dispossessed of her family home – all in all, not a good time.

By the time the process of re-survey for the 1:2500 reached its last county, one would think that the Survey had had time to smooth out all irregularities on what had become its flagship scale. It is not difficult, however, to criticise the Survey over the first edition 1:2500 of Yorkshire. Indeed much later, the Survey was to have reason itself for doing exactly that. As a cost-cutting measure, rather than carry out a survey specifically intended for sheets to be drawn to the large 1:2500 scale, as had been done elsewhere, it was decided to re-use the geodetic and planimetric data on which the six-inch had been based almost half a century before. This was a regrettable move. Yorkshire consequently became one of the 'replotted counties', and it was to be years before the errors induced at this time worked their way out of the OS large-scale mapping system.[25] At a micro-level, the representation of windmills on this 1:2500 edition is not without its inconsistencies. It is obvious that whatever protocols had evolved for the depiction of windmills in the thirty years before Yorkshire was started in 1887, when it came to depicting mills here at 1:2500, what was equally important was how they had been shown much, much earlier on the six-inch. In the intervening years, the practice of recognising redundant mill mounds, or even sites of one-time mills that had no visible remains, had predictably not really featured elsewhere. Logically, when a post mill was taken down but its roundhouse was left standing to be reused for some disconnected purpose, or a tower mill was reduced in height or converted into a house, then the circular ground plans of each would still appear on a subsequent edition, but equally sensibly there would be no annotation to signify the building's former purpose.[26]

However, one legacy of the initial Yorkshire six-inch that reveals itself much later on the initial 1:2500 is that a few windmills that had disappeared in the interim are cartographically resurrected, and other mills are given annotations of an ornate style that by the 1890s had long been discontinued elsewhere. To illustrate this we need look no further than the City of York itself. On the mid-century six-inch sheet, five windmills are recorded within what could then be construed as the city, three by annotations that include a place name – hence *Holgate Windmill*, *Nun Windmill* and *Heworth Windmill*. Two of these mills survived, later to be shown on the first edition 1:2500 quite normally. The other two windmills recognised on the six-inch had then been annotated as *Lady or Clifton Windmill (Corn)* and *Bootham Stray or Pepper Windmill*, which even by the standards of that edition were expansive. By the time of the 1:2500 edition, the first of these windmills – a post mill – had been photographed *circa* 1870 and pulled down *circa* 1879. Yet on the 1:2500 sheet, fifteen years after its destruction, this post mill is recorded, quite bizarrely by the new customs of the day, with an annotation

---

[25] The same was true for the other 'replotted county' – Lancashire. The treatment of these counties at this scale and time needs careful consideration. While the 1:2500 sheets for Yorkshire undoubtedly contain errors due to the re-use of earlier six-inch data, there is evidence to suggest that Lancashire fared much worse in this respect, possibly owing to its initial survey running a couple of years ahead of that of Yorkshire, so giving time for concerns over accuracy and improvements in procedure to benefit Yorkshire more so than Lancashire.

[26] One of many examples that can be cited to illustrate this is the roundhouse belonging to a one-time post mill at Fiskerton in Nottinghamshire. This is recognised simply by an unannotated open circle on the first edition 1:2500 dated 1885. On its subsequent survival to the second edition of 1900 the circle is then shaded. In fact, the roundhouse has survived to the present-day. Until comparatively recently any enquirer to the mill-house would be shown a small glass-plate photograph (black-backed so as to appear as a positive image) of the windmill as it appeared shortly after ceasing work. This photographic plate, kept just inside the front door to the house, had survived for well over a century when this writer first saw it and happily it has since been published. It is ironic that a windmill taken down before 1875 has survived photographically, yet so many that were still working in the twentieth century have not.

of *Lady or Clifton Windmill (Site of)*. (See Fig. 2.13.) The second of these mills survived (as a full-height tower) until 1898 but its earlier, somewhat florid, annotation did not. This mill appears on the 1:2500 as a distinct circle, but this time the accompanying annotation – *White Cross Mill (Corn)* – is more muted, though still generous for the period. Over the two next editions the depictions for this now truncated tower become more minimalist – the third edition six-inch is content simply to show a later large featureless rectangular building and annotate it as *Mill.* Reflecting the prolonged cartographic demise of Pepper Mill, a fragment of its brickwork survives. Other than pure chance deciding that these two windmills should be marked out for receiving ornate annotations, one characteristic shared by both mills was that the northern parliamentary boundary for York passed literally through each one. This would have been a big coincidence had it not surely been the case that each windmill had been chosen as a marker for that boundary. Under circumstances such as these it may have been thought appropriate to acknowledge the site of the original markers, even though one of them – Lady Mill – had gone.[27]

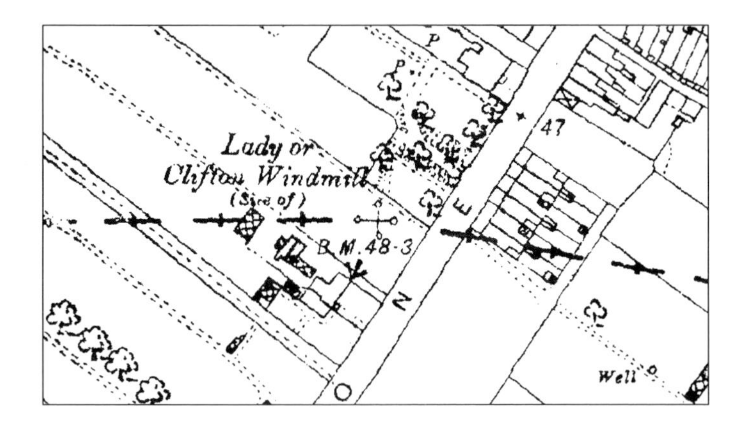

Figure 2.13. Lady (Post) Mill, Yorkshire: 1:2500 sheet 174.2 (1892).

[27] Some expansion of this point may be useful. The story of how the many types of boundary that needed to be, or could be, inserted on a map is a large study in itself. Technically, on the six-inch map this boundary had been a recognition of the division between the City of York liberty and the North Riding (the three ridings of Yorkshire all meeting around the 'hub' of York). In earlier centuries the city had developed two boundaries, the inner boundary of the liberty of York and an outer one within which the city exercised a reduced form of jurisdiction that centred mainly on the rights to pasturage. On the western side of the city this included Ainsty – an entire hundred (or, this being Yorkshire, a wapentake). These boundaries came close to one another at places, but not in the northen sector of the city on whose outer boundary the two windmills under discussion lay. It was this outer line that provided the boundary with the North Riding. In 1832, when the Boundary Commissioners drew up the parliamentary boundary in the wake of that year's Reform Act, they followed the ancient city liberty boundary for the most part, but where development was taking place they adopted the outer 'ridden' boundary – that of the annual mayoral perambulation around a town or city. The seven-storey Pepper (White Cross) Mill surmounted by a cap and sails would have been newly-built in 1832 and, together with Clifton (Lady) Mill, must have been the most obvious of landmarks to adopt as markers given that earlier city archives mention Clifton Mill and a White Stone Cross on the Haxby Road as perambulation markers. Even so, the OS surveyors found part of this boundary around York ill-defined when they arrived between 1846 and 1851. On 1 January 1836 the city had become a municipality, replacing the old liberty boundary with a new municipal boundary. It was not until 1884 that the municipal boundary in the northern sector of the city was extended as far out as the parliamentary boundary, and then it was again extended outwards in 1894. When the parliamentary boundary was eventually extended outwards to this new line as well, in 1918, the boundary that had been defined by Lady Mill and White Cross Mill became superfluous and was dropped from the map. The old post mill that had been demolished in 1879 and the tower mill, by now truncated to a stump, had truly had their day.

Remaining in Yorkshire, at Lings in Hatfield near Doncaster, two windmills had been shown on the early six-inch, one annotated as *Lings Windmill (Corn)* and, in what must have gone against the grain at the time but presumably as no alternative name was appropriate, the other simply by *Windmill.* By the time of the later 1:2500, the first mill has disappeared leaving an obvious mill mound and the other mill has inherited the same annotation that the first mill had had forty years before. At Beverley, windmills that had seemed abnormal in being given no names in 1855 remained equally so when they were later accorded some on the 1:2500, and at Market Weighton use is now made of a miller's name for a mill that had escaped disclosure earlier. Surprisingly this name survived, eventually to find its way onto the third edition six-inch even though by that time the windmill is also annotated as being disused. Even more surprising and decidedly bizarre, though not necessarily the fault of the Survey, is that census and other population statistics reveal no-one of that name (Stubthorn) to have existed at the time of the large-scale re-survey!

This review of the way that windmills were depicted at the two larger scales has routinely strayed into discussion of mapping from the second editions of these scales, so it might be prudent briefly to describe the changes to the 'look' of these later sheets to complement what has already been said in respect of the first edition sheets. Maps of the second edition six-inch were issued in quarter-sheets and have the unequivocal title 'Second Edition' placed prominently and centrally at the top of each sheet together with the date of publication. The date of revision is relegated, along with the original dates of survey, to the bottom margin of each sheet at its left hand side. For sheets that show urban areas, any element of re-survey that might have been prompted by significant alterations to the townscape, and which had taken place alongside revision, is also noted here. These sheets were coloured by the Survey to the extent that water appears in blue but, as with one-inch sheets, where other colour is found this will have been applied by the map-selling agencies or by the map buyer. Plate 24 is an extract from one of the two quarter-sheets that cover the Medway towns. As well as using additional colour to emphasise both the railway line and the principal streets here, the congested nature of any sheet at this derivative scale whose coverage is of an urban area becomes apparent – as indeed had also been the case for first edition sheets. Sheets at the 1:2500 scale too were transformed by changes in style and capacity for recording tiny detail. The early first edition sheets for Durham were not superseded until after 1895 and so had a currency of nearly forty years, but even where first edition sheets for other counties may have had a far shorter life before replacement by a second edition, they could well have been re-zincographed and reprinted many times, especially those of urban areas. Between printings, the size and placing of annotations may have altered or been deleted altogether and new ones inserted. The types of detail shown by first edition sheets had also changed, tending as time went on towards a loss of intricacy in the depiction of gardens, types of tree and a host of other minor features. As with other OS map editions, there is the possibility that a series of successive states, each different, exists for every sheet of the first edition 1:2500, and the depiction of windmills would not have been immune from this process of slight alteration between states. The chances of possible successive states offering radically different information for any individual windmill are reasonably remote, which is just as well given the impossibility of knowing just how many states may have been produced for each of the many thousands of 1:2500 sheets. The degree of variation in windmill depictions that would have occurred for any mill affected by this process is likely to have been no worse than that found for Lound Mill. Here, two slightly different depictions for this tower mill were offered by successive states of a sheet from the fairly short-lived first edition 1:2500 of Suffolk (Plates 25 and 26).

But the style, depiction of detail and colouring of 1:2500 sheets then dramatically altered with the advent of the second edition 1:2500. Sheets of this next edition were much bolder and crisper compared to before. Diagonal line shading replaced any earlier hand-colouring for buildings, so windmills were recorded, if considered to be roofed, by shaded circles, as in the case of Rainton Mill from County Durham (Plate 27). However, just as the number of mills recorded on any one-inch edition was less than the number recorded on a previous edition, so inevitably the numbers of windmills shown on second edition 1:2500 sheets are somewhat less than those seen on the first edition sheets. Nonetheless, a large number of mills did survive – albeit many were in a fairly parlous state – to be recorded for a second time by this safest of scales for ground-truth accuracy. But many windmills *had* been swept away in the interim, particularly those unfortunate enough to be standing in the way of civic growth. As well as schemes of suburban expansion, other building projects were to blame for the loss of mills such as the need for new reservoirs sited on higher ground. Contrasting any first edition 1:2500 sheet with its second edition equivalent will reveal the differences in style between them, as well as perhaps illustrating the loss of a mill. This is an exercise best undertaken in an area where urban development was underway and where the maps are able to show a lot of changing detail. Folkestone in Kent was just one of many, many expanding towns where a comparison of the two editions shows this well and, in also showing the loss of one of the local windmills, appears to deal a blow to the cosy notion that such mills were always allowed to subside into gentle dereliction (Plates 28 and 29).[28]

Turning to another, yet larger scale used by the Ordnance Survey: the town maps issued at a scale of 1:500 provide urban historians with a huge wealth of primary evidence. Limited only by a map coverage embracing what today would be regarded as a town or city centre, windmills are virtually always recorded, as would be expected, and stand out well as circular structures. The scaled and nearly always shaded diameter of a tower mill at 1:500 is much

[28] The extra layer of complexity due to a sheet of the first edition 1:2500 – or indeed a sheet from almost any edition of any scale – evolving maybe through successive states before being made obsolete by a replacement edition should not be underrated. A key to realising that this did happen to a 1:2500 sheet, and that one might not be dealing with an initial-state sheet, is when a 'print code' appears in its bottom left hand corner, as is the case with the sheet from which Plate 28 is taken. Alongside *Surveyed in 1871* the code *Reprint 25/97* appears, which tells us that demand for this sheet was such that another 25 copies had to be made as late as 1897. This does not imply that they necessarily comprised a later, or further, state – printings and revision states are not the same thing – indeed the dramatic difference between Plates 28 and 29 – the result of a revision ironically also dated to 1897 – should lead us to think otherwise. But, quite possibly, for this urban first edition 1:2500 sheet there *would* have been slightly altered states, with later states reflecting a 'lighter touch' as demonstrated by Plates 25 and 26, where the detail for Lound Mill was rearranged rather than revised. Certainly, changes to a townscape on the scale of those happening at Folkestone would have had to wait for a next edition. (Unless it would be misleading to do otherwise, where a later printing of a large-scale sheet is used as the basis of an illustration, the date cited for the sheet is its initial date of publication rather than that of the reprint). Turning to the circumstances of the mill recorded by Plate 28, but not then Plate 29: in such circumstances, one would assume from the evidence of any such pair of sheets that a windmill had been present and perhaps working at the time of the earlier sheet, but had then been demolished by the time of the later sheet. But in this instance the assumption would not be a safe one. This white smock mill at Folkestone (Dawson's Mill) did face being overwhelmed by new housing but, instead of being demolished, it changed hands *circa* 1886 and was taken by its new owner to Bethersden to start a new life, at which stage it was tarred and renamed as 'Black Mill'. Coles Finch, *Watermills and windmills*, 161, 205.

the size of a new penny and, together with an annotation, it thus becomes very noticeable. Yet, situations could occur that prejudiced the acknowledgement of windmills even at this scale. The centre of Newcastle upon Tyne offers an example in the shape of a smock mill that stood in a small street known as Darn Crook. The wooden body of this windmill had been erected on a substantial extension to the old city walls, probably late in the eighteenth century, to grind bark for nearby tanneries. Despite suffering fire damage early the following century, it survived sufficiently to be shown as carrying a set of sails in the early 1860s. How long it then continued to work and by what type of power is not known. But it did fall into severe disrepair and remained precariously perched on its base for decades until eventually, in 1896, it was considered to have become so dangerous that it had to be demolished (Plate 30). What had been the mill base lingered until it too was demolished years later, ironically leaving only the original adjacent section of the city walls still standing. Any thought that the OS surveyor might have had of conferring notifiable windmill status on either the derelict mill, or just its base after 1896, is academic. This windmill was missed not only by the one-inch Revised New Series – for which, in fairness, there was a problem of space – but also by the mid-century initial editions of the 1:2500 and the derived six-inch, together with each of their first revisions.[29] In recording a ground plan for the city wall extension that had been appropriated as a windmill base, and so whose shape did not conform to that of a normal mill base, and by providing an annotation that alluded only to the bark mill that was at work within that base, the 1:500 OS town sheet of 1861 did nothing at all to retrieve the situation. This windmill on Darn Crook was much in evidence almost to the end of the nineteenth century, even if it had lost its sails by this time. Despite the reality on the ground, though, this windmill was very poorly served by the Survey during the second half of the nineteenth century.

To summarise: the depictions of windmills by the Survey at its small one-inch scale and at all of its larger scales worked well for the great majority of English windmills. Each scale in its own distinctive way aimed to notify the map user of the prevalence of these mills with the worst instances of failing to do so occurring, unsurprisingly, at the smallest scale. The one-inch relied on using symbols: initially these were largely unannotated; then they went through a phase of disappearing from view altogether, though the significance and extent of this depended on the county being considered. The depiction of mills using symbols at this scale was reinstated towards the end of the nineteenth century when each symbol was now supplemented by one of a very small number of standard annotations. On the larger scales, it is the use of an annotation that best alerts us to the presence of a windmill. Ground plans of windmills are often nondescript – at best they arouse suspicion but at worst they remain innocuous. Undeniably on a six-inch sheet, a ground plan is insufficient by itself to ensure that a windmill is noted – all plans need annotating, and this rule starts with the very early six-inch sheets of Lancashire and Yorkshire. Luckily, annotations for these two counties are normally provided and they are usually also generous in their detail: otherwise, diligence and a very strong magnifying glass are one's only hope! This generosity continued to be applied to counties surveyed a lot later. Generally across the larger scales and the town mapping, an annotation normally gives a reasonable picture of a windmill's circumstances but exceptions

[29] The one-inch of this area – sheet 105SW of the Old Series, renumbered as sheet 20 of the New Series – was issued during the interval of time when windmills were not recorded.

can be found, such as for Draper's Mills in Kent. Far better to have a depiction such as the one used on the six-inch to describe the tower mill built, presumably just after the Battle of Waterloo, at Wavertree near Liverpool, but which then disappeared before 1890, than the one accorded Draper's Mills on the 1:2500 (Fig. 2.14 – this extract has been taken from a later state of the six-inch sheet and so shows a railway not present on the initial state). In using these two examples, a six-inch depiction has been shown to outclass one taken from a 1:2500 sheet but, when taking an interest in a particular windmill or group of mills, there is no restriction to having to use just one scale. Indeed, it is the ability we have to compare the depictions of any windmill across all scales and all editions that leads to the most insightful account of a mill's history.

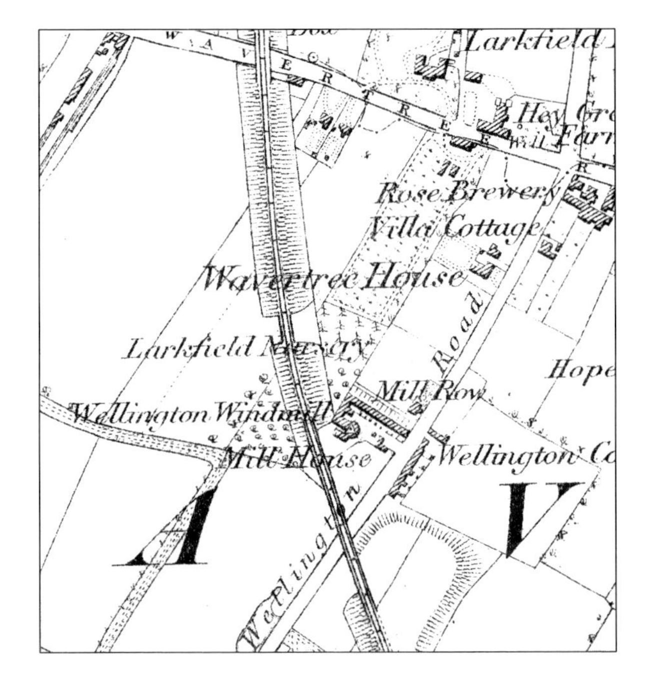

Figure 2.14. Wavertree (Tower) Mill, Lancashire: 1:10,560 (six-inch) sheet 106 (1851/reprint date of 1882).

Comparing depictions of windmills across the different scales

The question of whether or not windmills were coherently depicted by the Survey is clearly central to this study. It is quite clear that the Survey did attach importance to the recognition of windmills, and there is evidence that it sought to distinguish between types of mill, or at least it gave the appearance of wanting to do so. We can appreciate that windmills presented themselves to observers in a number of configurations that went beyond the basic tripartite typology, and there is further evidence that the Survey was at least partially aware of this. So, we should routinely expect to uncover evidence at each of the scales that the Survey made some attempt to do more than simply signify the presence of 'a windmill', whether through nuanced use of symbols or of ground plans. That it did not succeed in doing this evenly or, even more fundamentally, succeed in recognising all windmills, can be appreciated not so much by chancing upon isolated anomalies within one scale, but more by corroborating the evidence of one scale against that of another. Any mismatch encountered as the outcome of cross-scalar inspection reveals a potentially much more corrosive problem. The ambiguities and anomalies that arise from undertaking a comparison exercise of this kind clearly reflect a failing somewhere if the presumed remit for all scales is to depict all windmills. As the available space for detail within a map changes with each scale then, on the occasion of any anomaly, suspicion must naturally fall on the smaller of the scales being compared, and thus by extension the use of symbols for dealing with much of the detail at the one-inch scale.

A simple example of comparing the recognition and depiction of windmills on the one-inch and the 1:2500 quickly reveals the usefulness of this easy exercise. Consider the pair of post mills at Brill: both are recognised by the first edition 1:2500 and by the second edition too. The first edition sheets showing them – the mills were only yards apart but are shown on separate sheets – have survey dates of 1876-80. The second edition sheets were revised in 1898 and published in 1899. Usefully, something of the difference in condition between these mills and the potential durability of each can be inferred from the depictions that the second edition, in particular, accords them.[30] Importantly, the survey dates of the one-inch New Series and of the Revised New Series (1893) both predate the second edition 1:2500. Yet both one-inch editions, each in its own way, are content merely to show Nixey's Mill (Figs. 2.15.1, 2.15.2).[31] So, irrespective of any assumptions made about the transfer of data between the scales or their derivative or non-derivative status, on this occasion the one-inch sheets are clearly at fault, as is potentially the associated system of map symbols developed for them. Disagreement between the one-inch and the 1:2500 as on this occasion therefore suggests to us that care over the use of symbols is needed, though the possibility of a larger-scale sheet being at fault, rather than a smaller-scale one, can never be entirely discarded.

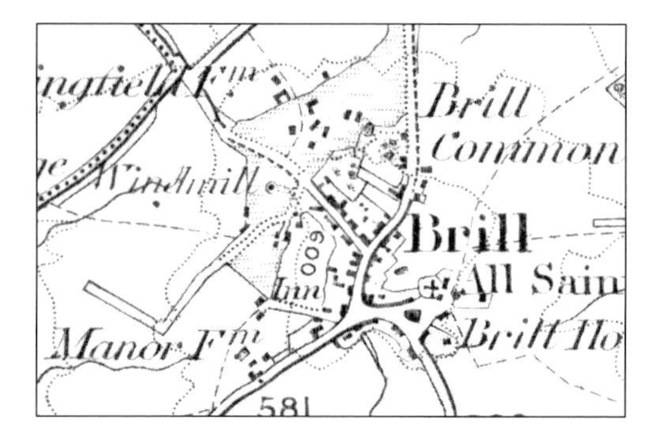

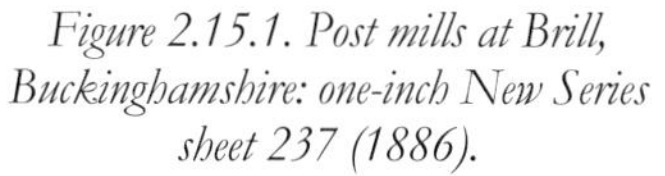
Figure 2.15.1. Post mills at Brill, Buckinghamshire: one-inch New Series sheet 237 (1886).

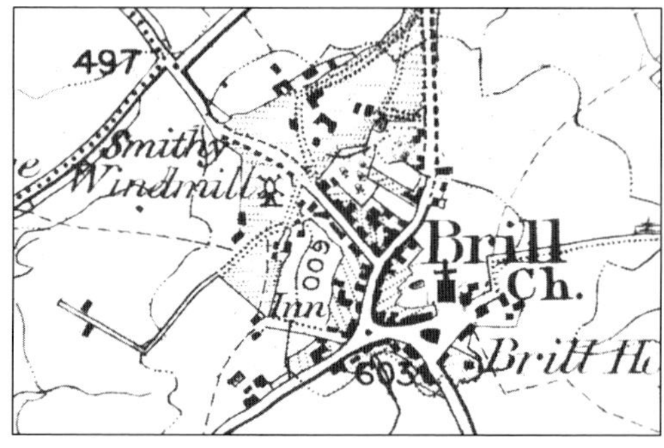

Figure 2.15.2. Post mills at Brill, Buckinghamshire: one-inch Revised New Series sheet 237 (1896).

If this idea of comparing the recognition of windmills across different scales is accepted as a sensible stratagem for assessing just how good the Survey was at recording windmills in the landscape, then, as suggested earlier, our ability to do this on those occasions when a one-inch map does not necessarily predate a contradictory larger-scale map will depend on

[30] The annotations accorded each mill by both editions is simply *Windmill*. The ground plans of both mills on the first edition 1:2500 are the same but, on the second edition 1:2500, Nixey's Mill is recognised by a shaded solid circle while Parson's Mill is recognised by a much less distinct unshaded, or open, pecked circle.

[31] The dates when each windmill disappeared are not crucial to this analysis since the contradiction highlighted by the different survey dates of the two scales needs no extra verification. In fact, Nixey's Mill survives as a preserved mill. Parson's Mill is known from safe evidence to have been pulled down in 1906.

one overriding factor. The rigour with which a one-inch sheet can be judged is dependent directly on being able to determine fairly accurately when individual windmills disappeared from the landscape. This may well turn out to be quite a daunting task even for those areas covered by one of the better county windmill books, but several kinds of primary material exist for providing the information being sought. Coupled with the complication of what actually is meant by a 'mill disappearance', the need to establish the date when a mill ceased to be a notifiable structure is such a significant part of being able to measure the accuracy of the representation of windmills by the Ordnance Survey, that detailed consideration of it will be made in due course.[32]

Taking a rather more ambitious example than the two post mills at Brill, the small town of Retford in Nottinghamshire is known to have had five active windmills in 1860.[33] Survey of the northern part of this county for the one-inch Old Series had taken place in 1837 and resulted in an orthodox two-inch scale manuscript drawing in the post-1836 regular-shape style.[34] This subsequently formed the basis of sheet 82NE, which carries a publication date of June 1840. The occasion of the 1837 survey was not, however, the first time that the OS surveyors had been to Retford. Located as it is on sheet 82NE within an inch of the border with sheet 83 – one of the eight Lincolnshire Old Series sheets nominally published in 1824 – Retford had come under the purview of Ordnance Surveyors' Drawing number 290 dated 1820.[35] On this earlier occasion three post mill symbols had been used to signify windmills at West Retford, South Retford and Spital Hill, only the last of which is known with any certainty – through a mill auction notice of 1850 – to have, indeed, been a post mill at that time. The later survey drawing shows the same three windmills together with a fourth, the Subscription Mill.[36] However, when sheet 82NE was published, not only is West Retford Mill not shown at all,[37] but Spital Hill Mill is mistakenly accorded a tower mill symbol and, in a reversal of fortunes, a post mill symbol is used for South Retford Mill earlier described in an 1827 mortgage deed as a tower mill. By the time of Margary state 5 (of 19) for this sheet dated *circa* 1852 and the arrival at Retford of the railway, this windmill had been erased from the Old Series, wrongly so since it was destined to become the town's last remaining

32 Later discussion of the secondary windmill literature will acknowledge the contemporary selection of county mill books to be the first source to approach for mill information. These are sometimes able to indicate the exact date being sought for a particular mill's demise but, other than the work of Farries already mentioned, most if not all fail to be comprehensive in either not covering all the windmills in their area or the collective nature of their decline, or they are are simply unable to provide well-referenced details for the precise date of any particular mill's demise. This is actually very understandable, as later discussion centred on Farries' study of Essex windmills will make clear.

33 B.J. Biggs, *The lost windmills of Retford,* Retford: Eaton Hall College, monograph 5, 1978, *passim.*

34 For further discussion on the procedure of sheet preparation at this time, see Harry Margary, *The Old Series Ordnance Survey maps of England and Wales,* volume VII, *North-central England,* Lympne Castle: Harry Margary, 1989, with *Introductory Essay* by J.B. Harley and R.R. Oliver, and *Cartobibliography* by Richard Oliver, pp viii-xxi.

35 Margary, *The Old Series Ordnance Survey maps of England and Wales,* V, pp x ff.

36 Documentary evidence for the construction of this tower mill survives in the form of a conveyance dated 1816 which refers to the mill as 'lately erected', thus discrediting the 1820 survey which does not show this windmill at all. Moreover, the facility for recording a tower mill by a tower mill symbol was introduced around 1820 but was, it seems, used sparingly to start with. One suspects that the 1820 survey, essentially made with Old Series sheet 83 in mind, should have noted Subscription Mill in Retford as being a tower mill given that on sheet 83 one of the new tower mill symbols is used for a mill at North Wheatley less than five miles from Retford, and also within the purview of OSD 290.

37 This mill is known to have survived until 1880 when it was dismantled.

windmill in spite of abutting onto the railway: it survived until 1937.

But all of this is symptomatic of the problems of the one-inch scale at this time. Of the five windmills in Retford in 1860 – and a sixth one built in 1866 – only two are recorded by later states of the Old Series. The family of New Series editions initially performed even less well, remaining inconclusive at best. The New Series sheet, although published in 1890, two years after the watershed for reintroducing the windmill symbol at the one-inch, shows no windmills. The Revised New Series sheet of 1897 recognises neither of the two tower mills known to have still been at work at that point, despite ample space being available in which to have done so. Switching to the other end of the scalar spectrum, the sheets of the 1:500 town mapping and of the 1:2500, all surveyed in 1884, depict the four tower mills then in working order, providing recognisably circular ground plans and *Windmill* annotations for all four at both scales. One looks in vain, however, at the derived six-inch sheet for matching information. At this scale, circular ground plans are reasonably discernible in the south-east quarter of the town, but only the annotated Storcroft Mill is noticeable for being a windmill (Plate 31). While the poor performance of the New Series probably stems from a derivative dependency on the six-inch, rather than on an absolute (*ie* the 1:2500) scale, the reason for the Revised New Series, as a supposedly non-derivative map, failing to show windmills on this occasion is harder to fathom. This brief analysis for the windmills of Retford highlights once again that presuming all scales to be of equal merit is dangerous: this time the six-inch sheet of Retford is evidently of less value than the 1:2500 sheets and, in likely consequence, the New Series has also been put into jeopardy. For convenience, the treatment of Retford's windmills by the Survey in the nineteenth-century is summarised in Table 2.1.

Name of Mill	Type	Built	Gone by	OSD (1820) and/or drawing (1837)	Old Series	New Series	Large Scales
West Retford	Post [1]	*pre* 1798	1880	✓	×	×	×
Spital Hill	Post	*pre* 1795	*circa* 1875	✓	✓	×	×
Subscription	Tower	*pre* 1816	*circa* 1886	✓	✓	×	✓
Thrumpton	Tower	1822-3	*circa* 1900	×	×	×	✓
South Retford	Tower [2]	*circa* 1827	1937	✓	✓[3]	×	✓
Storcroft	Tower	1866	1904	×	×	×	✓

Table 2.1. The nineteenth-century OS mapping of the windmills of Retford.

1. Widely assumed as such, but no documentary evidence cited.
2. Replacing an earlier post mill, which was probably removed elsewhere.
3. Erased from later states.

The confusion that can accompany a mapmaker's depiction of windmills – buildings that can be substantial yet are often vulnerable compared to other buildings, and that can also possess a variability of form and purpose – is manifest in the treatment here of the mills of Retford. The suspicion has to be that at the small one-inch scale, the pattern of inadequacy shown here is repeated elsewhere, despite the quite reasonable expectations that a map user might have of solid structures like windmills being sensibly represented. We can extend this line of thinking to argue that when contrasting a smaller-scale sheet that relies on the use of symbols with a larger-scale sheet that relies on ground plans, the suspicion for any errors found must lie with the one-inch sheet and its application of symbols. Map symbols are less 'of the map' than ground plans; they are more obviously a mental construct of some kind and because of that 'being at a remove' from the map they are more likely, for reasons discussed at the end of the last chapter, to be at fault in any cross-scalar comparison. Yet it was not only the one-inch that was faulty at Retford. The six-inch too was judged as remiss when compared to the 1:2500 and the 1:500. So, any confidence in the six inch that might have been gained through briefly considering the windmills of Belper and Shrewley, where in each case the detail of the 1:2500 was shown to have been replicated at the six-inch, may have to be reassessed. Another case study of a town or small area, where the mill depictions can be examined a little more closely, may help to decide whether or not the findings of Retford are likely to be widely endorsed by other settlements, urban or otherwise.

Of all the windmills that had once stood on Windmill Hills in Gateshead, the ones that survived long enough to be recorded by the first edition, or even the second edition, of the 1:2500 are of interest. The first edition records five mills (Fig. 2.16.1 shows three of them). Each was depicted conventionally and quite unambiguously by either a free-standing and circular ground plan, or one that is circular but assimilated into other buildings where the circularity of the plan nonetheless remains fairly conspicuous. Crucially, each is brought to the attention of the map user by an annotation of *Old Windmill* except for one where a clear annotation of *Gibbon's Mill* should be sufficient for the sensible user. So the ground plans in themselves are not needing to attract the user's attention. The annotations are presumed to be not incorrect since Gibbon's Mill is known to have been working until it burnt down dramatically *circa* 1858, since which time it has been cited as having been the last working mill of the area. A sixth windmill had stood on the mound annotated by the map as *Mill Hill* but had disappeared shortly before 1856 to be replaced by a monument to the first mayor of Gateshead; and this remained in place until fairly recently. The position by 1894 – the time of revision for the second edition – had radically changed in that three of the five mills had completely disappeared, and their sites obliterated by new housing development and an industrial school. The mill at Bensham Bank, the southernmost one of the group, remained. It kept its map annotation from the first edition even though it had been derelict for some time and the annotation was then repeated on the six-inch. The other mill to survive on the second edition was a large brick-built mill – the others had all been of stone construction – but by 1894 it had been converted into tenements for housing the poor and was therefore actively fulfilling a non-milling purpose at the time of revision. Under such circumstances, and particularly as an urban mill, any decision to record this windmill if using map symbols as the recording apparatus was bound to mean that it would disappear from the small-scale map record. At the large 1:2500 scale, however, the attribute of a map whereby it behaves as a palimpsest becomes very evident (Fig. 2.16.2). This one-time windmill structure survives cartographically, quite sensibly as a fully-circular ground plan, but one that is unannotated. This partial cartographic evanescence, understandably, then becomes more pronounced at the derivative six-inch scale. In any analysis of the cascading down of information between the large scales, the loss of map visibility for this mill can be viewed as a forfeiture that has arisen from a diminution of scale or, in its conversion to become tenements, from a shift to non-milling usage. This area of Windmill Hills shows well the styles of windmill depiction that can be expected at each of the larger scales where the six-inch representation is clearly a very much degraded version of the 1:2500. On the first edition of the six-inch dated 1862, it can be seen that the mill at Freemans Terrace is inexplicably no longer recognised as it had been at 1:2500 (it survived until *circa* 1879) and also that in the five years that it took for the six-inch to appear, Hood's Buildings Mill appears to have had something of a revival in no longer appearing to be *'Old'* (Fig. 2.16.3). Even if one were to give the Survey the benefit of the doubt by reckoning that it might have carried out a programme of field revision in that span of five years before publishing the six-inch sheets – which is very unlikely – it still managed to offer contradictory information at the two larger scales. To finish the story of these five surviving windmills on Windmill Hills *circa* 1860, Bensham Mill had held on to be recognised as the lone windmill in 1894 at 1:2500 as related above. It did not appear on the one-inch Revised New Series.

For the majority of counties, and especially for those south of the Preston-Hull line, this stratagem of comparing the two large scales improves the knowledge of windmills to be had, rather from just having the six-inch sheets to hand, but it adds little if anything to what can be achieved simply by using the 1:2500 at the outset – although the amount of mapping to be scrutinised does increase by a factor of sixteen! However, not for the first time, we are reminded that the early six-inch maps for Lancashire and Yorkshire, with the evidence they contain for windmills, differ from the later six-inch editions of elsewhere, which adhered to a newly-developed style. When, as can happen anywhere, a windmill's annotation does not actually incorporate the root word 'windmill', this can be more than sufficient to raise doubts in the mind of anybody trying to identify these mills on a map. This difficulty is exacerbated, it seems, on the first edition 1:2500 for Yorkshire, which has been seen not to acquiesce on every occasion to the late nineteenth-century coherent style of annotating windmills at this scale – but instead, of being influenced by the much earlier six-inch sheets. Accordingly, some depictions found for windmills on these sheets, which were among the last to be issued for the first edition 1:2500, went totally against the grain in reaching new levels of flamboyancy. These could easily lead to the word 'windmill' being excluded. For industrial windmills in the Holderness Road and Dansom Lane areas of Hull, these included such gems as *Craven Street Mill (Oil & Cake)*, *Oil Cake Mill*, *Anti Mill (Corn)*, *Subscription Mill (Corn)* and *Tower Mill (Oil)*, where obviously none of these mills is overtly declared as windmill structures, yet all annotations relate to circular ground plans. It is only through comparison with a much earlier six-inch sheet that all these structures are confirmed as having been windmills and, if they were not still fully-functioning mills with their sails by 1890, at least the mill towers with all their innards were probably still then intact, but now being powered by an alternative source. As later map editions came along these descriptions change, and the evidence of these buildings as having once been windmills becomes ever more diluted. The *Tower Mill (Oil)*, tellingly as a sign of changing times, becomes *Tower Mill (Brasso)* on the second edition 1:2500, at the same time losing its clear circular ground plan. Then, for the third edition, a lot more anonymously, the annotation changes to *Kingston Works (Metal Polish)* and the ground plan becomes a large featureless oblong, one of hundreds on the map. But at least for this industrial area of Hull, the first edition 1:2500 is offering some evidence of its former windmill population.

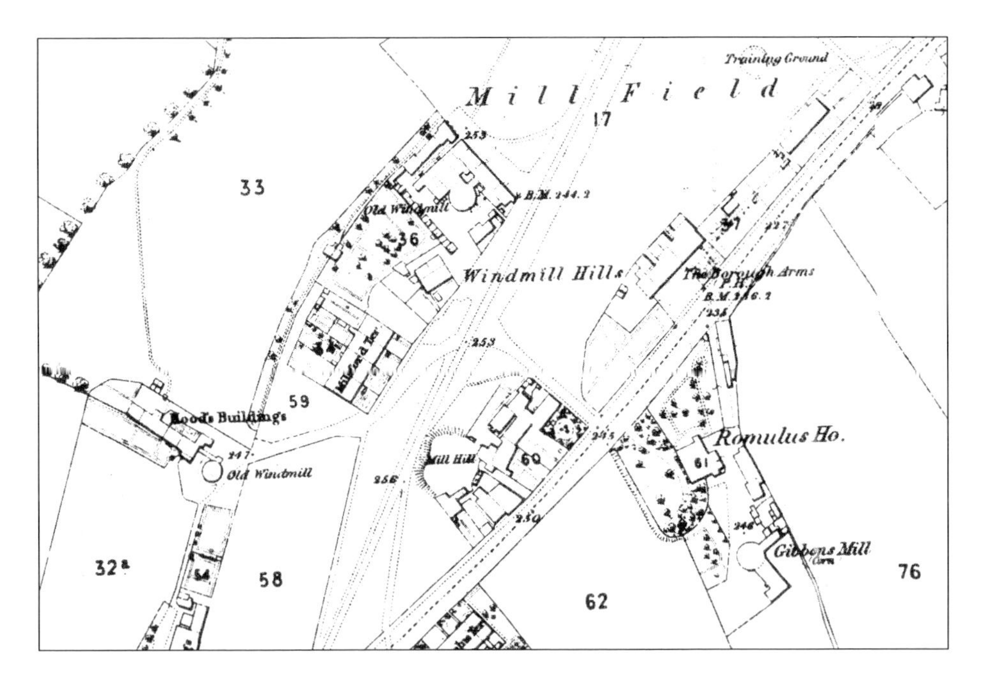

Figure 2.16.1. Tower mills on Windmill Hills, Gateshead, County Durham: 1:2500 sheet 2.16 (1858).

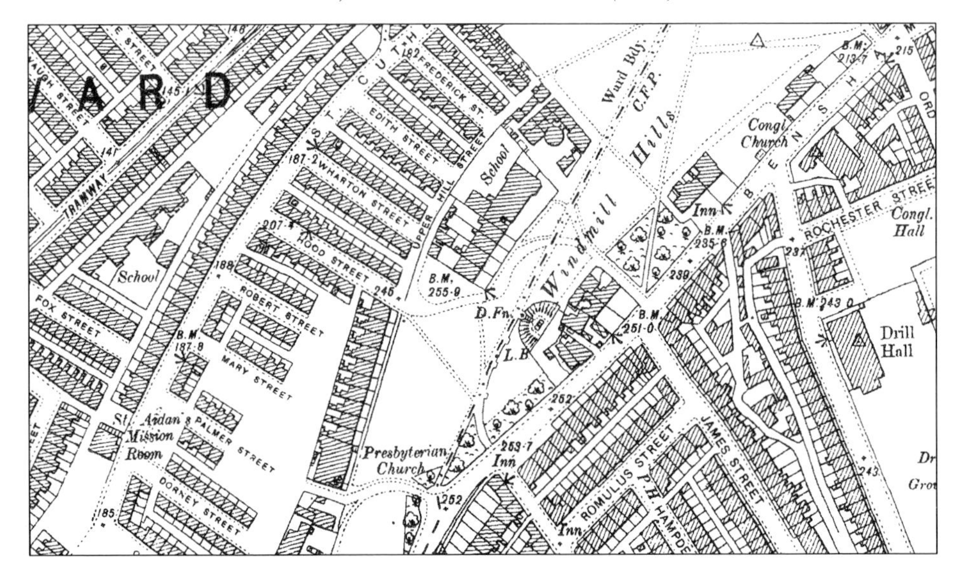

Figure 2.16.2. Tower mills on Windmill Hills, Gateshead, County Durham: 1:2500 sheet 2.16 (1894).

At Redcar, also in Yorkshire, two full-height tower mills were still in place at the time of the first edition 1:2500. The more easterly of the two, Redcar Mill, had previously been equipped with six sails and its cap was still intact at the time of survey, although the mill was unlikely to have been working at this point. A circular ground plan is discernible at 1:2500 but it has no annotation of any kind. The other windmill, Coatham Mill, had quite probably lost its sails and cap, had the curb of its tower castellated and been altered to non-milling use by the time of survey. A photograph taken of this mill tower before 1904 shows these alterations to have occurred by then, so providing a demonstration of just how quickly and easily a one-time tower mill could lose its visual essence as a windmill. The OS surveyor clearly must have succumbed to this sentiment as the only indication of this other full-height tower is a triangulation station symbol. This meagre recognition for the two mills in Redcar does not improve at either the six-inch or one-inch scales, where nothing at all is on offer for the closing decades of the nineteenth century. To cap it all, the Survey chose not to map Redcar at 1:500.

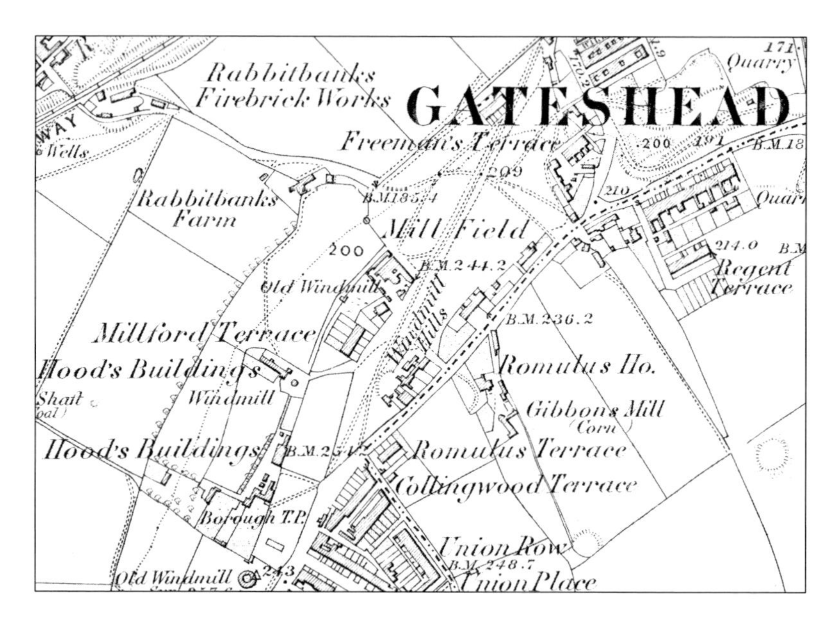

Figure 2.16.3. Tower mills on Windmill Hills, Gateshead, County Durham: 1:10,560 (six-inch) sheet 2 (1862).

To return, though, to our suggested convention of comparing the 1:2500 with the one-inch; the stratagem of cross-scalar examination, seen in two separate studies using Brill and Retford, proved to be most productive for revealing errors at the one-inch. Putting to one side sheets of the Old Series and of the first of the family of New Series editions as *possibly* flawed due to incompleteness in their recognition of windmills, the one-inch Revised New Series, produced as it was at a time of instability in the windmill population and supposedly a field-revised rather than a derivative edition, is acutely sensitive to being judged on how it depicted windmills. So, looking at this edition alongside an appropriate edition of the 1:2500 constitutes the strictest of tests for the one-inch, and for rigorously assessing the use of symbols by the Survey for windmill depiction towards the end of the nineteenth century. The study of Essex windmills by Farries has been mentioned, and it is suggested once more that this is without parallel in the secondary windmill literature. The degree of corroborative evidence for the survival patterns of each Essex windmill offered by Farries is sufficient to adjudicate in those cases of a mismatch between scales where, due to a difference in survey dates for the small and the large scales, the availability of evidence to determine the precise date for the demise of a windmill becomes necessary. The application of this stratagem – or more simply, comparison exercise – to the windmills of Essex (see Chapter Five) is thereby made comparatively easy, and can lead us from a path of looking at modes of depiction for mills to one of assessing maps for their completeness of windmill recognition. It is beyond the scope of this book to contemplate extending this comparison exercise across the whole country but, mindful of the categories of evidence signposted by Farries as being the best for deciding the date of a windmill's demise, their exploitation by others to achieve further county assessments for the Revised New Series will be eagerly awaited.

A large part of this chapter has been devoted to an understanding of how annotations were applied to ground plans at the larger scales; some closing remarks for that narrative may be helpful. It would seem that the collective justification for applying annotations to explicit, and often not so explicit, ground plans of windmills on early editions of the six-inch and 1:2500 is not a terribly easy one to characterise. Unless a particularly florid and apparently arbitrary choice of annotation is something we can merrily condone and put down to serendipity when it occurs, we have to look at the context in which each was provided to try and understand what possible purpose may have lain behind it. While the description of a mill as *Corn Mill (Steam & Wind)* as opposed to *Wind & Steam Mill (Corn)* can be seen as somewhat amusing, or even viewed with gratitude for contravening some all-embracing protocol, there were, fortunately, for a good majority of windmills, broader conventions for description that came to be applied later that are fairly easily discernible. Other factors are quite understandable too: the differences between the representation of isolated rural windmills and the more concentrated urban windmills owe something at least to the dramatic changes that were sweeping across the country towards the end of the nineteenth century. Mills working in rural Lincolnshire or Suffolk, for example, well away from the gaze of all but a small indigenous population, could routinely be annotated as *Windmill (Corn)* in the knowledge that their collective circumstances, though in decline, were not going to alter appreciably any time soon. But elsewhere things were changing. In 1870 two-thirds of all Britons still lived in the countryside or small towns. By 1914 that fraction had dropped to a quarter. In the large towns of England the developing industrial base was encouraging this change and tower mills that now found themselves robbed of their wind in the new manufacturing areas were either razed to the ground or stripped of their sails and adapted to house the new industries. In an urban landscape of tall chimneys and big industrial factories, these one-time windmills no longer stood out visually or culturally. Accordingly, their map annotations no longer needed to echo their past incarnations and they became lost to view as windmills.

Remaining with the city of Kingston upon Hull, the proliferation of new industries in the area of Stoneferry – adjacent to the River Hull – over the lifetime of three editions of the 1:2500 is very apparent. The first edition reveals clusters of small discrete buildings that includes an oblong ground plan with the distinctive circular shape of a maybe active or maybe defunct windmill embedded within it, annotated only as *Whiting Mill.* Thus, the windmill is already rendered anonymous. By the time we reach the third edition 1:2500, the ground plans in this area have become disproportionately large and densely-shaded leading one to suspect that the earlier practice of faithfully recording each and every type of building has been rationalised in favour of a 'cleaner look' to the map. This total lack of subtlety means that much less of the ground complexity is revealed; there is certainly no longer any indication of an old windmill here. To completely unravel the industrial story of Stoneferry and countless other areas like it, or even to understand on the largest of scales which buildings went with which annotations, needs a thorough examination not only of all the available mapping but also of other types of documentary source. This would give us a brilliant insight into the waning contribution of windmills to the urban industrial culture, but it would be one that looked very unlike the slow-changing agrarian culture of rural areas. If it seems that the last few pages have paid undue emphasis to the northern part of the country, one needs to remember that Lancashire and Yorkshire as well as the four counties further north were where the Survey first experimented with its larger scales. Consequently, this is where the foundations were laid for the consistency in windmill depictions that were later used to good effect for counties further south. But laying those foundations was not a smooth process; anomalies did occur. Later chapters, with their different remits, try to go some way to redressing the geographical balance and, hopefully, mollify those readers expecting a more countrywide account of how windmills were mapped by the Survey.

Before moving on to look at the one-inch scale in more depth in the next two chapters, one question that some readers may mentally have voiced to themselves will be whether or not there is a discernible link in the way that windmills were treated by the Survey, and the way that watermills were treated by it. It is a reasonable question. Both types of landscape feature were of equal merit in the work they did. In the range of applications to which each can be put, the watermill, if anything, ranks higher in the scale of usefulness, and in terms of the cultural identity of each, we can presume there to be parity between them. Nonetheless, there are real differences between them in the way they have been mapped by the Ordnance Survey. At the 1:2500 and six-inch scales, sensible annotations, applied to differently-sized and differently-shaped ground plans strung out along observable watercourses, achieve the same level of effectiveness in recognising watermills as we have seen realised for windmills. The huge difference, though, lies with the one-inch scale. The simple fact is that watermills were hardly ever acknowledged on OS one-inch sheets from the Old Series onwards. Two reasons can be offered for this. The first is complex and relates to the prehistory of the map symbol. Predating the windmill in the British Isles by centuries, the watermill had found widespread use as a facility in the community throughout medieval and early modern times, and its cartographic recognition was certainly equal to that of the windmill up to the end of the eighteenth century. But this acknowledgement of watermills by the one-inch and even smaller scales did not last much beyond 1825, and only then by private mapmakers such as Bryant and Greenwood. By the time the foundations of the Ordnance Survey were being laid at the end of the eighteenth century, a small number of variants of windmill pictogram had developed on European maps where these images were now 'standing for' windmills. This was a shift from an earlier position where intricate individual pictorial representations for each windmill had been used to reflect the reality on the ground. Unlike the windmill that offers distinctive visual parameters, the watermill would have appeared mundane and building-like and unsuited to the shift from pictorialism to increasing iconicity that was to mark out the development of the standardised map symbol. So, the fate of the watermill, cartographically speaking, on the small symbol-laden map scales, was sealed.[38] The second

[38] The development of the depiction of windmills – and also, at one time, of watermills – using highly iconic map symbols derives from before 1800, but this is a subject that still awaits definitive study.

reason for the poor showing of watermills on the one-inch relates to the idea that landmark status was a key ingredient for recognition by the Old Series and its successor one-inch series. Here, obviously, windmills scored heavily while watermills, often tucked away in the folds of deep valleys, did not.[39] Thus, a study of how windmills were mapped by the Survey can withstand separation from the parallel story of how (or if) watermills found themselves being mapped over the course of the nineteenth century and beyond.

[39] There were, of course, exceptions to this maxim. New Series sheet 273 shows the area around Faversham in Kent and has annotations of *P.Mills* or *Mills (Powder)*; at the larger scales these mills are explicitly annotated as *Mills (Gunpowder)* or described by name, such as *King's Mill (Gunpowder)*. These mills were all water-driven but their recognition would have been given added impetus by the military importance of their product. Sheet 273 later has improved annotations of *Powder Mills* as well as one *Mill* annotation accorded to a non-industrial watermill at Ospringe, but the same sheet misses two windmills in this village – Water Lane Mill survived until *circa* 1915 and the windmill which stood by the workhouse lasted until *circa* 1910. Also seen on the first edition 1:2500 near Faversham are many examples of *Pug Mill* in and around the clay pits and brickyards. These were small mills that ground lime for mortar, often horse-driven and too small to make it onto any sheet other than at the 1:2500 scale. The secondary mill literature for watermills makes rather better use of OS maps than does the windmill literature, possibly because the topographical setting of a watermill incorporates alterations to the natural passage of a watercourse so 'connecting' the watermill to the landscape much more than the setting of any windmill can claim to do. A good example of where extracts from OS large-scale sheets have been used in conjunction with schematic maps is Tony Bonson, *Driven by the Dane: nine centuries of waterpower in South Cheshire and North Staffordshire*, Congleton: The Wind and Water Mill Group, 2003.

3

The Old Series

The use of a pictogram or symbol on a one-inch map for denoting the ground presence of a windmill will now be studied in depth by taking a longer look at the Old Series in the earlier part of the nineteenth century. The Ordnance Survey emerged at the end of the eighteenth century as a state agency whose principal endeavour quickly became the provision of maps for the country at the one-inch scale. One strong incentive for doing this was the perceived possibility of an invasion by the French, but this need was only reinforcing an earlier notion that a set of homogenous maps for the kingdom should become the responsibility of a dedicated government organisation. So, for the subsequent half-century, production of Old Series one-inch sheets became the mainstay activity for the Survey in England and Wales, although only from about the beginning of 1826 with the abandonment of publication in 'parts' or instalments was all of this thought of as a national map and promoted as such.[1] In making maps that were contemporaries of the large-scale county maps published by private cartographers such as Bryant and Greenwood, the Survey was operating in a competitive market. This led to the wry and oft-cited situation of surveyors from the Survey as well as those of its rivals all working in the same county (Lincolnshire) at much the same time.

The northward sweep of the survey and the subsequent publication of Old Series sheets had begun, as discussed already, with the vulnerable southern coastline. After an extended period of time, measured in years, and after much in the way of revised mapmaking policy designed to enhance the utility of each sheet, the programme of survey and publication reached something of a marker for those interested in OS activities – a line drawn between Preston in Lancashire and Kingston upon Hull in the East Riding of Yorkshire. Passing beyond this line, the work eventually reached the northernmost counties and, as is shown in Figure 1.1 (see page 6), by *circa* 1865 the programme of work was nearing its completion.[2] Since the Old Series was to keep its currency for many years as the most up-to-date one-inch map available for much of the country – indeed, the currency of some sheets was to

---

[1] Margary, *The Old Series Ordnance Survey maps of England and Wales*, V, p.xxvii.

[2] Although the dates shown in Figure 1.1 corroborate the general idea of a northerly progress for the Old Series, there are caveats to this. As the inadequacies of windmill depiction on successive states of some of the sheets that make up the so-called Lincolnshire Map will be discussed, one such qualification can be recorded. Pressure from landed interests reminiscent of the system of patronage for early county mapping led to the survey of Lincolnshire being fast-tracked, which meant that this area – rich in windmills – was mapped ahead of turn. The eight sheets of the resulting county block – The Lincolnshire Map – each carry a publication date of 1824 though they were not issued until the next year.

last nearly to the end of the century – the way that it depicted windmills is obviously of huge importance to any study of windmill representation in the nineteenth century.[3] This need for a detailed consideration of the Old Series becomes even more acute in the light of an earlier suspicion that the revision of sheets to provide later states was flawed due to the woefully inadequate updating of windmill depictions.

At first sight, the depiction of windmills on early sheets of the Old Series is comparable to that of the private mapmakers with their large-scale county maps. On closer examination of these sheets, however, slight changes in form are found for the symbol resembling a post mill, and this is matched somewhat later for the symbol resembling a tower mill. The variety found for each of the two types of symbols, even on the same sheet, has been established by Rodney Fry – illustrator for the Margary volumes – who assembled the many variants of mill symbol that appear on the sheets covering Lincolnshire, Rutland and East Anglia (Fig. 3.1).[4] This variation on these and other sheets can be construed in one of two ways. Either it can be thought of as something of a stylish attempt to discriminate between different mill features or types of mill that existed or, more likely, it can simply be seen as whimsical by today's thinking – a reflection of the vagaries in the use of map symbols that were thought acceptable, and maybe even pleasing at the time, but not necessarily to be seen as useful for implying any diversity in the mapped feature.

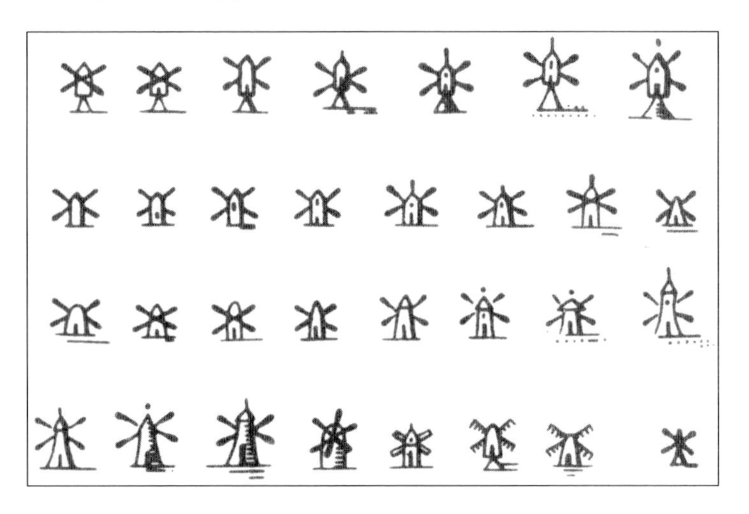

Figure 3.1. Windmill symbols used on Old Series sheets for Lincolnshire, Rutland and East Anglia (after Fry).

[3] As the last of the Old Series sheets were being replaced by the first few sheets of the New Series, a dilemma, typical of the difficulties that the Survey could get itself into, became apparent and served to keep the Old Series in commission longer than anticipated. By this time, late in the nineteenth century, the Survey was issuing sheets at larger scales and was now reliant on the survey information from those scales to prepare for the smaller-scale New Series, as will be discussed. The rolling programme of 1:2500 county surveys had not, however, dealt with Lancashire and Yorkshire by the time this information was needed for the New Series. In consequence, having decided to be consistent in its provision of information and not wanting to delay publication, the Survey left blank the areas of those two counties below the Preston-Hull line on its New Series sheets. Not until the Revised New Series was published in the 1890s, was one-inch mapping provided for these areas that could supersede the Old Series sheets dating from the 1840s, or even earlier.

[4] Margary, *The Old Series Ordnance Survey maps of England and Wales,* V, p.xix (drawings by Rodney Fry).

This is an important distinction since, either the need to discriminate between windmill features was seen by the Survey as being important at the time of initiating its Old Series, in which case a correlation should exist between the different forms of windmill symbol and what is known of the incidence on the ground of various kinds of mill feature. Or, perhaps more significantly, the issuing of these sheets came at the special moment in time when map symbols were changing from being unfettered pictograms – in this case possibly conveying some sense of a mill's individuality – to becoming increasingly uniform when, collectively, they became a mechanism for promoting and saying something else. As order and a sense of standardisation become much more noticeable in the make-up of OS maps – authorised descriptions of the landscape – then, by association, this may be seen as echoing a changing social order within the real landscape. Any contribution made by the windmill symbol to such a tacit process could hardly be expected to be a major one, and in any case this symbol was later to vanish from new one-inch publications for many years. But the sheer variety of symbols used for both types of mill on the Old Series before the 1840s does suggest this to be a time before views on rationalising its output had hardened, and before the Survey felt it should adopt a consistency of form for dealing not only with the windmill, but maybe other landscape features as well. Discrimination between different mill features had not occurred noticeably on previous mapping, and the obvious visual or functional differences between mills could not conceivably account for the large variety of symbols used on the Old Series. We are thus left with only one sensible inference to draw regarding the use of the windmill symbol on early sheets. However apparently quixotic it might seem, a culture of whimsical representation would appear to be the most credible explanation for the variation found in the earlier years of the century.[5]

On the other hand, just wanting to distinguish between post mills and tower mills was a far more sensible proposition, especially for the 1820s when an intensive programme of building tower mills was under way. Also, making this distinction would be following in the tradition of some of the earlier one-inch county mapping. So it was that the symbol visually akin to a tower mill was promptly introduced onto the Old Series to complement the very different symbol that resembled a post mill. It is a safe assumption that this newer symbol was designed to cater for the very different, modern type of windmill then starting to appear in large numbers. Like the post mill symbol, this symbol for the tower mill closely mimicked the visual reality of the feature it was meant to depict. But confusingly, as just seen, the two types of symbol came in a variety of forms with only minute differences to be discerned between them.[6] Possibly the real consequence from having this overwhelming variety of

---

[5] Moreover, extremely good eyesight is needed to take advantage of the subtlety of depiction engaged in by the Old Series. Any idea that this variation had any significance might be considered unlikely on this point alone. But this idea does need to be treated with caution since, as already seen, other OS maps, particularly those engraved at the six-inch scale for Lancashire and Yorkshire and largely of the 1840s, provide their ground-plan information for windmills at true scale and, therefore, a truly minuscule size.

[6] Vital to any discussion on the variety of forms that the two types of mill symbol could take, one possibility in particular has to be carefully considered. On sheets of the Old Series published before the introduction of the tower mill symbol – most likely done as one of Colby's revision measures in preparation for the Lincolnshire Map since they do not feature, for example, on sheet 7 of 1822 that covers an area where there would have been some of the new type of mill at this time – the Survey may have been distinguishing between post and tower mills using distinctive forms of the post mill symbol. But from what is known of the types of windmill in, say, Essex around this time, and also from the lack of two discernibly broad categories of symbol on the early sheets of Essex, this idea can strongly be refuted. This leaves us in a position where assessing the earliest Old Series sheets before 1824

symbols available to the engraver, was whether or not individual surveyors, faced with the knowledge that the engravers were capable of distinguishing between the two types of mill, felt the distinction between post and tower mills to be one of such importance that they all felt equally moved to make sure their field notes were detailed in this respect. The potential for the map record in respect of windmills to be severely distorted on account of this factor alone is one that may be hard to qualify.

The Old Series before 1830: the Lincolnshire Map

To address this concern further, and to examine these issues around the representation of windmills on the Old Series, some sample sheets can be chosen for closer examination. The standard reference sources for the Old Series – the individual volumes of Margary – have already been acclaimed. Their introductory essays provide, together with Seymour, the most accessible information on this map series.[7] For many years the maps of the Old Series were published as single sheets approximately 35 inches by 23 inches in size, and they had little useful information in the margins other than a scale bar and a nominal date of publication. The appearance of these sheets, their style of production and the type of information they provided were all to go through changes during the long life of the Series. This began with the very early sheets having only standards of accuracy needed to satisfy the exigencies of wartime. These standards were later deemed to be unacceptable by Major Colby when he assumed command of the Survey in 1820 and took responsibility for the then main Old Series priority – the Lincolnshire Map. The improved standards he set, and the measures he implemented to enforce them, though delaying publication of the sheets for Lincolnshire, set the tone for future OS map production until, again, a new set of imperatives emerged which led to the more finely-engraved and densely informative sheets of the 1830s.[8] Thus the initial states of sheets covering the south of England look very different to those for northern areas first published much later. Distinguishing between successive states of each individually numbered sheet is, however, a different matter. Revision carried out to provide a new state typically centred around railway construction but not much else, and the general initial style of a sheet was left untouched for succeeding states. Compounding an unhelpful situation – one where the value attaching to a less than carefully updated sheet was reduced – was the opaque way in which the later states were dated. The more years that had elapsed since the issue of a sheet's initial state – and so the corresponding lack of trust that could be placed in it – was disguised by retention of the original date of publication in its prominent position throughout the lifetime of that sheet. Some features within the marginalia of a map were subject to change a few times down the years, but the only hint of a revised date for a new state came when an unassuming note of an electrotyping date was provided and that, obviously, was only possible after the introduction of this process.[9]

---

for their windmill representation can only be based on slight evidence. For the first sheets, the four for Kent that made up the so-called Mudge Map, we are presented with a number of near-identical post mill symbols and little in the way of verification from other contemporary sources for them. Even other, commercial, one-inch county mapping tends to be scant between 1790 and 1815.

[7] Margary, *The Old Series Ordnance Survey maps of England and Wales,* I to VIII; Seymour, *A history of the Ordnance Survey.*

[8] Margary, *The Old Series Ordnance Survey maps of England and Wales,* V, pp xxviii-xxix.

[9] Here of course the Margary volumes come into their own. Together with defining the different states thus far

The choice of sheets from the Old Series needed for looking at all these issues does not have to be contentious, but simply has to accommodate each of the phases of production that the Old Series passed through. The earliest of the sheets – those issued up to the point of introduction of the tower mill symbol – all have the same simple characteristics of mill depiction, but from that point onwards the complications begin. So the eight sheets of the Lincolnshire Map that derive from the early 1820s provide a convenient group of sheets from which to make our first selection. In fact there is very good reason for sampling this group of sheets since, according to Harley, there are particular differences between these sheets which reflect the change in standards demanded by Colby on taking command of the Survey. While the southernmost four sheets had been surveyed before 1818, the manuscript OSDs (Ordnance Surveyors' Drawings) for the four northernmost sheets date from after 1820 and reflect the new measures for stringency in OS practice that Colby put in place, and which he continued to augment. In the words of Harley:

> 'The production of these two groups of sheets, the southern and the northern, as monitored in their associated manuscript drawings, marks a fundamental transition in the technical history of the Old Series mapping... [those for the southern sheets] belong technically to that phase of Old Series mapping which corresponded with the events of the Napoleonic war down to 1815. Under wartime conditions, the work was often undertaken in haste. Moreover, the simple methods of survey employed – including pacing and sketching – were open to different interpretation by individual surveyors... [those for the northern sheets] reflect the introduction of new procedures and new techniques within the topographical mapping. The comparison of any of the northern drawings with those surveyed to the south shows a marked improvement.'[10]

Given then that a divide exists between the northern and southern halves of the eight sheets making up the Lincolnshire Map, the middle four sheets (69, 70, 83 and 84) provide a good sample for exploring how the Survey was dealing with windmill depiction *circa* 1820. The legacy from this time of change, when the working practices from the very early days of the Survey were reformed, was that the improvements put in place were to have consequences for how effectively windmills were depicted up to and beyond the middle of the century. When the Lincolnshire Map was at last published, the accuracy of its detail depended not just on the initial surveys which had resulted in the OSDs – which could be more than a decade out of date – but also, at Colby's instigation, on a subsequent programme of revision that lasted right up to publication.[11] The emphasis placed on this revision may well have been different for separate areas, and perhaps markedly so between the northern and the southern areas bearing in mind the contrast seen by Harley in the manuscript mapping. But

---

uncovered for each sheet and providing a cartobibliographic account that details *some* of the revisions carried out for each state, facsimile copies of early states are given for each sheet. The cartobibliographic accounts make no claim to be definitive in the sense that not every change to a sheet is listed, and rarely are windmills mentioned – sheet 65 is one exception – but arguably enough detail is provided, either for the topographical content or for the sheet marginalia, for them to serve as diagnostic tools for the identification of any state that one might possess. By their own admission, expanding this knowledge base and discovering new states is acknowledged as possible.

[10] Margary, *The Old Series Ordnance Survey maps of England and Wales,* V, pp x ff.

[11] The last-minute attention that went into these sheets can also be inferred from the fact that, as already noted, though having a publication date of 1824, their sale to the public was delayed until the following year.

what is certain is that, like the OSDs, and despite a presumed desire for the published maps to be uniform as well as accurate, the published northern sheets are different in style from their southern counterparts. One marker of these differences is the depiction of windmills.

Looking at the sample of early-state sheets: sheets 69 and 70 between them are credited with recognising 298 mills, the majority of which appear on sheet 70. However, the pattern of distribution for these windmills, in reflecting the topography of the landscape, is typical of the pattern of mills on most sheets – it is anything but uniform (Fig. 3.2). More tellingly, all but three of these 298 windmills are unequivocally recognised with post mill symbols. Of the exceptions, one at Crane End is ambiguous. Another, at Gill Bridge, could be a post mill symbol overlain by a track, but only the third at Syderstone in Norfolk, close to the border with Sheet 68, is given a definite tower mill symbol. In contrast, sheets 83 and 84 between them can be credited with recognising 226 mills, with twenty of these accorded a tower mill symbol. Of the remainder, one symbol is ambiguous and the rest are depicted using post mill symbols. During this early period it was not just that the consistency of depiction had not evolved, but that sometimes the clarity of depiction was only poorly achieved. The mill

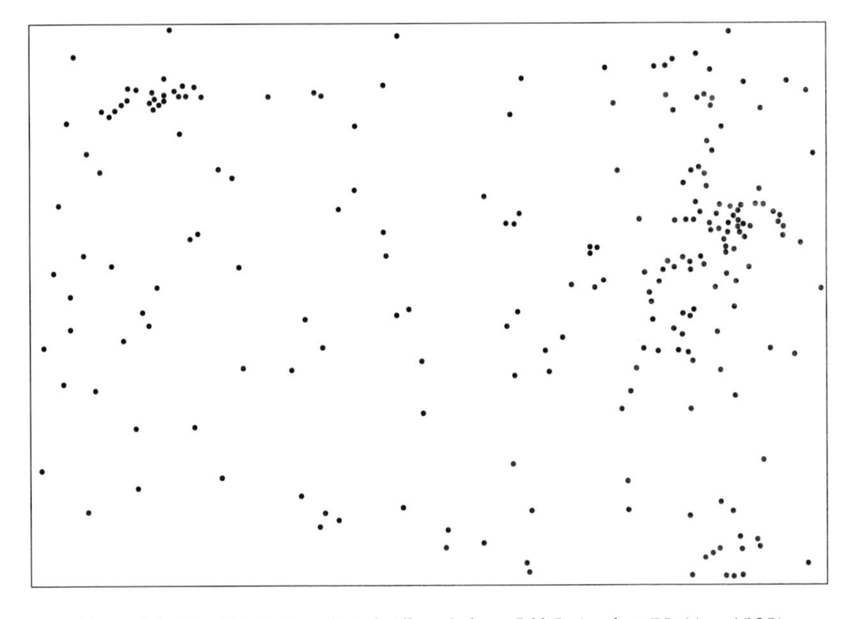

Figure 3.2. The distribution of windmill symbols on Old Series sheet 70 (circa 1828).

counts provided here are of course open to scrutiny, but any potential scrutineer would be advised certainly to use original sheets rather than modern copies – even the best facsimiles are unsatisfactory for this purpose – as well as a powerful magnifying glass. The problem posed to those looking for windmill symbols on these sheets is, in the worst of cases, the combination of a symbol's size, its delineation, and the existence of other map detail that might be obscuring it. For example, a mill symbol found on sheet 84 at Saltfleetby All Saints highlights the potential difficulty with which one can be faced (Fig. 3.3.1). Occasionally, the recording of two windmills that were close to each other – a not uncommon situation – can be made fussily intricate, as illustrated in the case of a couple of mills not from Lincolnshire, but from near Worthing in Sussex (Fig. 3.3.2). Certainly the integrity of mill depictions did vary, as mention of ambiguity in the case of some of the depictions noted above implies. But the early 1820s, when the sheets for Lincolnshire were being prepared, clearly marks a time when the tower mill symbol was making its debut, and when some attempt was being made to differentiate between the two types of windmill, at least on the northern sheets.[12] On the southern sheets, by way of contrast, any attempt at differentiation between types of mill was clearly half-hearted, if indeed it was made at all. We can gauge just how limited this was by having a knowledge of the dates of tower mill construction for the area.[13] The town of Boston, cited already for what its windmills can say about the use of forms of historical evidence, is easily capable of affirming or refuting the idea that post mill symbols and tower mills symbols were applied indifferently on sheets 69 and 70. Of Hanson and Waterfield's fifteen mill sites in Boston, nine are known to have been supporting tower mills in 1820.[14] They are all acknowledged using post mill symbols, suggesting that the appearance of the single tower mill symbol on sheet 69 is an eccentricity and that the southern sheets cannot be thought of as serious in their application of the newer symbol. But as far as the northern sheets are concerned – where a significant proportion of mill sites are provided with tower mill symbols – some analysis needs to be carried out on a sample of windmills to determine if the differentiation between mill types matches the differentiation between the two types of symbol.

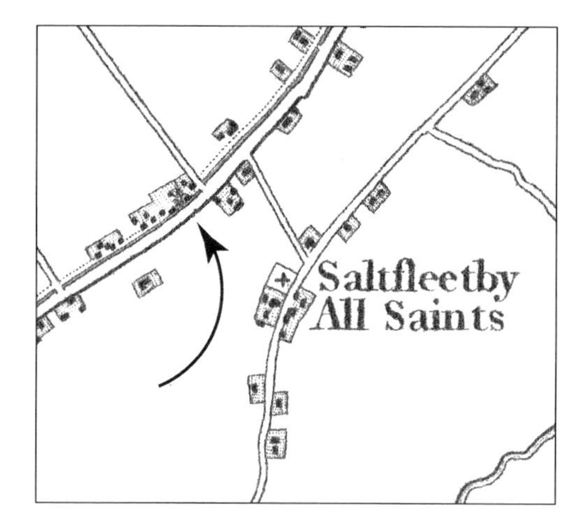

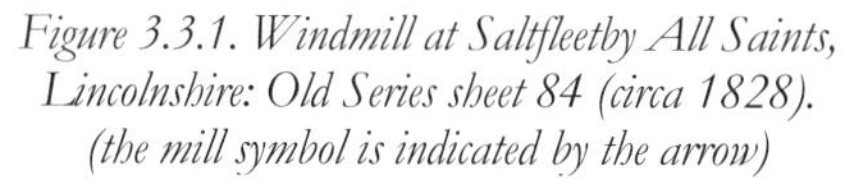
Figure 3.3.1. Windmill at Saltfleetby All Saints, Lincolnshire: Old Series sheet 84 (circa 1828). (the mill symbol is indicated by the arrow)

Figure 3.3.2. Windmills near Worthing, Sussex: Old Series sheet 9 (circa 1858). (one mill symbol nearly eclipses the other)

[12] The use of the tower mill symbol, while generally held in restraint for all northern sheets, is prolifically apparent on sheet 86 in the area of the Holderness Road on the eastern side of the city of Hull.

[13] The secondary windmill literature for Lincolnshire is not extensive. Studies offering a county-wide treatment are essentially limited to four: one delivered (as a paper in two parts) by Rex Wailes to the Newcomen Society that lists and describes only those mills which survived long enough (until 1923) for him to have examined them; later, a good account by Peter Dolman appeared for windmills surviving up to *circa* 1986. A more contemporary survey by David Jager of windmills that have made it into the twenty-first century was published in 2007 and, even more recently, we have Jon Sass, *Windmills of Lincolnshire*, Catrine: Stenlake, 2012. Not many texts of a more local nature have been published but the one cited already by Hanson and Waterfield for the mills of Boston is noteworthy.

[14] Hanson and Waterfield, *Boston windmills, passim.*

Selecting sheet 84 for this type of appraisal provides a population of 49 corn windmills to assess – disregarding the fifteen drainage mills powered by wind also found on this sheet. The dates when these 49 mills were built are not all known, but of the six mill sites that are given tower mill symbols, all are believed to have been supporting tower mills around 1820. Three of these mills – at Bilsby, Bolingbroke and Hagworthingham – survive today as tower mill shells, all strongly suspected of being built pre-1820.[15] A tower mill at Burgh-le-Marsh (Dobson's Mill), which had been built by 1810, and which also survives, was depicted using an ambiguous symbol that stands as an example of how difficult some interpretations can be for the map user. To assess the accuracy with which the remaining 42 sites were mapped all as supporting post mills – some refutation of the mills on these sites as being post mills needs to be sought. To help with this, recourse can be made to Rex Wailes' list of tower mill survivals to 1923 that includes, where known to him, dates for their construction.[16] Care needs to be exercised when using this list as some mills existing in 1923 were replacement mills for earlier post mills sited at the same location, such as Trusthorpe Mill built in 1880 or Alford (Myer's) Mill built in 1827.[17] Others would not necessarily have been replacing earlier mills, but were simply mills built too late to have been recorded by the initial survey for the sheets, or even by any revision for them. Many windmills of course from the list of 42 really were post mills, such as the one at Mumby or, if tower mills, then ones that quite reasonably disappeared in the long gap between being mistakenly recorded on sheet 84 as a post mill, and 1923. These latter two categories of mill obviously do not appear on the 1923 list; so all of the 42 mills that either do not appear on the list, or which do and are accorded a date for their construction later than 1825, are of no further use in this refutation analysis. If all such mills – a total of 29 – are excluded from our residual windmill population of 42, this leaves thirteen to be assessed. These thirteen mills are listed using the names given to them by Wailes, but still included among this number may be some, undated by him, where the construction of the tower mill may have postdated 1825 and been, like Trusthorpe Mill, a replacement for an earlier post mill (Table 3.1). Some additional information is cited that alleviates this difficulty.

So, these thirteen mills will include some that are definitely known to have been tower mills – those surviving long enough to be listed by Wailes, and for which he also felt able to furnish a date for the mill's construction. They are all depicted by post mill symbols on the Old Series and are thus collectively able to refute the idea that the tower mill symbol, on sheet 84 at least, was accurately applied. Those windmills for which Wailes provides no date of construction, and for which no credible evidence has come to light that places their construction before 1825, are, as the example of Theddlethorpe Mill must suggest, also now inadmissable as candidates for this appraisal. Even so, it can be seen that at least

[15] Peter Dolman, *Lincolnshire windmills: a contemporary survey,* Lincolnshire County Council, 1986, *passim.*

[16] Rex Wailes, 'Lincolnshire windmills: part 2 tower mills', *Transactions of the Newcomen Society* 29 (1955), 103-22. A small complication deserves mention here. This paper (together with its counterpart paper dealing with post mills) was subsequently reprinted as Rex Wailes, *Lincolnshire windmills*, Heckington: Friends of Heckington Mill, 1991. The reprint usefully contains corrections to Wailes' papers, particularly in respect of dates of mill construction as they became better known to us in the years between 1955 and 1991. Less meticulously, it is not apparent when a correction to Wailes' original material has been made, leaving us with a less than true impression of what Wailes wrote. The account here stays true to Wailes' original work but notes any later correction where it is appropriate.

[17] That Myer's Mill at Alford replaced a post mill is something of an assumption but, as the OSD indicates a windmill on this site, this had to have been the case, unless the tower mill construction date of 1827 is wrong.

eight tower mills standing by 1820 were each depicted using an inappropriate symbol – a greater number of mills than those recorded by the sheet, quite probably accurately, with the use of the tower mill symbol. This assessment of poor mill representation can only be made worse from realising that, of the windmills wrongly accorded a post mill symbol, a significant proportion will not have survived to 1923, thus escaping consideration in this simple exercise.

Name of Mill	Built (Wailes) ns: date not specified	Ordnance Surveyors Drawing		Shown on Bryant surveyed 1825 – 27	Shown on Greenwood surveyed 1827 – 28
		number	date		
Alford: Hoyle's Mill	1837 [1]	280	1819	✓	✓
Alford: Station Mill	ns [2]	280	1819	✓	✓
Burgh le Marsh	ns [3]	279/80	1819	✓	✓
East Kirkby	1820	279	1818	✓	✓
Grebby	1812	280	1819	✓	✓
Grimoldby	ns	284	1818	✓	×
Halton Holgate	1814	279	1819	✓	✓
Hogsthorpe	*circa* 1803	280	1819	✓	✓
Saltfleet	ns [4]	284	1818	✓	✓
Swaby	*circa* 1815 [5]	284	1818	✓	✓
Theddlethorpe All Saints	ns [6]	284	1818	✓	✓
Toynton All Saints	ns	279	1818	✓	✓
Wainfleet All Saints: Croft	1817 [7]	279	1819	✓	✓

Table 3.1. Selection of tower mills depicted as post mills by Old Series sheet 84 (circa 1828).

1. Subsequently corrected to 1813.

2. A date of 1789 is reliably provided for this tower mill in Gordon Willson (ed), *Recollections of a Lincolnshire miller: Robert Willson of Huttoft*, Louth Naturalists' Antiquarian and Literary Society, 1984.

3. Wailes may be less than reliable for this small town. This windmill is Hanson's Mill for whose site Dolman provides a date of 1842 for a post mill survival. The later tower mill was described as 'newly erected' in 1855. In passing, it can be noted that Wailes ascribes a date of 1844 for Dobson's Mill, the other tower mill in town. This was the mill that attracted the use of an ambiguous symbol in the early 1820s, so Dolman's date of 1813 is certainly much closer to the truth for when this mill – which still stands complete – was first built.

4. This mill is declared by Dolman to be 'one of the oldest in the county' and reputedly dated to *circa* 1770. Without corroboration for this early date, further consideration of this mill is felt to be unsafe.

5. Subsequently corrected to 1812.

6. A date of 1833 is provided by Dolman for this mill.

7. Subsequently corrected to 1814. From looking at the map, this mill appears to fall emphatically within the ambit of the village of Wainfleet All Saints, and Wailes' position on naming this mill as he did is understandable. However, the parish boundary between this village and the neighbouring village of Croft is such that the later description of this mill as Croft Mill is correct. Elsewhere, care is needed when comparing Old Series sheet 84 with the later maps that would have been used by Wailes in the early 1920s. For example, the mill referred to as Huttoft Mill on the Old Series, and the sole windmill in this village later shown on one-inch Third Edition sheet 104, were two different mills whose sites are almost a mile apart. But in other cases, the distances separating two such associated mills can be a lot less.

A pattern of misrepresentation revealed by a sample of just eight windmills is not going to convince everyone, but some further tentative observations can be made. Instances of correct and incorrect depiction occur alongside one another, as at Alford and Bilsby where two of the six or seven tower mill symbols on sheet 84 and also two of the proven 'errors' appear within a mile or so of one another (Plate 32). The other 'errors' are fairly widespread across sheet 84 and they offer no obvious reason for this inadequacy of representation. The position at Alford and Bilsby, however, can be illuminated by returning to the evidence of the manuscript drawing. OSD 280, surveyed in 1819 by H. Stevens, shows four windmills in this area. Three are at Alford, all of which are later recognised on the published sheet 84 by post mill symbols. Another mill is shown, more than a mile from Bilsby (and Alford), with the annotation *Bilsby Mill* (Plate 33). But on the published sheet this mill has gone, to be replaced by a 'new' mill in Bilsby. Both this mill and another 'new' mill at Alford are depicted by tower mill symbols. On the evidence of this small area, the suggestion clearly must be that the tower mill symbol was a favoured tool of the post-OSD revisers and kept for any windmill that was supposedly new. But this idea is refuted by the remaining four or five tower mill symbols on sheet 84, where these mills not only all appear on their respective OSDs but, certainly in one case, by doing so very explicitly as a post mill. The degree of clarity shown for this one mill is rare, however. That a mill can be determined as either a post mill or a tower mill from an OSD depiction is almost always difficult. Even on the better-quality OSDs, the way in which mills were depicted was certainly not done with any kind of differentiation in mind. Given the manuscript nature of an OSD, each mill depiction is obviously an individual iconic rendering in ink; some are intricate, even attractive, while others are scruffily drawn to a diminutive size. Duplication of the same mill on two adjacent OSDs, as happened to the two windmills at Burgh-le-Marsh, demonstrates that even allowing for the degradation of the images through the ageing of these manuscripts, any inference that can be drawn from one depiction can be discredited by the other. It is abundantly obvious that no attempt was made to discriminate between any two or more categories of windmill at the time of survey, and there is no record of any supporting documentation being made available to the engravers other than the pre-publication revision drawings. It is thus hard to understand how these sheets could have been expected to mirror ground-truth accuracy with respect to windmills, if an intention had been declared to represent these mills as something other than a single category of building or structure.

The only explanation left for an incongruous mix of symbols appearing on a published Old Series sheet at this time – a sheet seemingly trying to distinguish between the post mill population of an area, and that of tower mills – is that it is the outcome of the way in which revision was carried out between the OSD survey and publication. It must be the case that a difference in windmill counts between OSDs and published sheets will be as a result of this revision, which Harley informs us was occurring up to a late stage for the northern sheets.[18] The differentiation between the two types of windmill may have belatedly centred on very localised areas, except that the small number of tower mill symbols used are not clustered in any way and defy any attempt to put a pattern to them. Moreover, the fourteen tower mill symbols on sheet 83 provide no obvious arrangement that suggests the influence of either

[18] Margary, *The Old Series Ordnance Survey maps of England and Wales,* V, pp x ff.

the original OSD information, or the revision programme such as it is known to us. The revision drawings that survive are incomplete in their coverage, and this may well suggest that this process was only patchily applied in the first place. In addition, unlike the OSDs that are dated and carry the name of the surveyor, these drawings are fragmentary in the sense that each may only cover a small area of land and be untitled, undated and without a reviser's name. For sheet 84, the revision drawing survivals are minimal. Nevertheless, for the area around Louth, small hand-drawn revisions made on tracing paper do reveal the usefulness of the windmill to the surveyor and reviser alike. The detail of Louth itself is ignored, but five windmills in the neighbourhood are the single factor common to these drawings, each of which carries red-inked alterations in the manner described by Harley.[19] On these revision drawings the windmills of Louth are depicted in two different ways. First, a pictogram is used which has something of the shape of a post mill, but which differs from any published depiction; and second, a device is used where a cross on a stem surmounts a circle with a central dot (Plate 34). Elsewhere on the drawings, a legend shows a circle with a central dot as a way of indicating a triangulation station. One argument is that the pattern of individual windmills and their visibility will have served as a reference framework for the revisers of the Louth area and, for the revision drawings themselves, the pattern will have been used as a means of registration. Given the obvious and, by this stage, accepted use of windmills as triangulation stations, and now the similarity of their notations found on these examples of the revision drawings, it may be conjectured that the Survey will have wanted to highlight in some way those mills used in the geodetic survey, and hence made use of the tower mill symbol. To support this conjecture, Hagworthingham Mill is conspicuously annotated on OSD 281 as a triangulation station and on sheet 84 it is accorded one of the few tower mill symbols used on the sheet. However, remembering that at Alford the use of the tower mill symbol was reserved for late additions to the published windmill population, this thesis is badly compromised. The revision drawing survivals for sheet 83 are somewhat greater in number than for sheet 84, and something further of the process of revision as it directly related to windmills can be ascertained from them. As Harley relates, outline proofs of the prototype Lincolnshire sheets were used as templates for revision in the field.[20] One such proof survives for the area to the south and east of the town of Market Rasen. For the town itself, neither of the two mills eventually shown on the initial published state appears on this proof, but an amendment very crudely drawn in red crayon does acknowledge one of the mills, but in a way that again demonstrates a total lack of any attempt to discriminate between types of windmill. A note for this proof dates it as *circa* 1822, meaning that a later visit to the area, which must have been made to acknowledge the second mill, is in line with our understanding of the intense revision activity insisted upon by Colby before publication of the Lincolnshire Map.

From this limited analysis, provided by not much more than the evidence of one sheet, the suggestion by Harley that the more methodical programme of revision for the northern group of sheets contributed to their being a better topographical map than the southern sheets, is not overly endorsed by this embryonic move to record windmills by type. For the southern sheets, where the programme of revision appears to have been more flexible, any

---

[19] Margary, *The Old Series Ordnance Survey maps of England and Wales,* V, pp xiii ff.

[20] Margary, *The Old Series Ordnance Survey maps of England and Wales,* V, pp xiii ff.

provision for the introduction of the tower mill symbol, if it was considered, was obviously stymied in very large measure.[21] This is quite noticeable despite there having been obvious instances of late revision of sheets 69 and 70. One instance – not involving a windmill, but related – sees the annotation *Pode Hole Steam Mill* inserted close to Spalding at a very late stage without nearby annotation being moved. This mill was not an example of steam being used for the grinding of cereal, and a provincial successor, therefore, to the Albion Mill, but a very early example of steam being used for drainage purposes. In discussing the drainage of the fens, Hills records that the building of a pair of steam mills here was started in May 1824 though parliamentary approval had been given the year before.[22] The first mill was not completed until the following year, so the decision to include what must have been – or was going to become – something of an innovation in the neighbourhood would have come as an extremely late addition to the map.

The practice of recording the large numbers of wind-driven drainage structures by using the same symbol used to record grinding mills may reflect a less than purist approach to the notion of what constituted a 'mill' but, as discussed before, this was a mapping practice that the Survey was content to see extended to the end of the century. Such drainage structures that are depicted on the four sample sheets are shown either by unannotated symbols or, more helpfully, by symbols annotated with *Engine* or *Water Engine*, sometimes with a place name also added.[23] For those drainage mills depicted using unannotated symbols, the role of each can usually be inferred from a location that is adjacent to a water channel and not very close to any settlement, together with the use of one of the less bold symbol variants. This applies equally to the three cases on sheets 83 and 84 where tower mill symbols are used instead of the much more usual post mill symbols, even though the small smock-like structures then in use as drainage or marsh mills were, in reality, visually akin to neither post mills nor tower mills.

These first exercises using samples taken from the Lincolnshire sheets of 1824 provide, in short, a challenge to our thinking over the way in which the representation of windmills was handled, certainly with respect to any differentiation attempted between post mills and tower mills. This particular issue, and the map traits it led to, can possibly be put down to a lack of parity for mills in a hurried revision programme driven by a need for more cardinal features of the Lincolnshire Map to be correct on publication. But whatever the reasons, if indeed there ever were any, and whatever meaning the pattern of windmill depictions may once have held, it is obscure as far as today's audience is concerned. Moreover, it cannot really have been any more obvious to the map-buying public of the 1820s. For the historian of the Old Series, and those who seek to use the evidence of this long-lived and important map series, the enigmatic depictions for windmills that it used were hardly going to alter in the subsequent half-century.

The commendable rigour with which the Margary volumes seek to identify and define

---

[21] Margary, *The Old Series Ordnance Survey maps of England and Wales,* V, pp x ff.

[22] Richard L. Hills, *Machines, mills & uncountable costly necessities,* Norwich: Goose & Son, 1967, 57 ff. Hills makes a claim for Hatfield Chase as probably being the site of the first land-drainage steam engine used in England (1813). His Appendix 2 lists those steam engines erected in the fens from the date of the earliest (1817) onwards. Those at Pode Hole rank as fifth and sixth on this list, thus affirming their novelty value in 1825.

[23] Use of these expressions, which at this time had yet to change in connotation to imply the presence of steam power, can be found on not just OS maps but on other contemporary mapping as well.

successive states for each sheet of the Old Series can be gauged from examining the extent to which map details have been recorded for each sheet. Precise dating can be problematic for reasons already seen and very often the evidence of an embossed date of printing at the top of a sheet, or maybe of a watermark, is extremely useful in augmenting the evidence of any electrotyping dates. But it is largely the mapped detail itself that is of primary use in this process, and of that none more so than linking the developing railway network as it was mapped with known construction dates for each new railway line. For any sheet, comparing a copy of an early state with later states, it is readily apparent that little was done by way of additions or deletions to the mapped detail over the lifetime of that sheet. Such changes as were made centred around the need to accommodate new railway information, or were due to the coincidence of proximity to new railway lines. The few changes made to the windmill depictions on the four sample sheets from Lincolnshire support this assertion absolutely. The changes in windmill depiction on successive states of sheets 69, 70, 83 and 84 include simple deletions of a windmill symbol: these occur at Lincoln, Louth, Newark and South Leverton, where seemingly nothing took the place of the mills on the ground, with a further three drainage mills at Skirtbridge and Gibbet Hills also suffering the same fate.[24] The single replacement of a symbol by a new building occurs with the new workhouse at Spalding. Then there are the more complex cases. Of these, South Collingham was to suffer the worst indignity: initially, one of the two post mill symbols south of this village had been coupled with an annotation *South Collingham Mill* while further north, beyond the adjoining village of North Collingham, another mill, in a seemingly logical arrangement, was annotated as *North Collingham Mill.* With the arrival of the Nottingham and Lincoln Branch Railway in 1846, however, priorities in the vicinity were obviously deemed to have changed when one of the annotated symbols was erased, though not completely, and its annotation requisitioned to denote instead an architectural feature of the new railway. This was a windmill that truly suffered a cartographic fall from grace (Figs. 3.4.1, 3.4.2). At Spalding, a windmill symbol was untidily moved to create room for the railway and a symbol depicting a drainage mill at Great Hale Fen, instead of being removed, was simply left to be semi-obliterated by a new railway line. Very noticeably, *all* of these sites were less than half a mile, and in most cases a lot less, from one of the new railway lines spreading out across the country by the 1840s, or from schemes of building work associated with them such as stations and new streets.

As well as a few windmill symbol deletions, which would not in any way have reflected the true incidence of windmill removal from the landscape over a period of more than half a century, there were also mill symbol additions to the map. Such cases, fewer even than the number of deletions, were again linked to a revision programme centred on the railways. A tower mill at North Leverton that had been built in 1813 is belatedly recorded, though by a post mill symbol, on Margary state 7 (of 17) of sheet 83 dated *circa* 1853. On the same state of sheet 83, and even closer to a railway first recorded by this state, a 'new' windmill can be seen at Market Rasen next to the Manchester, Sheffield and Lincolnshire Railway that had reached the town in 1848. This windmill, which may or may not have been built after 1824, was subsequently to have a history of disappearing and then reappearing on the one-inch scale. It was also, incidentally, very close to another of the small number of mills later fitted

[24] South Leverton Mill was to survive despite this treatment and today remains a tower mill capable of working. A symbol depicting this windmill reappeared much later on the Revised New Series.

with eight sails. But, unlike at Wisbech, and due as much to a lack of space as the fact that the time for fanciful annotations was largely over when the multi-sail arrangement of this second mill was raised *circa* 1868 (see page 28), its new look seems not to have excited the reviser's annotative flair. Other windmills fared worse: any mill built after 1825 – and there were many – naturally did not appear on the initial state of the Lincolnshire Map. If these mills were built, as the great majority obviously were, well away from the railway reviser's gaze, then they were simply ignored by later states. Even some that were built very close to the new railways were missed. One windmill, later to become known as Pocklington's Mill, was built at Heckington in 1830 within fifty yards of where in 1859 the Boston, Sleaford & Midland Counties Railway built Heckington Station. This mill – still standing in a preserved condition – was resolutely ignored by the Old Series.[25] All of these additions and deletions to the windmill count for the four sample sheets are set out below (Table 3.2).

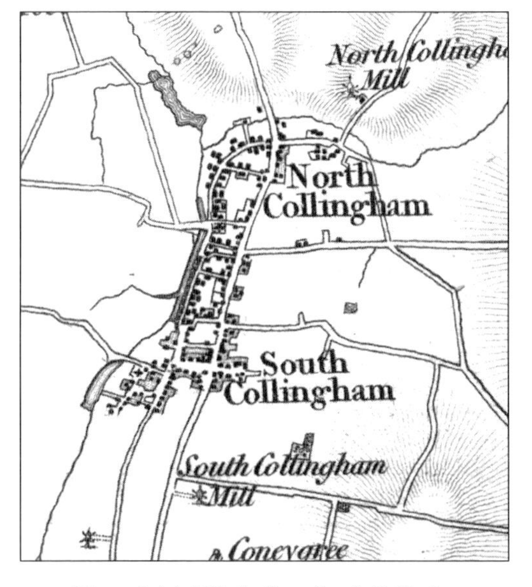

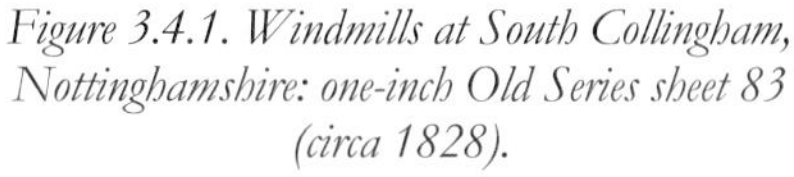

Figure 3.4.1. Windmills at South Collingham, Nottinghamshire: one-inch Old Series sheet 83 (circa 1828).

Figure 3.4.2. Windmill(s) at South Collingham, Nottinghamshire: one-inch Old Series sheet 83 (circa 1863).

All that we can say for certain after analysing the Lincolnshire Map is that it embraced a new facility for discriminating between different types of windmill presence. Irrespective of whether or not this was successfully used in some way that eludes us – though doubt must remain for any sort of discrimination even being feasible in light of the regime for gathering field data – it would have been disingenuous to have used the facility of two broad families of mill symbol to do anything other than to distinguish between post mills and tower mills. But the evidence that this, indeed, is what was done is not terribly persuasive, and certainly it was not done with any rigour. From the time of preparation for the Lincolnshire Map –

[25] This feature of the Old Series, where the revision of its sheets over the lifetime of each was very selective, has been commented upon by earlier writers. See particularly, Richard Oliver, 'What the bird doesn't necessarily see', in Peter Barber and Christopher Board, *Tales from the Map Room: fact and fiction about maps and their makers,* London: BBC Books, 1993, 32-3.

circa 1820 onwards – a mixture of post and tower mill symbols appears on each one of the newly-issued sheets where windmills were present in the landscape. One might think, with the regularising of procedures introduced by Colby for dealing with the revision of earlier OSDs, that the altogether more accountable system for ensuring the topographic accuracy of later sheets would lead to an improvement in the correlation of type of mill with type of symbol. But, with accuracy in this respect found wanting for the sheets of Lincolnshire, the later gathering of momentum for issuing better and more finely-detailed sheets in the years up to 1840 does not, unhappily, seem to have led to windmills being recorded with greater accuracy.[26]

Name of Mill (D) signifies a drainage mill	Old Series sheet [1]	Railway Line & date of opening		Earliest Margary state to show revision (& approximate date)	
Deletions:					
East Louth	84	EL	1848	8 (of 15)	1849
Lincoln [2]	83	M, S & L	1848	7 (of 17)	1853
Spalding	70	GN	1848	9 (of 18)	1853
South Leverton	83	GN	1850	7 (of 17)	1853
Newark	70	GN	1852	11 (of 18)	1857
Great Hale Fen (D)	70	B, S & MC	1859	12 (of 18)	1859
Skirtbridge: 1 (D)	70	B, S & MC	1859	12 (of 18)	1859
Skirtbridge: 2 (D)	70	B, S & MC	1859	12 (of 18)	1859
Three Gibbet Hills (D)	70	B, S & MC	1859	12 (of 18)	1859
Additions:					
Market Rasen	83	M, S & L	1848	7 (of 17)	1853
North Leverton	83	GN	1850	7 (of 17)	1853
Other Changes:					
South Collingham	83	N & LB	1846	7 (of 17)	1853
Spalding	70	GN	1848	9 (of 18)	1853

Table 3.2. Changes to windmill depictions on Old Series sheets 69, 70, 83 & 84.

Railway lines: EL: East Lincolnshire; M, S & L: Manchester, Sheffield & Lincolnshire; GN: Great Northern; B, S & MC: Boston, Sleaford & Midland Counties; N & LB: Nottingham & Lincoln Branch.

1. For this analysis not every state of the four sheets could sensibly be consulted, but nor did they need to be. Access to original copies in the public domain is reasonable: BLML, for example, holds on its open shelves three states of each sheet. These are selections of early, middle and late states, and are referred to as such. The Margary volumes provide ready access to the railway histories of each sheet. The original sheet states that were consulted are listed below. The dates of each (using Margary) are approximate:

[26] Between OS concern for the worth of its Lincolnshire sheets and, only five years later, the handling of survey notes for sheets covering central England, Harley states that 'the system of revision – and in turn the techniques used for new surveys – had been revolutionised by Colby'. Harry Margary, *The Old Series Ordnance Survey maps of England and Wales,* volume IV, *Central England,* Lympne Castle: Harry Margary, 1986, with *Introductory Essay* by J.B. Harley, and *Cartobibliography* by J.B. and B.A.D. Manterfield, pp xvii ff.

sheet 69: states 3 (1829) and 9c (1870 – with an Embossed Printing Date (EPD) of 1886)
sheet 70: states 2 (1828), 9 (1853) and 12 (1863)
sheet 83: states 2 (1828), 8 (1855) and 10 (1863)
sheet 84: states 3 (1829), 9 (1859 – with an EPD of 1862) and a hybrid state with an electrotype and price characteristics of state 13, but only the railways of state 12 (*circa* 1873).

2. Of the windmills listed, this was the furthest away from the railway associated with its revision – 650 yards. All the others were considerably closer to a new railway line, in some cases perilously so.

The middle years of the Old Series

To judge the truth of this assertion, assessments need to be made of Old Series sheets that were published in the years after 1825, but before the point in time when their production became linked to the introduction and preparation of the larger scales. So, from the broad swathe of mapping of southern-central England, a modest sample of four adjoining quarter-sheets can be the focus for a first scrutiny into how windmills were treated during the 1830s and early 1840s.[27] Sheets 53NW, 54NE, 62SE and 63SW were all published between May 1831 and June 1835, and so provide us with an insight into some of the OS practices of the early 1830s. Analysis of these sheets will be followed by an equally straightforward scrutiny and narrative for the treatment of some of the windmills of East Anglia during the late 1830s. Then, moving much further northwards, the sheet covering Liverpool and virtually all of the Wirral will be the next to be studied. This sheet was published in 1840 just before success of the six-inch mapping in Ireland prompted the short-lived adoption of that scale for the then remaining counties of northern England. To finish this section, one of the few complete revisions of an early sheet – sheet 1, issued in 1805 and later republished in 1843-4 – will be examined to see how the OS one-inch was handling the recognition of windmills in the mid-1840s, shortly before it became a derivative scale.

First, though, the study for *circa* 1832 examines the composite sheet made up from four adjoining quarter-sheets (53NW, 54NE, 62SE and 63SW), and therefore limits itself to the equivalent of just one full sheet.[28] These sheets cover a central area of England that includes much of northern Warwickshire together with parts of Leicestershire, Staffordshire and Worcestershire. In all, 65 windmills are depicted in this mill-rich area.[29] Their representation shows that use of the tower mill symbol had become proportionately greater in the decade after the Lincolnshire Map was published. To assess once again the correctness of its use requires, as before, that the type of each mill be known. Clouding the issue somewhat is the fact that the Survey chose this time to make experimental use of other symbols to signify windmills, but this was done only to a very limited extent. A crude type of pictogram not

[27] Since 1831 sheets had not only been published as quarter-sheets in one of Colby's measures to streamline map production but, to accommodate the greater amount of surveyed detail generated by the new regime, sheets from this time attain 'the achievement of both a more standardised and a more delicate style of engraving... amounting to a *new look*'. See Margary, *The Old Series Ordnance Survey maps of England and Wales,* IV, pp xxviii-xxix.

[28] A composite sheet is simply a group of sheets – full sheets or quarter-sheets, but normally rather more than just the four quarters of a full sheet – either mentally put together for consideration, or literally pasted together for sale to the public.

[29] The arrangement of counties assumed to be in force here, and the one used throughout this study, is the one commonly thought of as having been in place before the administrative changes of 1974, except that care needs to be exercised over the plethora of mid-nineteenth century county detachments – parts of one county embedded within another – that endured, in some cases, until nigh on the end of the century. More will be said on this later, but this proviso is particularly relevant to the area considered here.

unlike that often used on an OSD to signify a windmill – and reminiscent of the pictorial representation of a gibbet seen on older mapping – is sometimes used, as is also a symbol that consists of a circle overlain by a cross, but these and lesser variants obviously did not meet with lasting approval and were quietly dropped.[30] As an improvement on Lincolnshire, studies of windmills from the counties of Warwickshire and Leicestershire exist in a form where awareness of their entire nineteenth-century mill populations has been sought, and where reasonable documentary evidence for the type of windmill has been unearthed for about a half of all sites.[31] These sources and others for neighbouring counties suggest, a few complications aside, that the incidence of misrepresentation of windmills is still significant. The numbers of windmills depicted by type of symbol, and the proportions of each that are correctly depicted, as far as can be known, are set out below (Table 3.3). Of the numbers of mills known by type, something in the order of a third are depicted using a symbol that is inappropriate. This continuing failure to signify correctly each mill by type is not obscured here by the relatively high numbers of mills for which knowledge of their type is less than certain. To help readers examine for themselves the conclusions that are being drawn here, a list of the individual mills contributing to the figures in Table 3.3 is also given (Table 3.4).

Types of Mill & Symbol used	Post Mills	Tower Mills	Windmills of uncertain type
Post Mill symbol	23 correct recognition	11 inexact recognition	13
Tower Mill symbol	3 inexact recognition	8 correct recognition	3

Table 3.3. Accuracy of mill depiction by type of symbol on a composite Old Series sheet for central England (circa 1835).

1. The single mill deleted from later states of this composite sheet is excluded here. The post mill at Marton found itself, cartographically at least, too close to the Rugby and Leamington Branch Railway when the railway arrived in 1851. The mill survived north of the line until *circa* 1907.

2. As an example of the continuing difficulty over revision of much earlier OSDs and the provision of up-to-date information even on quarter-sheets, the post mill moved to Shrewley Common in 1832 is not shown by quarter-sheet 54NE. But another mill at Chilvers Coton is shown, only it is thought to have ceased work by the end of the eighteenth century and must therefore have been quite derelict by the 1830s.

---

[30] The illustations by Rodney Fry of depictions used on Old Series sheets are instructive here. See Margary, IV, p.xxi; VIII, p.xix.

[31] Wilfred A. Seaby, *Warwickshire windmills,* Museum abstract 1, 1979 and Nigel Moon, *The windmills of Leicestershire and Rutland,* Wymondham: Sycamore Press, 1981 are useful here. Also helpful is Barry Job, *Staffordshire windmills,* Birmingham: Midland Wind and Watermills Group, 1985 and Joseph McKenna, *Windmills of Birmingham and the Black Country,* Studley: K.A.F. Brewin Books, 1986. Worcestershire is less well served by the windmill literature for the period under discussion.

Names of Post Mills	Names of Tower Mills	Names of mills of uncertain type
23 mills provided with correct recognition by use of post mill symbols Baddesley Ensor Barwell: West Mill Bascote Baxterley Birdingbury Botley Chilvers Coton Corley Moor Cubbington Earl Shilton: Cooper's Mill Earlswood Lakes Haseley Hinckley: Mill View (Mill 1) Kenilworth: Common Mill Lode Heath Monks Kirby Pailton: North Mill Pailton: South Mill Pinwall Rowington (Mill 1) Sharnford Witherley Wolvey Heath *3 mills provided with inexact recognition by use of tower mill symbols* Foleshill Henley Stockton	*11 mills provided with inexact recognition by use of post mill symbols* Atherstone Attleborough: North Mill Coleshill Fillongley Kenilworth: Tainters Hill Mill Norton Lindsey Packwood Towers Rowington (Mill 2) Sheldon Shrewley Thurlaston *8 mills provided with correct recognition by use of tower mill symbols* Balsall Common Earl Shilton: Union Mill Exhall Hinckley: Brick Kiln Street Mill Sapcote: Granitethorpe Mill Solihull Lodge Tuttle Hill Weatheroak Hill Note: This list of tower mills correctly recognised includes Weatheroak Hill Mill for which the symbol given is ambiguous – on this occasion the Survey has been given the benefit of the doubt.	*13 additional mills recognised by use of post mill symbols* Attleborough: South Mill Barwell: Red Hall Mill Bleaks Hill Bulkington Dunchurch Earl Shilton: North Mill Hartshill Headley Heath Higham on the Hill Hinckley: Mill Hill Hinckley: Mill View (Mill 2) Perry Bar Stapleton *3 additional mills recognised by use of tower mill symbols* Alton End Sapcote: Lodge Barn Mill Shenton Note: This list excludes three mills: Bentley Heath, Copt Heath and Lapworth – all on sheet 53NE. Each is depicted by the type of crude pictogram, totally unable to distinguish between types of mill, noted by Harley/Fry as featuring occasionally in this area. See Margary, IV, p.xxi.

Table 3.4. List of windmills shown on a composite Old Series sheet for central England (circa 1835).

It might be thought that any plan to differentiate between types of windmill, seemingly newly applied as a feature on the Lincolnshire Map, would have been consolidated in the decade after 1825. So, with the representation of windmills remaining problematic in 1835, and mindful of the unsettling imponderables raised by the study of Alford for 1825, it needs to be considered whether this continuing failure to distinguish accurately between post and tower mills is suggesting something else; that another type of duality of windmill presence is being signified. If this is the case, the use of two different families of windmill symbol, each of which clearly corresponds visually to one of the two dominant types of windmill, may be an unfortunate coincidence and one that, inexcusably, will have caused some confusion. At this time of prolific tower mill construction and for years afterwards, the advertising of such mills normally centred around the concept of their being 'modern' compared to the post mill, though this varied across the country. Hence, the types of depiction used by the Survey on the one-inch may instead have been meant to provide some indication, if not of a mill's modernity, then at least of its commercial worth based maybe on the number of pairs of stones it operated, or perhaps on its age, or on some other non-visual attribute. It seems hard to imagine, though, that this would be done for a non-cadastral map where confusion would predictably be caused. The much more likely explanation remains that the surveyors were simply not sufficiently determined in their resolve to get this distinction between post mills and tower mills correct – if indeed there had ever been an explicit policy decision that they should – and that this issue continued to be unaddressed, even under the strictures of Colby's new regime of better fieldwork.

If a premise of laxity on the part of the surveyors is taken as the reason for a seemingly indiscriminate use of the different windmill symbols, then a further trait that could perhaps be expected is a clustering of each type of symbol, where the preferred option of a surveyor was holding sway within his area of surveying responsibility. This hypothesis can modestly be explored for part of the broad area that was next on the Old Series survey agenda. That the position for windmill representation had not improved by the late 1830s is confirmed merely by glancing at the sheets published for East Anglia. Virtually all the mills shown, for example, within a mile or so of the centres of Norwich and Great Yarmouth are shown as post mills – fifteen out of seventeen and thirteen out of fourteen respectively – whereas in reality this was far from the case. In both centres of commercial activity, state-of-the-art tower mills had by this time long since outstripped the capacity of the few post mills that remained, and these tower mills were active in a variety of industrial roles.[32] The two very imposing mills already referred to at Southtown in Yarmouth had been erected in 1812 and 1815 with the earlier of the two being the one that probably held the record for the largest windmill ever to be built in the country. No-one could have been in any doubt as to the scale of windmilling operations, and the predominance of the tower mill, in both of these places by this time.[33]

Discussing areas of responsibility for individual surveyors brings this narrative back to the subject of OSDs – each one originally overseen by a named surveyor. Our interest now lies in whether or not the incidence of error in recognising types of windmill by symbol matches the tracts of land covered by each OSD. The processing of windmill data from any OSD – with all its revisions – so as to provide a reliable published sheet was demonstrably problematic in the years *circa* 1830. This position was not set to improve quickly. Down the

---

[32] Across the country, tower mills were being used in roles other than grinding cereal, as already noted. Apling provides details of the particular uses to which some Norfolk windmills were put, noting that some in Norwich were used to grind bark and snuff as well as to crush bone; they worked also as saw mills and were used to drive cotton-mill machinery. Harry Apling, *Norfolk corn windmills,* volume 1, Norwich: Norfolk Windmills Trust, 1984.

[33] Other tower mills had been erected in the built-up part of town just inshore from the beach; these replaced mills shown by earlier maps as having been on the beach. There was probably only a single post mill surviving in Yarmouth at the time of survey for the Old Series.

coast from Yarmouth, an anomaly can be found for Lowestoft where the published sheet is at odds, in respect of a mill, with the manuscript OSD even though this had been corrected. The care taken to create this OSD – drawn to the standard two-inch scale and conforming to the newly-introduced more regular-shaped style for these drawings – as well as the gains achieved from it being corrected, are largely reflected in the final map. But one exception to the correspondence between the two is the depiction of a tower mill that is known to have survived until after 1890. Elsewhere in the vicinity of Lowestoft, other discrepancies can be found between the corrected OSD and the published sheet.

Moving slightly inland from Lowestoft, the south-eastern quarter-sheet of sheet 50 was published in 1837. At first sight, the distributions of both post mill symbols and tower mill symbols appear to be highly clustered, supporting the conjecture that individual surveyors favoured a standard symbol for their areas of responsibility. The many symbols found in the north-eastern quadrant of this quarter-sheet are almost all tower mill symbols, and those in the south-west are very nearly all of post mills. Moreover, superimposing the boundaries of the individual OSDs used in the compilation of this quarter-sheet onto these mill symbol distributions separates the symbols into clusters where the predominance of one type of symbol over another is absolute in one instance, and in others very nearly so. Within the area of OSD 315^{B}, for example, 32 windmills are recorded, all of them by use of a post mill symbol. Conversely, on another part of the same quarter-sheet for which no earlier OSDs are thought to have been compiled – meaning that the first time this particular plot of land was surveyed was as late as 1836-7 – there is a mix of ten post mill and eighteen tower mill symbols.[34] So, where OSDs were compiled for this quarter-sheet, each appears very heavily to have influenced the type of windmill symbol later used, but it should be remembered that unequivocal findings similar to these found for OSD 315^{B} were not immediately obvious for Lincolnshire or central England. In any case, this finding for a small corner of Suffolk needs to be endorsed, or perhaps negated, by knowing the types of mill that were present on the ground, just as with previous studies.

East Suffolk was an area where post mills were still very much in evidence during the 1830s, and many were to endure to become the sophisticated sort of mill seen at Friston. Tower mills, and smock mills too, were also built here in reasonable numbers, leading to a balanced mix for the three types of windmill that was matched by only a few other counties. Returning to the exclusive use of 32 post mill symbols on OSD 315^{B}, only five of these are known with certainty to have been wrongly used as the mills being recorded were definitely tower mills.[35] Another three symbols were recording (wooden) smock mills and the type of mill is unknown for a further five sites. This would seem to absolve the survey team from complete blame, as it could claim to have applied the correct symbol in over eighty per cent of sites for this area. Not so for the area to the north that was first surveyed in 1836-7. Here fourteen of the eighteen tower symbols applied were definitely used in error as the sites recorded are known to have then been supporting post mills. Of the remaining four mills, one was a smock mill and one was of unknown type, suggesting that maybe no more than two tower mills were correctly recorded. But then the ten post mill symbols also seen here

[34] The OS did have earlier knowledge of this last area but this seems to have depended upon a map unusual in that, though of Board of Ordnance provenance, it was not a standard OS-derived map: TNA PRO MR 1/1415.

[35] A good source for use here is Brian Flint, *Suffolk windmills*, Woodbridge: Boydell Press, 1979.

were correctly applied since what they were recording were, indeed, all post mills apart from one smock mill.[36] This contrast of fairly-accurate and severely-inaccurate sets of depictions – where the sets are based on knowing the actual types of windmill – is severely out of kilter with the patterns of symbol-type clusters for which at least some sort of rationale, related to the OSDs, seems to have been found. Arguably, this reveals as much about the *ad hoc* nature of the agenda for OSD revision undertaken just prior to publication, as anything else. But it also raises the more absorbing question as to where the authority lay for the production of a completed sheet. Based on the modest amount of evidence so far assembled for the years before the larger-scale OS maps made their appearance, it is hard to avoid the conclusion that ground-truth accuracy in respect of the two main types of windmill was not accurately mirrored on the Old Series. This may mean that the field data from the survey teams was routinely lacking in this detail; that the draughtsmen were often unable to make informed choices as to the type of mill symbol they needed to instruct the engraver to apply, thus obliging them to lend some artistic licence to the situation. In the meantime, the name of the engraver appears at the bottom of each sheet so that he was, in modern-day parlance, a stakeholder in the final outcome. One can imagine, for an engraver, that the completed map would have been a object of pride but, in following a draughtsman's drawing, he would not have been able to exercise any discretion as to the content of the sheet. Whether, when bringing meticulous care to other aspects of their work, the suspicions of any engraver was ever aroused by the engraved types of windmill symbol appearing to depend on constituent OSD boundaries, has to be an open question.

By the beginning of the 1840s – the decade that saw the introduction of the larger-scale sheets with their serious effect on the Old Series and the later story of one-inch mapping – the advance of the Old Series northwards had reached the line on the map joining Preston with Kingston upon Hull. Sheet 79, published in 1840, straddles the English-Welsh border and covers much of the Wirral as well as Liverpool. The evidence of this sheet, although modest in terms of the number of windmills that it shows, is nonetheless interesting. There are just 24 mills recorded on this sheet for the part of England it shows, and though at least two sited on the Wirral were post mills – at Burton and Irby – only the tower mill symbol is used for the peninsula. Elsewhere on the sheet, in the Liverpool area, all the symbols except one are also of the tower mill type. Any speculation that this may reflect a considered move towards wanting to use only this one type of symbol in the future is weakened, however, by realising that at least one later quarter-sheet – sheet 87SE, published in 1841 – is still happy to display a mix of symbols. Moreover, on this quarter-sheet of the area around Doncaster, at least one post mill – that at Stainforth soon to be given an expansive six-inch annotation acknowledging it to be a post mill – is recognised by a tower mill symbol (Fig. 3.5), and a newly-built (1836) tower mill at Hatfield Woodhouse is recognised by a post mill symbol. But another trend in windmill depiction can be seen as having its origins on sheet 79, one that would outweigh the continued use of either post or tower mill symbols. Only eighteen symbols appear on sheet 79, while six mills are recorded instead by annotated ground plans where, quite obviously, the annotation provided takes up considerably more space than the

[36] The confusion over the status of the smock mill is understandable. They were often described as tower mills, to which they were similar in having a cap that turned to face the wind. Moreover, they could possess substantial brickwork bases. But otherwise their structural essence was indisputably one of wood rather than brick or stone.

use of a symbol would have done. Had there been more windmills to record on this sheet, this style of depiction may not have been considered, but with a lot of the Old Series sheets for the north of England faced with only having to record similarly small numbers of mills, this was perhaps seen as a viable alternative for the future acknowledgement of windmills. It may even be that using symbols to record this variable landscape feature had become such a recognised problem that this new way of depicting mills offered a much-needed solution.[37]

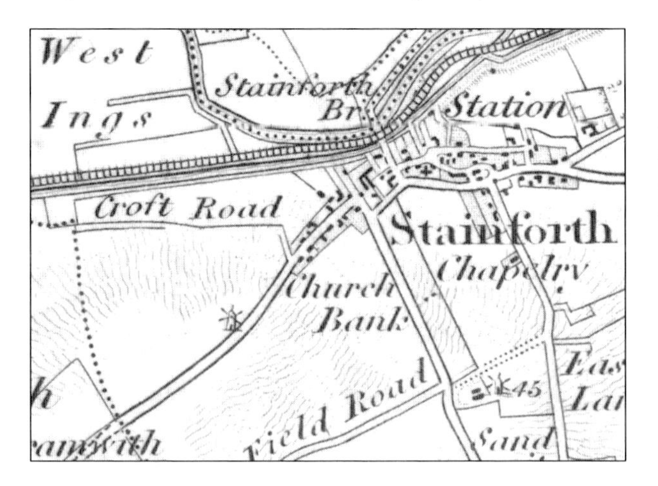

Figure 3.5. Windmills at Stainforth, Yorkshire: one-inch Old Series sheet 87SE (circa 1856).

Having by now got as far north as Lancashire and Yorkshire, and with approval given for surveying these two counties under a regime that would reduce the importance of the Old Series, the Survey needed to review what it had achieved in the previous half-century. Several of the early Old Series sheets were considerably out of date topographically and, by the standards of 1840, were also woefully lacking in planimetric accuracy and style. Earlier, in 1834, problems with a whole nexus of issues surrounding the operations of the Survey had led to an internal investigation. Some outcomes from the unease of around this time have already been touched upon: the introduction of quarter-sheets in 1831; the change to regular-shaped OSDs and the need for 'Hill Sketches' and 'Revision Drawings', all of which contributed to Colby's 'new look' for his Old Series sheets. Undoubtedly by the late 1830s, the newer sheets were an improvement on those of fifteen years before, but there was still the problem of the earliest sheets. Looking at sheet 1 – which functioned as the south-west sheet of the Essex Map when it was first published in 1805 – its style alone was enough to condemn it by the mid-1830s, irrespective of its content. So it was decided to overhaul this sheet completely, thus bringing it up to the standard of the prevailing model for sheet style and depiction. Following complete revision to one of the two northern sheets of Essex and partial revision to the other, the south-west sheet – sheet 1 – was completely re-surveyed and re-published in four quarters in 1843-4.

[37] It has to be considered too that an element of doubt creeps in over just what sort of mill is being recorded when the annotation of only *Mill* is applied, as is done here for six mills. Luckily, the near-contemporary map of Cheshire by Bryant (1831) confirms that these mills – distributed across the Wirral – were windmills rather than watermills, apart from at one site where both were present. The manuscript OSDs shed no light on why these six windmills should later be recorded in this way, and why the other nine on the Wirral were not, but they do record all of them as windmills. OSD 344, for example, records Eastham Mill by using a full *Windmill* annotation, while Willaston Mill is instead clearly recorded by symbol.

The two features of windmill depiction on the Old Series discussed up to this point – differentiation between type of mill and revision of the status of mills on successive states – leave one further feature, arguably the most important, left to consider. The issue of how comprehensive, or all-embracing, the recognition of windmills was on each sheet of the Old Series has yet to be discussed. For this to be assessed, knowledge is required of all the mills in an area, and moreover a knowledge that has not been map-derived. For Lincolnshire and central England, the information on windmills provided by the secondary literature is not sufficiently complete to have permitted rigorous analyses to have been done in this regard. Suffolk fares somewhat better with enough knowledge readily available to declare that three mills at least – post mills at Friston, Leiston and Parham – were omitted from sheet 50SE. But it is only when Essex comes to be considered that the published evidence for windmills can be seen as sufficient for this more demanding type of scrutiny. This is entirely due to the rigour with which a study of the windmills of Essex was conducted by Kenneth Farries. His study satisfies the simple, yet demanding requirement; that of being able to assess the comprehensiveness of representation for windmills on an Old Series sheet by considering only the sources of evidence that, transparently, do not depend on the evidence of this map series.[38] Indeed, because of the calibre of his work, Farries could be critical of various maps drawn for Essex. This includes the Survey's early county map of 1805 that, together with its manuscript OSD, he dismisses as:

> 'Both defective and inconsistent in their windmill representation... some mills were shown by rectangles and are indistinguishable from houses... on the 1805 map, taken alone, there are 43 known windmill omissions... ten mills did not graduate as symbols or in a recognisable form from the preliminary OSD to the publication, though all, except the mill at Scarletts, Colchester, remained active.'[39]

When speaking of the revisions made in the early 1840s to the four Old Series sheets that cover Essex, Farries is cautious with his comments, declining to provide a mill count as '[it] would have little meaning and is not, therefore, offered'.[40] Instead, he provides an analysis for the north-western sheet of Essex that diagrammatically shows all mill omissions and incorrect non-deletions – in other words, windmills not shown that should have been, and mills that are shown, but which had disappeared by the early 1840s. Together with justified anxieties about the one-inch map of Essex published by Greenwood in 1825, little doubt is left as to Farries' scepticism about the ability of the cartographic record alone to provide reliable evidence for windmills. That the revised OS sheet 1 published in 1843-4 is fallible in its acknowledgement of ground-truth accuracy for windmills is taken as given, but such is the work of Farries that the extent of this fallibility can be examined. As well as Essex, parts of Hertfordshire, Kent, Middlesex and Surrey also appear on this sheet and, in the analysis that follows for the completeness of windmill recognition by sheet 1 in the early 1840s, the mills of these other counties have to be included. It needs to be appreciated, though, that the findings of this analysis may be distorted owing to the uneven rigour on offer from the sources for these other counties. But the number of windmills known from documentary

---

[38] Kenneth G. Farries, *Essex windmills, millers & millwrights,* volumes 1-5, London: Charles Skilton, 1981-8.

[39] Farries, *Essex windmills, millers & millwrights,* 1, 42.

[40] Farries, *Essex windmills, millers & millwrights,* 1, 45.

sources other than maps to have been on the ground in the part of Essex covered by sheet 1, and which are at least strongly suspected of having been at work during the years 1835-45, is 95. This compares with 31 for all the other counties combined, as best as the easily available mill literature is able to judge. The windmills that are of unknown type from these 31 non-Essex mills represent a disproportionately high share of all the mills that are of unknown type – a consequence of the differing rigour in mill evidence between counties. We should, in any case, therefore look to pay particular heed to the 95 windmills of Essex for the most significant findings. The distribution pattern, meanwhile, of all the windmills revealed by this one-inch sheet is shown below (Fig. 3.6).

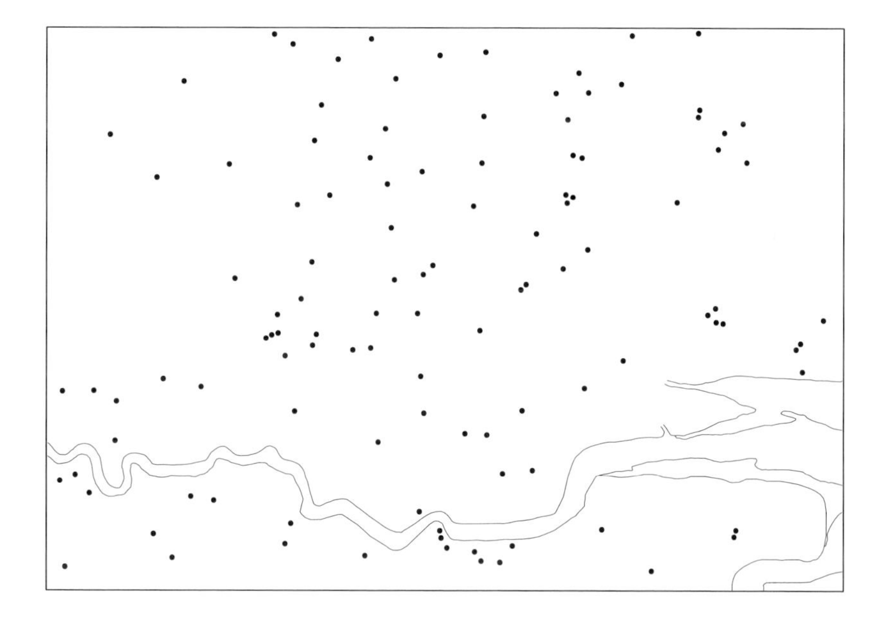

Figure 3.6. The distribution of windmills in the area covered by Old Series sheet 1 (circa 1845).

But first, the numbers of windmills for all five counties can be grouped by type – where known – and symbol accorded (Table 3.5). The figures demonstrate that, for counties other than Essex, the declaration of mill omissions is fairly minimal, which probably reflects the dependence of the secondary literature for these counties on OS mapping in the first place or, for those absentee mills, that the degree of congestion in places like Millwall was already making it awkward to insert a mill symbol. But also for Essex, the number of mill omissions is not large, with half of them forming a small cluster in West Ham. Realising that the total windmill population for the part of Essex covered by sheet 1 at this time was as high as 95, the slightly less than complete recognition may be thought not unreasonable. This certainly goes some way to salvage some credibility for the Old Series and its representation of mills even though on this sheet, and as late as 1840, any ambition for correctly applying the two types of windmill symbol still seems unapparent, though the relative scarcity of tower mills here, and indecision over which symbol to use for recording smock mills, may account for this. So, in an apparent reversal of the policy supposedly used on sheet 79 for the Wirral, a single tower mill symbol appears on the Essex part of the revised sheet 1 alongside 86 post mill symbols. This represents, more or less at the end of the era of autonomy for the Old Series, a return to its practice of depicting windmills of forty years before, though in fairness it does have to be stressed that the number of brick-built tower mills in Essex at this time was small. Again, to help readers examine for themselves the conclusions being drawn here for this next population of windmills, this time those that were recorded by the overhaul of sheet 1, or which should have been, a list of mills by name is provided (Table 3.6).

Map Symbol used [1]	Windmills of Essex (95 in total)	Windmills of Hertfordshire (1), Kent (21), Middlesex (7) & Surrey (2)
Post Mill Symbol	55 post mills [2] 24 smock mills 3 tower mills 4 mills of unknown type	5 post mills 6 smock mills 3 tower mills 9 mills of unknown type
Tower Mill Symbol	0 post mills 1 smock mill 0 tower mills 0 mills of unknown type	0 post mills 0 smock mills 1 tower mill 3 mills of unknown type
Mills that are unrecognised	7 post mills 1 smock mill 0 tower mills 0 mills of unknown type	1 post mill 1 smock mill 0 tower mills 2 mills of unknown type

Table 3.5. Numbers of windmills by type shown on re-surveyed Old Series sheet 1 (circa 1840).

1. The Margary volume (1) that provides a cartobibliography for sheet 1 of the Old Series chose, for reasons that need not delay us, only to consider the states of this sheet that appeared before railways were introduced to the area, and certainly before the wholesale revision of the early 1840s. A cartobibliography of later states for this sheet was subsequently provided in Guy Messenger, *The sheet histories of the Ordnance Survey one-inch Old Series maps of Essex and Kent*, London: Charles Close Society, 1991. The fuller picture revealed by this source is that there are now seven known states for sheet 1 as a full sheet – an increase of two on the five states acknowledged by Margary – and for the quarter-sheets dating from 1843-4 onwards, anything then from a further nine to thirteen states depending on the quarter under consideration. The Old Series sheet 1 used for this analysis is a composite of two Messenger state 3 quarter-sheets – the northern ones – and two state 1 quarter-sheets, suggesting it can safely be dated to *circa* 1849. The actual date is not of huge concern as the extent of any windmill revision done after 1843-4 on these four quarter-sheets will no doubt mirror the pattern of that done for other sheet numbers subsequent to their initial state, *ie* not very much. In support of this, it can be pointed out that Messenger state 7 for sheet 1 – still as a full sheet and carrying a publication date of 1805, yet dated to *circa* 1838 – preserves the same windmill representation as was offered in 1805.

2. Included in this subtotal is the composite mill at Little Laver. A composite mill was a hybrid mill that looked like a post mill, but which had had its post and supporting trestle removed, and had rollers attached to the underside of the mill body. These rollers sat on a curb on top of the roundhouse wall in much the same way that the cap of a tower mill sits on a curb placed on top of the brickwork of the tower. This was a relatively uncommon type of windmill.

Name of Mill	Type	Old Series sheet 1 (1843-4)	Name of Mill	Type	Old Series sheet 1 (1843-4)
Essex windmills [1]			Ingatestone	Post	Post
Aveley	Post	Post	Kelvedon	Smock	Post
Barking	Smock	Post	Little Laver [7]	Comp	Post
Blackmore	Post	Post	Little Stambridge	Post	Post
Bobbingworth	Post	Post	Little Totham	Smock	Post
Boreham	Post	Post	Little Warley	Smock	Post
Buttsbury [2]	nk	Post	Moreton	Post	Post
Chelmsford:			Mountnessing	Post	Post
Moulsham [3]	Post	Post	Nazeing	Post	Post
Rainford	Smock	Post	North Ockendon [8]	Post	Post
Widford	Post	Post	Orsett:		
Chigwell Row [4]	Post	Post	Baker Street	Smock	Post
Dagenham:			Mill Lane	Post	Post
Beam River	Smock	Post	Pitsea	Post	Post
Becontree Heath	Smock	Post	Prittlewell:		
Chadwell Heath: 1	Post	Post	Hamlet	Post	Post
Chadwell Heath: 2	Post	Post	West Street: North	Smock	Post
Chadwell Heath: 3	Post	Post	West Street: South	Post	Post
Marks Gate	Smock	Post	Purleigh:		
Danbury: North	Post	Post	Cocks Clarks	Smock	Post
Danbury: South	nk	Post	Raven's Mill	Smock	Post
East Horndon	Post	Post	Ramsden Bellhouse	Post	Post
East Tilbury [5]	nk	Post	Rayleigh:		
Epping	Post	Post	Castle Farm	Tower	Post
Fobbing	Post	Post	Eastwood Road: East	Post	Post
Fyfield [6]	Post	Post	Eastwood Road: West	Post	Post
Great Baddow:			Ruffle's Mill	Post	Post
Barnes	Post	×	Rettendon	Post	Post
Galleywood	Smock	Post	Romford:		
Steven's Mill	Smock	Post	Collier Row	Post	Post
Great Burstead:			Edward Collier's Mill	Post	Post
Bell Hill: East	Post	Post	Pratt Collier's Mill	Post	×
Bell Hill: West	Post	Post	Rising Sun	Post	Post
Outwood Farm	Post	Post	Roxwell	Post	Post
Harlow:			Shenfield: East	Smock	Post
Foster Street	Post	Post	Shenfield: West	Post	Post
Potter Street	Post	Post	South Ockenden	Smock	Post
Hatfield Peverel	Smock	Post	South Weald:		
Hazeleigh	Post	Post	Bentley	Post	Post
High Ongar	Smock	Post	Brook Street	Tower	Post
Hornchurch:			Springfield	Smock	Post
Bush Elms	Post	×	Stanford Rivers:		
Howard's Mill	Post	Post	Littlebury	Post	Post
Horndon-on-the-Hill	Post	Post	Shanks	Post	Post

Name of Mill	Type	Old Series sheet 1 (1843-4)	Name of Mill	Type	Old Series sheet 1 (1843-4)
Toothill	Post	Post	Eltham	nk	Post
Stapleford Abbots	Post	Post	Erith	Post	Post
Stapleford Tawney	nk	Post	Gravesend:		
Stock:			Denton	nk	Post
Common: North	Tower	Post	Windmill Hill	Post	Post
Common: South	Post	Post	Hoo Common	Smock	Post
Threadgold's Mill	Post	Post	Lee	nk	Post
Upminster:			Lower Stoke: North	*(note 17)*	Post
Abraham's Mill	Smock	Post	Lower Stoke: South		Post
Common	Post	Post	Northfleet:		
Waltham Holy Cross	Post	Post	Perry Street [18]	Post	Post
West Ham:			Rosherville	Tower	Post
Abbey Mills [9]	Smock	×	Stone Bridge: North	nk	Post
Nobshill	Post	×	Stone Bridge: South	nk	Post
Pudding [10]	Post	×	The Hive	nk	Post
Romford Road	Smock	Tower	Plumstead	Tower	Post
Stent's Mill [11]	Post	×	Sydenham	nk	Post
West Thurrock [12]	Smock	Post	Woolwich	Tower	Tower
West Tilbury	Smock	Post	*Middlesex windmills* [19]		
Willingale Doe	Smock	Post	Bow:		
Woodham Ferrers [13]	Post	×	Old Ford New Town	nk	Post
Woodham Mortimer	Smock	Post	Wrexham Road	nk	Tower
Writtle:			Hackney:		
Southgate's Mill	Post	Post	Stoke Newington [20]	nk	×
Warren's Mill	Post	Post	Union Mill	nk	Tower
Hertfordshire windmills [14]			Poplar:		
Cheshunt	Smock	Post	Blackwall [21]	nk	×
Kent windmills [15]			Kerby Street	nk	Tower
Bexley Heath	Post	Post	Millwall [22]	Smock	×
Cliffe	Smock	Post	*Surrey windmills* [23]		
Dartford	Smock	Post	Bermondsey &		
Deptford:			Rotherhithe:		
Black Horse Road	Smock	Post	Dale's Mill	Tower	Post
Tanners Hill [16]	Post	×	Poupart's Mill [24]	Post	Post

Table 3.6. Recognition of windmills on re-surveyed Old Series sheet 1 (circa 1840).

The notes below provide details, where known, of any significant changes to windmills that occurred within the time-span 1835-45 together with any peculiarities of mill depiction on the sheet. Also noted are the main county sources, definitive and not so definitive, for individual mill histories. The use of × indicates an omission of a windmill on the map, and nk (not known) indicates that any declaration of the type of mill would be unsafe from the evidence provided by the source quoted, or from other information easily available.

1. Farries, *Essex windmills, millers & millwrights*, 1-5. This authoritative text for the windmills of Essex can be considered definitive. Volumes 3, 4 and 5 provide individual mill histories by parish.

2. The only evidence for this windmill is this particular OS map. It does not otherwise appear in the historical record. A small farm mill is strongly suspected here.

3. Mill auctioned and probably demolished in 1844.

4. This mill burnt down in 1842.

5. Like Buttsbury Mill, this mill is shown only on this map. A small farm structure is suspected.

6. What looks like a tower mill symbol on the map is more likely to be a post mill symbol that has had its lower part over-engraved by a double-pecked line indicating the track leading to the mill – a style of depiction occasionally found across the country.

7. An instance of the very rarely found composite mill.

8. This mill worked until *circa* 1841 – no mention of it in later tax returns. On the map it is uniquely annotated as *Old Mill*. This is a good example of where the Survey got its detail right.

9. No symbol is provided for this mill but an annotation of *Windmill* is used. This obviously takes up more space than would have conventionally been the case in what is already a congested part of the map. No reason for this variation is offered other than, perhaps, a lingering influence suggested by the name of the mill.

10. This mill was auctioned in 1843-4 after which it fades from view, presumed dismantled.

11. This mill was severely damaged in 1834 and left as derelict. It remained in this condition for some years.

12. Another case of a probable post mill symbol truncated to appear as a tower mill symbol by over-engraving.

13. No symbol is provided for this mill but a simple *Mill* annotation is offered.

14. The source for use here is Cyril Moore, *Hertfordshire windmills & windmillers*, Sawbridgeworth: Windsup, 1999.

15. The principal source for the windmills of Kent is William Coles Finch, *Watermills and windmills*, London: C. W. Daniels, 1933. This source is clearly dated if judged by modern criteria for scholarly inquiry, but it has the advantage of being written by someone who started his fieldwork in the early years of the twentieth century, and who was thus closer in time to the period of interest than more modern authors. The evidence cited within the individual mill histories varies from comprehensive to minimal.

16. This mill survived until 1849 when it was pulled down.

17. One of these two mills was a smock mill, the other was of unknown type. Coles Finch is silent as to which of the two sites supported the smock mill – which he had visited.

18. That this was a post mill is known from photographic evidence rather than Coles Finch.

19. The source here is Guy Blythman, *Watermills & windmills of Middlesex*, Baron Birch, 1996.

20. This mill is declared to have been in ruins in 1852 but, with no precise location offered for it, this parish mill may well have been outside the sheet neat line (or map border).

21. This mill is known to have been working in 1840.

22. One windmill from the row of up to ten mills survived here until at least the mid-1850s. Another mill – a post mill thought to be working in 1830 – was probably missed by the Survey.

23. The source here is K.G. Farries and M.T. Mason, *The windmills of Surrey and inner London*, London: Charles Skilton, 1966. Co-authored by the later author for Essex windmills, this text is considered to be rigorous in outlook.

24. The symbol used on the map is an amalgam of what appears to be a post mill body sitting on top of a tower mill stem.

By the time that Old Series sheet 1 was undergoing its complete revision, the survey was also underway – it had started in 1842 – for the new six-inch sheets of Lancashire. From the mid-1840s onwards, the story of windmill depiction at the one-inch scale follows a radical new trajectory, departing from its former course and having now to adjust to the primacy of the new larger scales. This might be a convenient place to remind readers that the OS one-inch mapping of this period did not exist in a cartographic vacuum: there were comparable mapmaking ventures that included the work of private cartographers. Two major players here were Bryant and Greenwood who were publishing maps of a similar scale that made use of representational styles not dissimilar to those used by the Survey. They each recorded mills on county maps and each has been mentioned in passing for his ability to corroborate, or not, the evidence of OS one-inch sheets. Other types of contemporary maps also deserve at least a mention: estate maps and tithe maps from this period are all capable of providing information about windmills to those willing to search for it, and privately-published town maps too are of value.[41]

The Old Series as derivative mapping

The new regime of surveying for, and publishing maps at, much larger scales assumes a pivotal role in any understanding of the further progress of the Old Series. A lull in issuing Old Series sheets is linked with the decision to map the next two counties to be surveyed – Lancashire and Yorkshire – at the six-inch scale. Subsequently, the 'Battle of the Scales' lent further flux to the situation. Emerging from this period of uncertainty in the late 1850s, the Survey had continued to provide Old Series sheets in the interim, but these were derivative – reduced from the surveys which had provided the two counties with their six-inch sheets. Later, for the remaining unmapped areas of northern England, it was to do the same from

---

[41] Tithe maps are particularly useful. In 1836 the Tithe Commutation Act was passed, the aim of which, in the wake of the 1832 Reform Act, was to pacify grievances stemming from a system which required payment to be made by the agricultural community to the church, and which over time had become iniquitous. Future payments were to be made fairer by linking the apportionment of liability to land holdings. This meant that reasonably-scaled maps were needed for indicating boundaries and acreages, and also to show land usage, thus becoming cadastral maps similar to those available elsewhere in Europe. The creation of these maps began in 1836, went through peak years of 1839-40 and was largely over by the early 1850s. A standard characteristic sheet was made available which included two rather unglamorous symbols for depicting windmills. Their designations, *Wooden Mill* and *Stone Mill*, were clearly intended to distinguish between the constructional nature of post mills and tower mills. Even so, this does seems rather a throwback to a time when tower mills could be primitive and before their widespread use of brickwork, wrought-iron galleries and cast-iron external fittings. But the time was approaching when windmills were starting to fall out of commercial favour, and they were heading also for cartographic near-oblivion, certainly as far as the Survey's one-inch scale was concerned. Their visibility and treatment by symbol or annotation on the tithe maps is correspondingly uneven with the definitive work on the subject being content to leave its principal mention of mills to the section dealing with decoration within the tithe-map *corpus*. Very likely, this is because pictorial images are also used on tithe maps to signify windmills. Cumulatively, all this is a reminder of much earlier mapping practices for depicting windmills, and of those times when a windmill might even appear within the cartouche of an eighteenth-century county map. With this slightly archaic feel about them, the styles of depiction recommended for windmills on tithe maps only serve to reinforce the idea of a diminishing status in the landscape, and it is unlikely that tithe maps were operating to a different dual-meaning for their basic two symbols than the Survey was for its maps. See Roger J.P. Kain and Richard R. Oliver, *The tithe maps of England and Wales*, Cambridge University Press, 1995, 789; Roger J.P. Kain and Hugh C. Prince, *The tithe surveys of England and Wales*, Cambridge University Press, Studies in historical geography 6, 1985; Geraldine Beech and Rose Mitchell, *Maps for family and local history: the records of the Tithe, Valuation Office and National Farm Surveys of England and Wales, 1836-1943*, London: The National Archives, Readers Guide 26, 2004.

the surveys which provided mapping at the new prime scale – the 1:2500. So it might be assumed that the accuracy of the Old Series as it progressed northwards from the Preston-Hull line with sheets 91-110 would improve, since it was now, after all, the outcome of a much more demanding survey process. But the trend towards the Survey being disinclined to depict windmills at the smaller scale was gathering pace. Its failure to replicate on the Old Series the recognition of windmills that it made available on its Lancashire six-inch sheets has been commented upon by Oliver, and an inspection of corresponding Old Series sheets for Yorkshire reveals a similar story.[42] Examination of Old Series sheet 94, published by 1858, reveals an attitude to depicting windmills that seriously hinders any understanding of the true circumstances of windmill presence in this part of the East Riding. Only three symbols are seen, and they are dispersed across the sheet; all are post mill symbols where two of the mills they acknowledge are safely known to have been tower mills. Additionally, only one of the symbols – that for Mappleton Mill – is unannotated. All other disclosures of windmills are by annotated ground plan where the annotation can vary between the full word *Windmill*, sometimes with a name, to use of simply *Mill* or, for one of the annotated symbols, to use of the abbreviation *W.Mill*. The possibility of a windmill not having any annotation at all and therefore going completely unrecognised becomes a matter of some concern. But any suggestion that the Survey purely had it in for windmills at this time would be entirely misplaced. This failure to disclose the presence of windmills has to be seen in the context of the wider implementation of a more planimetric or 'scientific' approach to the depiction of detail that was introduced around this time, and which fitted the ethos of the larger scales. This new approach devalued the tendency to have pictorial detailing on a map, and this disapproval extended to the use of some symbols. Presumably, how this approach might affect what a map could 'say' will have been considered on a feature-by-feature basis and so, equally presumably, the loss of windmill-related information at the derivative one-inch scale was not felt to be too much of a problem.[43]

The great achievement of the six-inch stems from the engraving at this scale being able to show much finer detail than had been possible on one-inch sheets, owing simply to the increase in map space for any given ground area. This became apparent in the last chapter, of course, when examples of individual windmill depictions were discussed. No longer was there any need at this new scale to embellish sheets with a pictorial symbology. It allowed maps to be good at getting minor topographical detail correct, and this included the precise location of windmills. The depiction of a windmill at near-true-scale becomes much harder to see than when the deliberately enhanced mill symbol device is used, but gone at the six-inch scale is the need to be so selective of landscape detail because of restrictions of space, or because the cartographer maybe feels that the artificiality of symbols permits their more subjective use. Hence, the representation of windmills on the six-inch sheets of Lancashire and Yorkshire can probably be accepted as near-definitive, and certainly as good as that of the better sheets of the Old Series. This realisation can be put to good use in our continuing effort to consider sheets of the Old Series, since we now have a relatively simple means of assessing the one-inch sheets of these two counties. If the six-inch sheets of Yorkshire that cover the area of Old Series sheet 94 are carefully studied for their recognition of windmills,

---

[42] Hellyer and Oliver, *One-inch engraved maps of the Ordnance Survey from 1847*, 27.

[43] Hellyer and Oliver, *One-inch engraved maps of the Ordnance Survey from 1847*, 26 ff.

this will result in knowing the population of mills that, ideally, should have been cascaded down to the one-inch sheet.[44] But, manifestly, this cascade did not happen. Three symbols and 30 ground plans with various annotations, each denoting a single mill, can be found on sheet 94, but as many as 35 windmills are totally ignored by it.

Returning to the dichotomy of post and tower mill differentiation: one possible pointer to a rule of disclosure conceivably being used on sheet 94 is that the only absentee from the group of mills on Beverley Westwood sheet is the single post mill of the group – Fishwick's Mill – dismantled in 1861. Certainly, for the mix of post mills and tower mills still available to be recorded by sheet 94, the minority number of post mills assumes a disproportionate share of those not recognised. Yet three such mills are recognised – one of them vividly and unambiguously annotated as *The Old Post Mill.* But, given that many tower mills are also not recorded – present-day survivors include mills at Hornsea, Lelley and Skidby – it seems that, once again, a clear-cut explanation underpinning the rules of disclosure – if one ever existed – is as elusive as ever. Following usual practice, a summary for the depiction of windmills by Old Series sheet 94 and a listing of the mills included appear below (Tables 3.7 and 3.8).

Type of Mill	Mills recognised by Old Series sheet 94 with annotated ground plans	Mills not recognised by Old Series sheet 94
Post Mill	2	6
Tower Mill	19 (two by symbols instead)	19
Unknown type	12 (one by symbol instead)	10

Table 3.7. Recognition of windmills by type on Old Series sheet 94 (circa 1850).

[44] The East Riding of Yorkshire is an area where valiant attempts to map both its windmills and watermills have been made. A good schematic map showing its windmills can be found in Alan Whitworth, *Yorkshire windmills,* Leeds: M.T.D. Rigg, 1991. This map has almost certainly been influenced by K.J. Allinson, *East Riding watermills,* York: East Yorkshire Local History Society, 1970. The only information on windmills contained in this earlier source is their inclusion on a fairly rudimentary map that principally shows watermills and their watercourses. But, in purporting to show all windmills of *circa* 1850, Allinson's map is quite clearly compiled from the evidence of the OS six-inch sheets, though it misses at least five windmills – one of the trio of mills east of Howden, one of the two mills at Market Weighton, one of the mills at Leven, and post mills at South Skirlaugh and Wetwang – a fault attributable, no doubt, to the small size of engraved detail. Allinson, lately editor of the Victoria County History for the East Riding, treats all his material in scholarly fashion, but rather than try to capture the complexity of the six-inch sheets which depict and annotate, for example – and as already noted – redundant mill mounds, his basic map merely acknowledges those windmills presumed to have been working. Whitworth, in repeating the same oversights of Allinson's map, does at least contribute a far more stylish map for the distribution of windmills of this area. A good study of the windmills of the East Riding had been published in the gap between Allinson and Whitworth (Roy Gregory, *East Yorkshire windmills,* London: Charles Skilton, 1985) and this includes a small map showing the distribution of mills. The number of mills shown is an improvement on Allinson, but the sad part is that it carries no notes of explanation and the author, or his publisher, has chosen to relegate it to the inside back flap of the dust cover. After all of this, no guarantee is offered by this writer that he has spotted every diminutive windmill depiction provided by the six-inch for use in the next Old Series sheet analysis!

Name of Mill	Type	Old Series sheet 94	Name of Mill	Type	Old Series sheet 94
Aldbrough: East [1]	nk	✓	Hornsea	Tower	×
Aldbrough: West	nk	×	Huggate	nk	✓
Atwick	Post	✓	Hutton Cranswick *	Tower	✓
Bainton	Tower	✓	Kilham	Tower	×
Beeford	Tower	×	Langtoft	Tower	✓
Beverley Westwood:			Lelley	Tower	×
Black Mill	Tower	✓	Leven:		
Fishwick's Mill	Post	×	Canal Head	Tower	×
Hither Mill [2]	Tower	✓	New Mill	Tower	✓
Lowson's Mill	Tower	✓	Wright's Mill	Tower	×
Union Mill	Tower	✓	Lockington	Tower	✓
Beverley East: North	Tower	×	Mappleton *	Tower	✓
Beverley East: South	nk	×	Market Weighton: East	Tower	×
Bishop Burton	Tower	×	Market Weighton: West	Tower	×
Bridlington:			Marton	nk	×
Black Mill	Tower	✓	Middleton	nk	✓
Duke's Mill	Tower	×	Nafferton	Tower	✓
Folly Mill	nk	×	North Dalton: North	nk	✓
Forty Foot Lane	Tower	×	North Dalton: South 1	Post	✓
Spink's Mill	Tower	✓	North Dalton: South 2	Tower	✓
Spring Mill	Tower	×	North Driffield: North	Tower	✓
Union Mill	Tower	×	North Driffield: South	nk	×
Catwick	Post	×	North Frodingham	nk	✓
Cottingham:			North Newbald: North	nk	×
Low Mill	Tower	×	North Newbald: South	nk	✓
North Mills	Tower	×	Rise	nk	×
Danthorpe	nk	✓	Salts House	nk	×
Dringhoe	nk	✓	Skidby	Tower	×
East Lutton *	nk	✓	Skipsea	nk	✓
Ellerby	nk	✓	Sledmere	nk	×
Etton	Tower	✓	South Skirlaugh	Post	×
Everthorpe	nk	×	Sutton	nk	✓
Foston	Tower	✓	Thearne	Tower	✓
Fridaythorpe	Tower	✓	Walkington	Tower	×
Ganstead	Post	×	Wetwang	Post	×
Garton	Tower	×	Withernwick	Post	×

Table 3.8. Windmills shown at the 1:10,560 (six-inch) scale circa 1850 for area of Old Series sheet 94.

* indicates recognition of a windmill to be by symbol rather than by annotated ground plan.

1. This village had both a post mill and a tower mill: one mill lay to the north-east and the other to the west of the village centre. Which was which is unclear, although an annotation of *Old Mill* for the easterly one almost certainly establishes this as the post mill.
2. An anomalous case for this list of mills: Yorkshire six-inch sheet 210 was surveyed in 1851-2

and published in 1855. Old Series sheet 94 was issued in 1858. This postdated the demolition of the windmill, *circa* 1856.

All windmills listed are those recorded by the six-inch map for the area covered by Old Series sheet 94. This recognition is normally by a distinctive, circular or square, shaded or unshaded, ground plan that is occasionally integrated into another building. Normally also there is an adjoining cross, and – apart from Market Weighton (East Mill) and South Skirlaugh Mill – this will be annotated with either *Windmill* or just *Mill*. In some instances, place names that it would appear sensible from the map as the obvious ones to have used, have been replaced by alternative ones for the six-inch annotation. This occurs, for example, at Elstronwick *(Danthorpe Windmill)*, Skipsea Brough *(Dringhoe Windmill)* and at Morton for one of its two mills *(Ellerby Windmill)*. In addition, for some mills the accompaniment to the annotation is a ground plan superimposed onto a triangulation-station symbol (or *vice versa*), and sometimes it is just the triangular symbol in isolation. Lastly, what must have undoubtedly been an open-trestle post mill at Withernwick is denoted by four small black dots and a cross. Our knowledge of the types of each mill is only as good as the East Riding of Yorkshire regional windmill studies – as part of a secondary mill literature that can easily extend to not confirming mill types – allows. Here, probably all the windmills whose type is declared as 'nk' (not known) would have been tower mills. In all instances of a mill being recognised on the one-inch scale, the depiction found on the initial state of Old Series sheet 94 remained the same up to the last known states of New Series sheets 64, 65, 72 and 73 – remembering that this was one of the Old Series full sheets that was later adopted to become New Series sheets. Ironically, anyone wishing to buy a copy of sheet 94, first published 1857-8, could do so more than a century later as the New Series subsequently provided base maps for overprinting with geological information. In this manner, sheet 94SW was effectively being reprinted as a 1:50,000 enlargement at least as late as 1990!

That the Survey had not quite forsaken using symbols, or even pictorial depictions, in its recording of windmills on the one-inch *circa* 1855 can be seen from other Yorkshire sheets. On quarter-sheet 93NE (adjacent to sheet 94) two windmill depictions can each be construed as being more than just a symbol. At Sheriff Hutton, a square, solid ground plan has a detached cross alongside it – the squareness probably indicating a post mill – while at Dunnington, a cross abutting a circular, solid plan suggests either a tower mill or a post-mill roundhouse. This type of image, anomalous on the one-inch scale by its very limited use, is the same 'flattened-cross' style of depiction discussed in earlier chapters as being extensively used on the six-inch and later, more sparingly, at 1:2500. Finding this type of depiction used on a one-inch sheet reminds us of the question raised in the last chapter concerning the blending of ground plans with symbols. A workable response here is that the mere handful of times this sort of large-scale depiction is found on the one-inch points to its use as having been experimental, and if, through oversight, it was not subsequently removed, it became a flaw. This possible explanation chimes well with the moves taking place to produce a more planimetric style of mapping, which may have led to it seeming a good idea to replicate a standard six-inch depiction on the one-inch. If so, it was an idea that did not catch on, but this was clearly a time when changes were in the offing – associated, undoubtedly, with a new Director-General – and of key interest here are not so much the dates of either survey or publication, but the exact timing and supervision of a sheet's compilation. What remains certain is that the loss of visibility for windmills was happening on the one-inch scale by 1860. This, as we have already seen, was to lead to a drawn-out period of time when windmills were largely ignored by all but the largest scales.

As the Old Series moved towards completion, the depiction of windmills got no better than it had been for Lancashire and Yorkshire, a state of affairs probably exacerbated by the sparser mill populations in the rural parts of the northern counties, together with the earlier onset of windmill decline in their urban areas. Continuing northwards from Yorkshire and crossing the River Tees, the mills of Durham have already been the subject of discussion. Forty active windmills are recorded for this county at around this time, mostly located on the coastal plain.[45] The depiction of them on three quarter-sheets – sheets 103NE, 105NE and 105SE – all of the early 1860s, is a bleakly meagre, and familiar, story where only eleven windmills are recognised through use of an annotated ground plan.[46] Durham had been the first county to be mapped at 1:2500 with both survey and publication of sheets taking place from the mid-1850s onwards while the 'Battle of the Scales' was still unfolding. Eventually vindicated in providing these sheets – a total of 705 to cover this county – the information they contained now became the basis of the Old Series sheets issued for Durham. But the earlier use of the older circle-and-cross style of depiction occasionally found on these large-scale sheets had been discontinued, and any 'mistakes' such as those seen on quarter-sheet 93NE were not repeated. Crossing the Pennines and coming forward in time to 1869, we arrive at almost the last of the Old Series sheets to be published.[47] Hughes provides a figure of just sixteen for the total number of windmills in Cumberland known definitely to have survived until the time of the large-scale survey here in 1859-65. Of these, four were sited in the area covered by sheet 107. But the three quarter-sheets making up this coastal sheet only record one of these four windmills, together with a second mill for which Hughes' evidence for survival to that time is inconclusive.[48]

In conclusion, it has been shown that for its first half-century and more, the Ordnance Survey's recognition of windmills by the use of map symbols was achieved with only partial success. As only one factor in the principal task that the Survey set itself during this time – the survey and publication of sheets for its one-inch Old Series – its policy concerning the recognition of windmills was subject to change. For those sheets whose initial states appear to have been comprehensive in their windmill coverages, the fact that they were then only updated in response to new railway construction dilutes any acclaim that can be given these sheets. In its treatment of windmills, an added lack of credibility derives from realising that

---

[45] Wiggen, *Esh leaves, passim.*

[46] Familiar also is the need to apply a minor corrective to this fragment of the secondary windmill literature. In the absence of any other ready evidence, the account of 40 active windmills from Wiggen is useful but probably not definitive. One of his windmills was actually in Yorkshire (Mandale, Thornaby) though the others would appear to be *bona fide*. The evidence of the maps, in particular the 1:2500 sheets, suggests a small increase to his count of active windmills. Some of these extra mills such as Whitburn Mill, West Hartlepool Dock Mill and three others feature on the Old Series. Their recognition by the one-inch scale is in addition to that of another six mills which *do* appear on the Wiggen list, but from that list as many as 22 windmills from an area covered by just three quarter-sheets are not recorded by the Old Series.

[47] The final sheet to be published was that covering the Isle of Man: this finished the series in 1874.

[48] J. Hughes, 'Cumberland windmills', in *Transactions of the Cumberland & Westmoreland Antiquarian & Archaeological Society,* 72 (1972), 112-41. On Hughes' evidence, the windmills that could be expected to appear on sheet 107 are those at Bowness-on-Solway, Cardewlees, Laythes and Monkhill. But of these, only the mill at Laythes is shown together with another mill at Holmelow (see page 157). There are other signs that might be taken to imply a windmill is occupying a site, but these can be misleading. An annotation of *High Mill* that in many cases across the country would be signifying an exceptionally tall tower mill – or perhaps the tallest in the area – appears at Thurstonfield on sheet 107, but here this annotation is recognising the upper of two watermills on a very minor watercourse.

the symbolic notation used to distinguish between the two main types of mill is seriously at fault. The analyses done for a chronological range of sheets confirm that the representation of windmills on the Old Series is flawed, and that any understanding of mills to be had from its sheets will be limited to an overall picture of mill distributions. Realising this, it could be thought that the use of the windmill symbol as an ornamental device by prior generations of cartographers may have left such a lingering impression that a precision in attributing types of mill symbol was not felt to be necessary, or even appropriate, at the one-inch scale. Yet, at the end of this half-century, an expansion of the Survey's activities led to the emergence of larger scales that were able to offer an improved capacity for recognising windmills. The promise of a better accommodation of the many sorts of windmill presence did not prove to be a disappointing one in the following half-century with, eventually, all scales improving on their representation of windmills.

4

The New Series

By the early 1860s, when preparation of the Old Series sheets for England and Wales was nearing completion, the earliest sheets – other than those revised in a wholesale manner as sheet 1 had been – were out-of-date by anything up to half a century. In England especially, where the growth of industrial centres, and urbanisation in general, were comprehensively altering the look of the landscape, the criticism levelled at the Ordnance Survey over the outdatedness of many of its one-inch sheets was becoming something of an embarrassment to Sir Henry James, its then Director-General. However, by this time also, the programme of re-survey of the southern counties to produce larger-scale mapping was fully underway. Safe in the knowledge that Treasury approval for this was settled, this programme of rolling survey would lead to sheets at the 1:2500 scale becoming available for all habitable areas of the country, a process destined to finish in England and Wales with the sheets of Lancashire and Yorkshire. The solution to the problem of the one-inch scale was obvious. Using the new survey data as it became available from what was now the major OS venture, derivative smaller-scale maps at both the six-inch and one-inch scales could be provided. Indeed, this course of action had formed part of James' scheme for ensuring continuity of funding for the large-scale surveys. This least-cost solution to the problem of the one-inch later received further impetus when it became understood in the War Office that the French Army had been seriously disadvantaged in its engagements with the Prussian Army in 1870, owing to a similar lack of current maps.[1] In applying this solution, the second generation of one-inch maps that the Survey now thought to provide would be the genesis of what later came to be known as the New Series.

While no doubt striving to make the later Old Series sheets better than the earlier ones, it was more the bold decision to initiate a replacement scheme for all Old Series sheets that allowed James to silence his critics more effectively than if he had merely used the new data to update the Old Series more convincingly than had happened before. This was particularly so since it was more than topographic detail that was to be enhanced. The source material for these newer sheets meant that their geodetic accuracy could be improved. Their format, too, was to become clearer, and this can be seen in the change in style introduced onto Old

[1] Not that the War Office had to be concerned any longer with the costs incurred by OS activities. It had been responsible for the Survey since 1855 but, following an initiative from the Treasury, it was presumably happy to agree with James' view that as the one-inch 'Military Map' was finished, and as the large-scale surveys were seen as being conducted for civilian purposes, it could relinquish that responsibility and fiscal burden. This it did, handing over responsibility to the Office of Works in 1870.

Series sheets north of the Preston-Hull line. The precepts of mapping projection used as an earlier basis for drawing sheets of the Old Series had meant that, as one got further north, the more non-rectangular, or trapezoidal, the shape of some of the sheets became. To avoid this visual peculiarity, the projection was altered, which allowed the remaining Old Series sheets each to assume a strict rectangular shape. This change in style is evident on sheet 91, and from this point onwards each quarter-sheet represents a tract of land measuring exactly eighteen miles by twelve miles.[2]

The sheet lines of the Old Series north of the Preston-Hull line were thus regularised, and since these sheets were also derivative – first, from six-inch Lancashire and Yorkshire sheets and later from 1:2500 Cumberland, Durham, Northumberland and Westmorland sheets – and thus both accurate and up-to-date, they were suitable for adoption by the New Series when it finally came to be sanctioned. Extending the rationale of these sheet lines southwards, and retaining the eighteen-inch-by-twelve-inch format, the configuration and sheet lines of the 360 sheets that constituted the New Series were created, all drawn to the same meridian (see page 8). The quarter-sheets making up sheets 91-110 of the Old Series were simply renumbered, thus becoming sheets 1-73 of the New Series. The numbering sequence now started in the north of England – Berwick upon Tweed, for example, appears on sheet 2 – and progressed southwards to end with sheet 360, which was one of the two sheets covering the Isles of Scilly. But it was not for a couple of years after the New Series had been sanctioned by the Treasury, in July 1872, that standard sheets for it were published for anywhere south of the Preston-Hull line, though at least one prototype map, numbered as a quarter-sheet, is known to have existed a little earlier.[3]

During the later years of the Old Series, when sheets above the Preston-Hull line were being prepared, it had been decided to issue two versions of each. In a departure from what was then normal, the indication of relief with the use of hachures was omitted from the new alternative version and contours were used instead. When, sometime later, preparation of the New Series sheets for southern England was underway, it was again decided to provide two versions of each. The first was partly of customary design with relief being shown by hachuring but augmented by contours, and the second, in what turned out to be a much more viable and popular version, was one where contour lines alone were provided. These two versions, the 'hills' and 'outline' versions respectively, would be followed in the long history of the New Series, lasting until relatively recent times, by other variants that by the end of the nineteenth century were making use of multi-colour printing.[4] As just implied, the New Series came to be published in several editions of which only the first two are decidedly nineteenth century in origin. The third is associated with the first decade of the twentieth century and remained current up to the outbreak of the Great War. Each of these engraved editions had its own identifying hallmarks and features, but changes in style and

---

[2] For detailed discussion concerning the later sheets of the Old Series and their relevance for the inception of the New Series see Harry Margary, *The Old Series Ordnance Survey maps of England and Wales,* volume VIII, *Northern England and the Isle of Man,* Lympne Castle: Harry Margary, 1991, with *Introductory Essay* by J.B. Harley and R.R. Oliver, and *Cartobibliography* by Richard Oliver, *passim*, particularly pp viii, xxii ff.

[3] Hellyer and Oliver, *One-inch engraved maps of the Ordnance Survey from 1847,* 43-5.

[4] Seymour, *A history of the Ordnance Survey,* 164; Tim Nicholson, *The birth of the modern Ordnance Survey small-scale map: the Revised New Series colour printed one-inch map of England and Wales 1897-1914,* London: Charles Close Society, 2002; Hellyer and Oliver, *One-inch engraved maps of the Ordnance Survey from 1847,* 63-4.

policy for depicting topographic detail or the use of annotations occurred throughout the lifetimes of all three. The sheets of each of these first three editions – or, as they have been described: the New Series, the Revised New Series and the Third Edition – were issued in a number of states. But, unlike sheets of the Old Series where evidence of changes in style for the mapping of detail went in phases, the sheets of this family of New Series editions reveal changes in style showing them to have been constantly evolving. Unlike the Old Series, and in recognition of the need for revised editions, across-the-board upgrading of topographic detail on New Series sheets took place, not just once – to provide the Revised New Series – but for a second time too, resulting in the Third Edition, almost within twenty years of the sluggish introduction of this series to southern England. Again in contrast to the Old Series, the use of symbols for showing ground detail was expanding, requiring the use of a legend to assist the map-user, and this too evolved with successive editions.

The early sheets of the New Series

But in contrast to this later burst of energy, early progress on the southern sheets for the New Series was muted, so much so that only 24 sheets had been published by 1880; all were for the south-east of the country and, with a single exception, only as yet issued in outline version.[5] However, the pace of production eventually quickened, reaching a peak *circa* 1889, and all sheets for England below the Preston-Hull line were published by March 1895, with the remaining Welsh ones by the end of the year. This meant that the New Series had been completed over not much more than two decades – contrasting starkly with the protracted progress of the Old Series. For its early coverage of the south of England, the manner of depicting topographic detail on what retrospectively came to be seen as just the first of a sequence of New Series editions – but usually abbreviated, as here, to 'the New Series' – not unnaturally kept the conventions of the later Old Series sheets that had become New Series sheets by adoption. As will be appreciated, this was a time of silence for the depiction of the windmill on the one-inch. This meant that the iniquities of disregarding the sparse windmill populations in the north of England were now about to be repeated with the much larger windmill population of Kent.[6]

Showing the Medway towns, the outline version of New Series sheet 272 – or, if we use the Oliver notation (see page 7) to create a shorthand description for it: sheet NS-1-O:272 – is known to have been issued in at least ten states, where not only was the railway account regularly updated, but changing thoughts were also incorporated on how best not to show the naval dockyard at Chatham and its protective ring of army fortifications.[7] This militarily very important sheet had been among the earliest of the southern sheets to be published – its first state is dated 1876 – and, as further replacements for the most outdated of the Old Series sheets were issued, Kent inevitably benefited particularly well from the small number of new sheets published for the south-east before 1880. But in line with the new thinking,

[5] The outline version of each sheet quickly became the standard sought-for product, so much so that for many sheets, the hills version – a much more difficult map to read – never made it to publication in this edition.

[6] As suggested already, the first priority of the Survey in returning south was to make new maps available for key military areas. Primarily, that meant the naval stations. So it was that Hampshire and Kent became two of the first counties to be surveyed under this regime. Of these, Kent was (and is) much the more prolific for windmills.

[7] The evidence for military establishments was often suppressed, simply leading to blank spaces on maps.

the invisibility of windmills on these sheets is near-complete.[8] On the eastern side of Thanet – a part of Kent centred around Margate, Ramsgate and Broadstairs – the Survey had much earlier in the century recognised a windmill population of seven: in three groups of two for each of these towns, together with a solitary mill at St. Lawrence. This position was updated on the occasion of a later visit from the OS railway reviser, with three mills being added at Drapers, Newington and Northwood, though at the same time the St. Lawrence windmill disappears.[9] Later again, the loss of a mill at Broadstairs – seemingly due to the building of the Kent Coast Branch Railway – was acknowledged, and even though by *circa* 1873 when a later state of Old Series sheet 3 showed eight windmills left in the area – a reality that was certainly flawed – this had been a not unreasonable overall account of the windmills here, even though it may not have been a definitive one.[10] In embracing the norms of windmill depiction for the Old Series, this account for the eastern part of Thanet epitomises the type of accuracy that the Old Series had broadly been achieving in its recording of windmills, especially those close to railway lines. But, in sharp contrast to the recognition of these mills offered by the Old Series, the depiction of windmills in this area on the New Series is dire. Thanet is shown on sheet 274, which was based on large-scale survey data of 1871-2 and first published in 1878. How later states of sheet 274 portrayed Thanet allows features of New Series sheets to be noted. Contours, for example, appear as streams of small, very faint dots and, after 1881, parish names appear in open-shaded lettering.[11] In addition, sheets are no longer just numbered but each is now given the name of a prominent place shown on it. Hence sheet 274 is perhaps more easily and better known as the *Ramsgate* sheet, which can result in it being more fully labelled as NS-1-O:274 *Ramsgate*, though in most contexts sheet 274 *Ramsgate* should be sufficiently unambiguous. More noticeable, however, is the silence of this particular sheet with respect to the windmills in its area. This is all but complete; only the pair of mills outside Margate at Drapers is recorded, and then only with an annotation of *Mills*. This observation echoes what the careful reader may have gleaned from looking at

[8] Although the date of initial-state publication was provided on each New Series sheet, the year(s) of survey for its constituent 1:2500 sheets initially were not, thus leaving the casual map user unsure as to the datedness of the one-inch sheet. This was later remedied when survey details were added but this was not done until the early summer of 1891, and then only for newly-published sheets. Sheets of the Revised New Series carried the survey and publication dates of their earlier New Series counterparts and also their own dates of revision and publication. Discrepancies can often occur between New Series and Revised New Series sheets over some of this information. This is the result of a change in procedure introduced in 1888 whereby the date of survey was now to be taken as the date of final examination on the ground, whereas previously it had been the date of parent large-scale plans being passed fit for publication.

[9] All the symbols seen are of the post mill type as would be expected for a pre-1820s sheet – in this case a sheet published in 1819. This is all the more surprising as Old Series sheet 3 was extensively re-engraved *circa* 1840, and furthermore also ironic since Northwood Mill was a tower mill and the rest probably all smock mills. The mill at St Lawrence had been moved by the railway company who needed to build a station on its site. In all likelihood, it was moved to become the windmill later recorded at Newington.

[10] The authority for the windmills of Kent is William Coles Finch, *Watermills and windmills*, London: C.W. Daniels, 1933. This volume deals less than completely with the precise fate of all these windmills but considers that of the eight, one had certainly disappeared by 1870. However, another three mills – at Margate, Ramsgate and a second mill at Drapers – are not shown on the Old Series even though all are safely known to have been working at this time. Of these, though, two were sited in areas of close detail and hence not easily shown.

[11] The lettering style introduced in 1881 followed earlier Old Series usage. It subsequently appears on some initial states of sheets as well as on later states of the earlier pre-1879 sheets (including sheet 274). But it was very soon phased out with no sheets first published after 1883 having it; see Hellyer and Oliver, *One-inch engraved maps of the Ordnance Survey from 1847*, 48.

the extract from an adjacent sheet (see Fig. 2.5.2). In providing a rather odd depiction for Borstal Hill Mill, and in failing also to record any other mills in the vicinity of Whitstable, sheet 273 *Faversham* is doubly remiss in its windmill representation.[12]

This deficiency of the New Series quickly repeated itself throughout much of Kent, so that areas poorly served in the previous half-century because the Old Series did not provide up-to-date accounts of their mills continued to suffer. Taking the village of Headcorn as the hub of a small circular area – five miles or so in diameter – just large enough to include two windmills (one at Smarden, the other at Staplehurst) that were recognised by the Old Series, neither of them is later shown by the New Series. Instead, just a single *Windmill* annotation is provided for this small area: this acknowledges a previously unrecorded mill at Headcorn. This record of two windmills, and then just one, by the Old and New Series respectively falls hopelessly short of the mark since at least half-a-dozen windmills are known to have been working in this area at the times of survey for each of the two series. During the entire period of large-scale survey here (1864-71) this tally of windmills had included not one mill at Headcorn, but two that stood beside each other next to the South Eastern Railway.[13] One would think that if any windmills were going to find themselves recorded, then these two in such a prominent position next to a railway would so qualify; it is certainly surprising to discover that one mill could be recorded without the other.

Thanet and the small area centred on Headcorn both exemplify the policy for depicting windmills on the New Series which was in force up to *circa* 1880, but thereafter the policy of suppressing knowledge of windmills seems gradually to have been relaxed. So, although it is possible to come across sheets for the south coast where one may suspect the recording of mills has been seriously blighted, New Series sheets first published in the 1880s fare better. As already seen in the case of sheet 272 *Chatham*, New Series sheets, like their predecessors, could be the subject of incremental ongoing revision over a number of successive states, the exact number depending on how liable an area was to change and, particularly, if there was a strong railway presence in the area. The number of states issued was also likely to be more, the earlier the initial date of publication had been. Thus, for example, more than a dozen different states have been identified for sheet 270 *South London* reflecting, just as for the Old Series, a continual desire to update information about the railways. But for other sheets, far fewer states were generated; we find sheet 304 *Tenterden*, for example, to have been issued in only five identifiable states over a span of fourteen years, starting in 1879.[14] But, this sheet is

[12] Figure 2.5.2 usefully also shows the form of weak contouring used on these New Series sheets, as well as some hand-colouring, seen here highlighting the road next to the mill but used also for drawing attention to parks, railway lines and water. This was an optional extra applied by the map-selling agencies rather than the Survey, and was clearly not part of official OS practice. By the same token, hand-washes were applied across each sheet on a county-colour-coded basis where boundaries with other counties or the sea were emphasised by a heavier wash. Most counties kept their stock colour throughout the later period of the Old Series and into the twentieth century before the practice was dropped. Kent, whose colour had still been red at the time of the early New Series sheets, was one of the few counties to later change its colour – in this case, to sienna. Lincolnshire was another county whose colour changed at the same time, this time from sienna to green.

[13] This area around Headcorn is mainly covered by sheet 288 *Maidstone* that, like sheet 274 *Ramsgate*, was initially published in 1878. Other windmills in use at the time of the 1:2500 survey were at Frittenden (Sinkhurst Green), Smarden (two mills) and Staplehurst. Together with the pair at Headcorn, all six mills were later recorded by the Revised New Series.

[14] Hellyer and Oliver, *One-inch engraved maps of the Ordnance Survey from 1847*, 385. The cartobibliography that forms part of this book was prepared by Roger Hellyer. So, in much the same way that different states of Old

interesting for what it says about how windmills were handled during that time. The initial state carries no windmill information at all, but the second state, which has an electrotyping note of 1879 though it was probably issued *circa* 1882, has had *Windmill* annotations added in respect of twelve mills. No discernible ground plan is present for nine of these mills but, for Goodshill and Peasmarsh Mills, small dots within small circles indicate precise locations for these windmills. Silver Hill Mill, in Sussex, is marked by a small dot within a triangle, so this could be referring to the triangulation station that lay a few yards from the mill. The last three Hellyer states all have open-shaded parish lettering and, by the time of the fifth state of *circa* 1892, another four *Windmill* annotations have been added to the sheet including one at Woodchurch. The date of this fifth state of sheet 304 *Tenterden* is critical since it postdates the reintroduction of the windmill symbol *circa* 1888. It would be very convenient indeed to assume that these last four annotations had been added by 1888, otherwise it would mean that not only were later states for a New Series sheet being issued with enhanced evidence for windmills, but that this was being done using a, by now, outmoded means of depiction rather than the newly-reinstated recording device of symbols.[15] What may not be assumed, though, is that the representation of mills on the last state of the *Tenterden* sheet was, at last, correct. Many prominent mills still escaped disclosure, among them two large smock mills at Sandhurst and Cranbrook (Union Mill). Interestingly, rather than Woodchurch having just one mill, there were two smock mills working here in close proximity to one another, though each was distinctive in its name and paintwork – the White Mill and the Black Mill. But, in only adding a single annotation to the last state, the associated ground plan, which had been altered and moved between states, maybe to accommodate the parish writing, fits the true position of neither mill. Other changes to the small ground plans at this scale, or additions to them, are found elsewhere such as at Benenden Mill, but the motives for these, and indeed for the partial windmill recognition on the *Tenterden* sheet, are hard to fathom. Whatever they were, they could have had no basis in any structural or visual differences, or in operating differences between the windmills then active in this part of Kent and Sussex.

The evidence of the *Tenterden* sheet state of *circa* 1882 may be a signal that the eventual decision to reintroduce the mill symbol was the result of a gradual move to rehabilitate the windmill rather than this arising from any sudden change in policy. But once it is realised that many other New Series sheets of the time were disinclined to recognise windmills at all, and that no other later states for a sheet appear to have been upgraded in respect of mills in the same partially-considered way that the *Tenterden* sheet was, it surely has to be concluded that sheet 304 *Tenterden*, if judged on the changing depiction of its windmills, is something

---

Series sheets can be identified as Margary states, different states of New Series sheets can be acknowledged as Hellyer states.

[15] This assumption is perhaps endorsed rather than refuted by a general absence of windmill-related evidence in Hellyer's account of New Series sheet states. Mills are certainly mentioned by him but only in the sense that they can be helpful in acting as one single diagnostic tool among many for distinguishing between states. No claim is made for offering an all-inclusive treatment of mills between sheet states. Even so, it is significant that the remarks for windmills made by Hellyer, beyond his mentioning sheet 304 *Tenterden*, are as few as two. He notes that state 6 (of 9) of sheet 286 *Reigate* – a sheet whose Revised New Series edition would recognise more than a dozen windmills, five of which were totally ignored by this state 6 of the New Series – added one mill symbol *circa* 1889. This was recording an old post mill thought not to have worked since 1875. The suggestion is also made that sheet 79 *Goole*, issued in only one outline state, was maybe the first to make use of windmill symbols – in October 1889. Hellyer and Oliver, *One-inch engraved maps of the Ordnance Survey from 1847*, 213, 369, 385.

of an aberration rather than the outcome of considered policy. One explanation for this, it has been suggested, could lay with an officer whose tour of duty was not sufficiently long for him to become familiar with all aspects of how map detail was to be shown; except that this was an anomaly that had a lifetime of nearly a decade.

The reinstatement of the windmill symbol onto the New Series

Publication of the early New Series sheets in the south was followed in the fullness of time by provision of all the rest up to the start point of Lancashire and Yorkshire – all reliant on the rolling programme of large-scale county surveys. Broadly, the programme for publishing the derivative New Series sheets shadowed that of the large-scales. One objective may have been that the datedness of Old Series sheets would decide the order in which counties were re-surveyed, but other considerations played an important part here. Among these was the priority accorded the 'mineral districts'. These primarily meant the coalfields, but included the tin-mining areas of Cornwall as well. Also, the War Office had voiced its concern that the eastern counties be considered urgently. This led to the surveys of Norfolk and Suffolk, for example, taking precedence over that of Lincolnshire, which was thus late (1883-8) in an overall programme that lasted from 1854 to 1893. This contrasted unhappily with the earlier Old Series experience when this county had been mapped ahead of turn. The upshot was that the New Series sheets for Lincolnshire, badly needed to replace Old Series sheets of the mid-1820s, were not issued until 1890-2. On the other hand, publication of northern New Series sheets of Lincolnshire, such as that of sheet 89 *Brigg* in 1890, still predated some of the southern sheets of the West Midlands and parts of East Anglia.[16] So, the broad plan for publishing New Series sheets in a general sweep from south-eastern counties progressively northwards and westwards does come with complications.[17] After Hampshire and Kent, it was the turn of Sussex and the home counties to be re-surveyed. Maps for these counties continued the practice of depicting windmills in a desultory fashion by supplying a modest number of annotations, sometimes with an identifiable ground plan. But by the mid-1880s, some time before the final state of the *Tenterden* sheet, but approaching 1888, this was being done in an increasingly enthusiastic manner.

It is not difficult to assess how coherent this application of windmill-related annotations was becoming, and how comprehensive New Series mill counts became on ensuing sheets. The utter inadequacy of the New Series in Kent was revealed by looking at two small areas in conjunction with a good (but less than definitive) sourcebook for the county's windmills – Coles Finch.[18] For many counties, there are no books that can compare with this one. Yet even the most insubstantive and mediocre listing for the former mills of a county can offer a level of knowledge useful for affirming, or not, the poor standing of New Series sheets. But to be able to assess all sheets fairly and uniformly, what is needed is a way of selecting unbiased, manageable samples of windmills from all those that existed in an area at the time

---

[16] Devon provides an even more extreme case of the large-scale survey leading belatedly to a replacement for the Old Series. The Old Series sheets for this county had been published in 1809 yet under the rolling programme of large-scale county surveys, it was the last county (excluding the two re-plotted ones) to be finished, and this was not until 1889.

[17] Hellyer and Oliver, *One-inch engraved maps of the Ordnance Survey from 1847*, 45.

[18] Coles Finch, *Watermills and windmills*.

of re-survey. One way of choosing such a sample is to consider only those mills that have survived – however that may be defined – to the modern day, since the survival of any one windmill through a century's worth of uncertainty can have little, if any, association with its chances of having been chosen by the Survey, whatever the criteria, to feature on the New Series more than a century ago.[19] Moreover, as the windmill was soon to be rehabilitated onto the New Series with the reinstatement of mill symbols in 1888, this way of selecting a sample can equally be applied to gauging how inclusive the New Series was of windmill populations after that point when, seemingly, it was more prolific in its depiction of mills.

Following the deficiencies in representation of Kent's windmills and undoubtedly those of other southern counties, the publication dates of New Series sheets for Buckinghamshire straddle that time when the windmill symbol was being reintroduced. The result of this is a combination on these sheets of some mills going unrecognised, while others are recognised in different ways, either by an annotated ground plan of varying explicitness, or by a symbol that is unannotated. The secondary windmill literature for Buckinghamshire can safely be described as lightweight. Aside from many generalist, county-magazine types of article, it is practically non-existent apart from a slim volume that is one of a series of county surveys of surviving windmills undertaken in the 1970s, primarily by Arthur Smith.[20] The format Smith introduced remained a popular one for quite some time and was adopted by other authors for similar work, the thorough study of the windmills of Lincolnshire by Dolman being just one example.[21] In these surveys, the amount of historical knowledge assembled for each mill is minimal; essentially their importance is as gazetteers. The sixteen windmills noted by Smith as surviving in Buckinghamshire *circa* 1976, as well as an indication of how the New Series recorded them *circa* 1887, are listed below (Table 4.1).

The way in which these windmills are either recognised or not is instructive. From a small sample of just sixteen mills, half are recognised with the familiar New Series device of an annotated ground plan. Indeed, the plans of the two post mills, one of which is Nixey's Mill at Brill, each consists of a circular solid spot encircled by a pecked line, which can be imagined as the line traced out by the foot of the stairs to the body of the mill when they are turned, or luffed, to ensure the sails face into the wind.[22] The tower mill plans lack these pecked lines, which is to be expected if the conjecture just offered is correct. For the eight

[19] This is, for the time being, setting aside the issue of whether or not a windmill was in working order when it was surveyed. Obviously, a mill in a state of near-collapse at the end of the nineteenth century is less likely to have survived to modern times than one which was then still working. Essentially, the argument here is that the many factors affecting the occasion of a subsequent mill clearance could not have been apparent to the surveyors in the case of any individual windmill. Thus, this way of sampling a population of windmills can justly be considered an unbiased one.

[20] Arthur C. Smith, *Windmills in Buckinghamshire and Oxfordshire,* Stevenage Museum Publications, 1976. Although sole author for the survey of surviving windmills in Buckinghamshire, Smith also collaborated with other authors to provide comparable surveys for other counties.

[21] Dolman, *Lincolnshire windmills*. This review of one county's windmills conforms to the format developed for the series, but works harder than most to provide some historical context for each windmill in the county still around at the time of publication. This text can be regarded as representing a second-generation model for the series.

[22] On a technical note (though not applicable to the comparatively modest post mills of Buckinghamshire) the weight of the stairs to the body of a large post mill, which typically would have had a fantail and been self-luffing, was conveyed to the ground through heavy iron wheels running on a set of, usually, flat wooden boards laid as a permanent feature on the ground in a circular pattern around the mill. Though neither a plateway nor a tramway in the true technical sense, this set-up for a large post mill is often referred to as its 'tram'.

annotated ground plans, the annotation *Windmill* is used with the single exception of the mill at Coleshill, where *Mill* is used instead. In another solitary instance, an unannotated windmill symbol is used without any ground plan recognisable as being that of a mill, but this appears on the single sheet from the group that was published in 1889. But for each of the seven windmills recognised in the then normal manner of a ground plan accompanied by a *Windmill* annotation, another windmill goes entirely unrecognised. What is more, the geographic spread of each group of similarly recorded mills offers no easy reason as to why these differences may have occurred. This sample of windmills, chosen with methodological concern, and illustrative of the divisions in windmill depiction that were occurring late in the 1880s, is nonetheless a very small one. Moving ahead in time and hence ever northwards, the same type of exercise can be undertaken for other counties – preferably ones with a larger surviving windmill population – where the New Series sheets were published after the reintroduction of the windmill symbol. The requirement for a statistically useful sample size severely limits the candidacy of most counties, since most from the north and the Midlands can usually only offer the sorts of mill numbers that Buckinghamshire was able to provide. To illustrate what is meant here; Derbyshire is another county where the windmill literature was for many years almost non-existent. This gap in the literature was eventually filled by Gifford who provides details of the meagre twelve windmills that remained in one form or another in Derbyshire in the mid-1990s.[23] Their details of depiction on the New Series are again listed below (Table 4.2).

Name of Mill & type		New Series depiction, sheet number & date of publication		Name of Mill & type		New Series depiction, sheet number & date of publication	
Beaconsfield	T	×	255 1886	Ibstone	S	*Windmill*	254 1887
Bradwell	T	×	203 1887	Lacey Green	S	*Windmill*	238 1887
Brill	P	*Windmill*	237 1886	North Marston	T	×	219 1887
Cholesbury	T	×	238 1887	Pitstone Green	P	*Windmill*	238 1887
Coleshill	T	*Mill*	255 1886	Quainton	T	×	219 1887
Edlesborough	T	×	220 1889	Stewkley	S	✓	220 1889
Fulmer	T	×	255 1886	Thornborough	T	*Windmill*	219 1887
Great Horwood	T	*Windmill*	219 1887	Wendover	T	*Windmill*	238 1887

Table 4.1. Recognition of a sample of Buckinghamshire windmills by the New Series (circa 1887).

Key: ✓ indicates the use of an unannotated symbol to denote this windmill
Windmill / *Mill* indicates the use of this annotation with a ground plan
× indicates that this windmill is not recognised at all by the New Series

The depictions of these windmills on the New Series, published just a couple of years after those of Buckinghamshire, present a very different picture. Here, the sole attribute of recognition for each mill is whether or not a symbol is used. From an even smaller sample than before, eight mills are accorded a symbol while only four are not, which, the question

[23] Alan Gifford, *Derbyshire windmills,* second edition, Birmingham: Midland Wind & Water Mills Group, 2003.

of sample size aside, does suggest that the New Series recognition of windmills is improving with the move northwards. This suggestion is maybe supported by the realisation that one of the unrecognised mills is Belper Mill, which was shown earlier (see page 20) as probably having been in a parlous state *circa* 1890. The possibility of extending our interpretation of these findings is limited as the absentee windmills show no apparent common grounds of dereliction, remoteness or map congestion to explain their non-disclosure. In any case, the small sample size clearly dilutes the value of whatever meaning is imputed to these figures.[24]

Name of Mill & type		New Series depiction, sheet number & date of publication		Name of Mill & type		New Series depiction, sheet number & date of publication	
Belper	T	×	125 1890	Nether Heage	T	✓	125 1890
Bolsover	T	✓	112 1889	Normanton:			
Carsington	T	✓	124 1889	Fordbridge	T	×	112 1889
Dale Abbey	P	✓	125 1890	Common	T	×	112 1889
Findern	T	✓	141 1890	Ockbrook	T	✓	125 1890
Fritchley	P	✓	125 1890	Spancarr	T	×	112 1889
Melbourne	T	✓	141 1890				

Table 4.2. Recognition of a sample of Derbyshire windmills by the New Series (circa 1889).

Key: ✓ indicates the use of an unannotated symbol to denote this windmill
× indicates that this windmill is not recognised at all by the New Series

Plainly, what is needed to advance this analysis of the New Series is a county-sized area where the windmill symbol was the only type of depiction considered *ab initio* for depicting windmills, and where a sizeable portion from a large one-time mill population has survived. Then, with a good sample size of mills to hand, an indication can be sought for just how comprehensively windmills were recognised in the years after 1888. (It is assumed here that since the Survey made a decision to change the nature of its depiction for windmills, then it would have been serious about its later recording of them...!) Norfolk offers one such area where the towers of grinding and drainage windmills, both found in large numbers, were sufficiently resilient to have endured *en masse*. Accordingly, the survey of 'corn windmills' of Norfolk made by Smith might be used to better effect than some of his others. Better still is the availability of Apling's definitive study, mentioned in the last chapter, of all the grinding and other industrial windmills of Norfolk that survived in one form or another to 1984.[25]

---

[24] What can helpfully be added here is that the Survey embarked on a new regime for field-checking its one-inch scale *circa* 1887-8. See Hellyer and Oliver, *One-inch engraved maps of the Ordnance Survey from 1847,* 49.

[25] Apling, *Norfolk corn windmills.* As the title of the book implies, its remit is limited to considering cereal mills and, though industrial grinding mills are also included, drainage mills most certainly are not. The inclusion of drainage mills under the general umbrella of windmills has already been discussed and, in the context of surveyors making their field notes and seeing much in common between the two types of structure, to the Survey this inclusiveness made a lot of sense. Nonetheless, there are those who make a clear distinction between them on the basis of their having very different functions, and this remains a strong tradition among modern-day writers. As an example of the meticulousness that Apling applied to his definition of the continuing existence of mill remains, he indicates for the tower mill at Swanton Abbott that 'in 1926 the tower was still standing as an empty shell, but it was later

Apling's recognition of industrial windmills – the fact that mills powered by wind could be used in roles other than just the grinding of cereals – is particularly useful. The secondary mill literature is often guilty of overlooking this, and it may also have been a grey area for surveyors when needing to decide what constituted a notifiable windmill structure. The restored waterside Berney Arms Mill at Reedham in Norfolk, for example, had been used for grinding cement as part of the nearby Berney Arms Works before being converted to a drainage mill in 1883 after the works closed. The New Series chose not to record this mill.[26]

Apling deemed that the remains of 123 Norfolk windmills were still identifiable in 1984. How those 123 mills were recognised on the New Series can be categorised by type of mill and type of depiction (Table 4.3). More than it just being the case that a greater number of windmills are involved, the findings from Norfolk are much more gratifying than those for either Buckinghamshire or Derbyshire, in that a solid majority of the sample windmills are now recognised. But the geographical distributions of mills that are recognised and of mills that are not offer no easy explanation for this differentiation. In the event, from the sample of 123 mills (where the population of windmills in Norfolk in the 1890s would have been considerably bigger) a few retain use of an annotated ground plan. One possible reason for this is the association of each of these mills with a more prosperous watermill – putatively in all four cases. Otherwise, the numbers of windmills recognised to those unrecognised, rests at 95 to 24, suggesting that very nearly one fifth of the sample windmills were ignored, which, although something of an improvement over the two earlier samples, is still less than admirable given the derivative nature of this mapping and the presumed revival of intent to recognise all mills. Comparing the small-scale New Series sheets with those of a larger scale is one obvious step to help us decide if particular mills were deliberately overlooked, or if in some way their presence on the ground was a problem for the surveyor. Otherwise, reading Apling's description for each sample mill might isolate a common feature for the absentee mills but, if not, then the approach of appealing to the larger scales to find a reason for this significant shortcoming is our best option.

As might be expected, the larger, yet still derivative, six-inch map offers some increase in the number of sample windmills recognised. One standard type of depiction found at this scale – a circular ground plan with an annotation that includes the word 'Windmill' – is used for depicting many windmills on the Norfolk six-inch, including fifteen of the sample mills not recorded on the one-inch. For other sample mills, either where their six-inch depictions are ambiguous, or where there appears to be no recognition at all for windmills at this scale, then further recourse can be made to the 1:2500. Often, as explained in Chapter Two, the

demolished to ground level. Only grassed-over foundations remain to mark the spot'. In truth, any assessment of numbers of surviving windmills is prone to being flawed, not only for reasons of definition, but for the fact that the destruction or dubious refurbishment of incomplete remains is an ongoing activity, and any assessment that is published inevitably misrepresents the up-to-the-minute situation. But more importantly, Apling appears to put minimal value on the evidence of maps for his individual mill biographies, instead placing more emphasis on, for example, auction notices and land use returns. It had been intended by Apling that his book would be the first in a number of volumes that would cover all Norfolk windmills, those remaining and those that had disappeared. Sadly, his death thwarted the publication of any further work beyond volume one.

[26] Apling, *Norfolk corn windmills,* 45. This mill seems to have been built around 1865 to continue the work of an earlier mill that had at one time been a saw mill, but which by the 1860s was grinding cement clinker. This gave rise to the unusual Old Series annotation of *Sawing Mill*, but the New Series was initially not nearly so helpful as far as the replacement mill was concerned.

way of recording the presence of a windmill at 1:2500 was simply replicated on the six-inch, but this did not always happen. But first, the evidence of the six-inch scale for all those windmills that went unrecognised on the New Series should be considered (Table 4.4). The diversity of annotations given at this scale owing to the different sets of circumstances that the mills presented, and which the Survey clearly thought it ought to disclose, is noticeable. Though discussion of this diversity might be interesting, it could say nothing to excuse the inadequacy of the one-inch scale. For a significant minority of sample mills to be of normal appearance and working conventionally, yet be disregarded on the one-inch even though the six-inch sheets, and thus the surveyors' field notes, do recognise them, seems irrational and inexcusable.

Type of Mill	Mills depicted by the New Series using an unannotated symbol	Mills depicted by the New Series using an annotated ground plan	Mills not recognised by the New Series
Post Mill	15	0	5
Tower Mill	76	4	17
Smock Mill	2	0	2
Composite Mill	2	0	0

Table 4.3. Summary of recognition of a sample of Norfolk windmills by the New Series (circa 1890).

Of the 24 mills not recognised on the New Series, eleven – almost half – are identified on the six-inch by ground plans, where each is unequivocally annotated with *Windmill (Corn)*. From the main sample of 123 windmills, three mills are known to have combined the roles of grinding and pumping – Catfield (Swim Coots), Oby and Smallburgh Mills. In addition, two grinding mills are known to have later been converted into drainage mills – Horning and Reedham Mills. Three of these five windmills are included in the list of 24 that went unrecorded on the New Series, and all three have helpful annotations on the six-inch scale. A further five windmills missing from the New Series have six-inch annotations that either make use of the word *Windmill* or, in the cases of Harpley Mill and West Walton (Highway) Mill, indicate the presence of a mill where the topographical circumstances safely point to a windmill. The five other windmills for which evidence from the six-inch is being sought to offset their lack of recognition at the one-inch, are each portrayed on the six-inch in a way that is much less positive. Close examination reveals that these five mills are all but invisible on the six-inch: they are accorded ambiguous unannotated ground plans. We are therefore obliged to resort to sheets of the 1:2500 to see what evidence they can offer for these five mills (Table 4.5).

1. *(right)* Post mill at Mumby in Lincolnshire *circa* 1910.

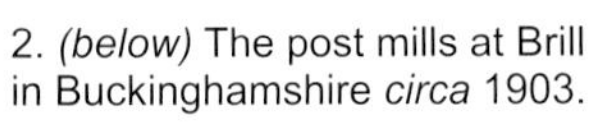

2. *(below)* The post mills at Brill in Buckinghamshire *circa* 1903.

3. Barrington Mill, Cambridgeshire *circa* 1920.

4. Greatham Mill, County Durham *circa* 1907.

5. Friston Mill, Suffolk *circa* 1925.

6. Earnley Mill, Sussex *circa* 1906.

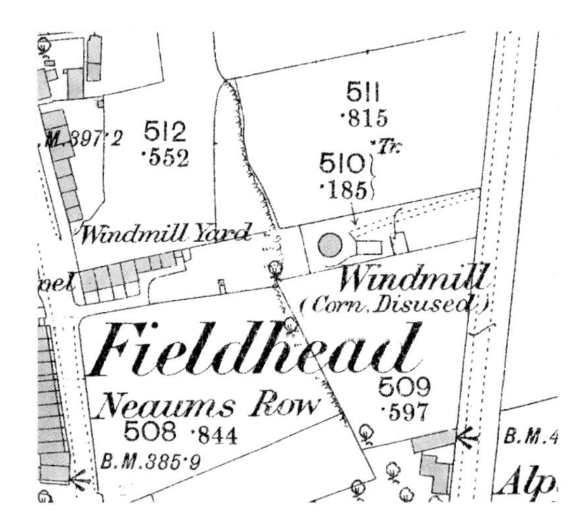

7. Belper Tower Mill, Derbyshire depicted at 1:2500 *circa* 1880.

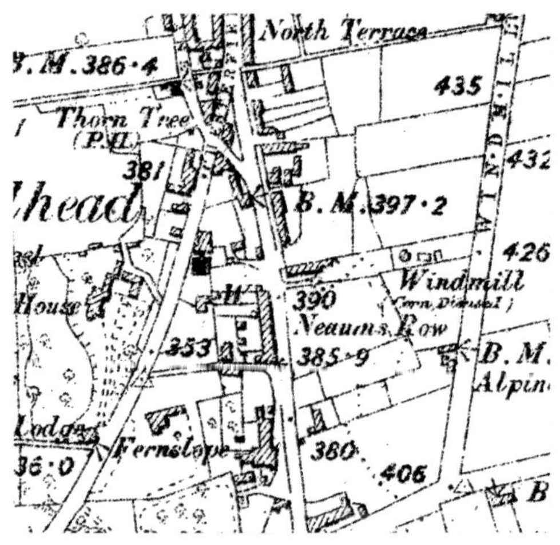

8. Belper Tower Mill, Derbyshire at 1:10,560 (six-inch) *circa* 1884.

9. *(left)* The demolition of Aldbourne Tower Mill, Wiltshire in 1900. *(Image courtesy of English Heritage)*

10. *(above)* The post mill at Sullington, Sussex *circa* 1912.

11. Thearne Mill, Yorkshire *circa* 1910.

12. Thurne Dyke Pumping Mill, Norfolk *circa* 1920.

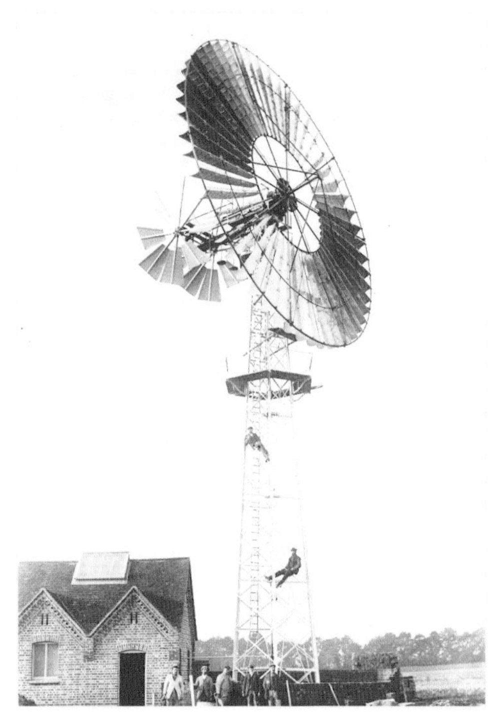

13. Wind Engine at Reading, Berkshire *circa* 1895.

14. Post mill at Thorne, Yorkshire *circa* 1904.

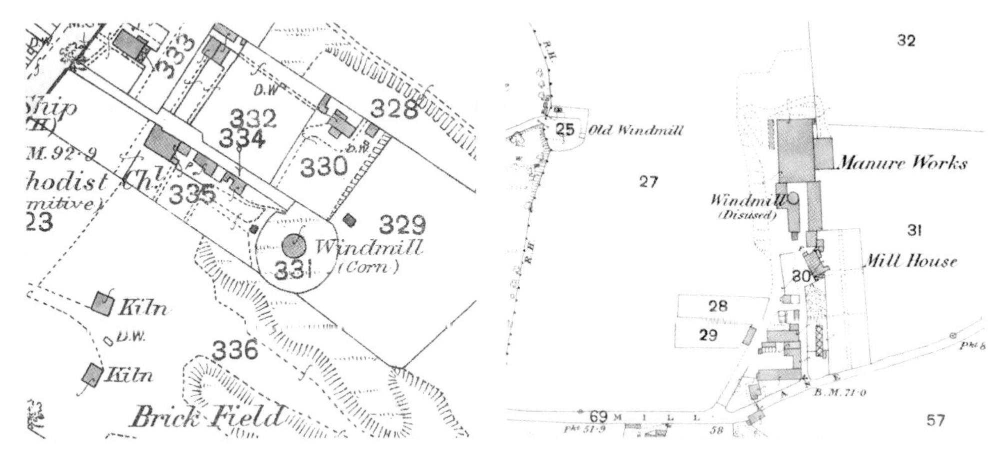

15. Windmill in Sprowston, Norfolk at 1:2500 *circa* 1884.

16. Two windmills near Norwich at 1:2500 *circa* 1886.

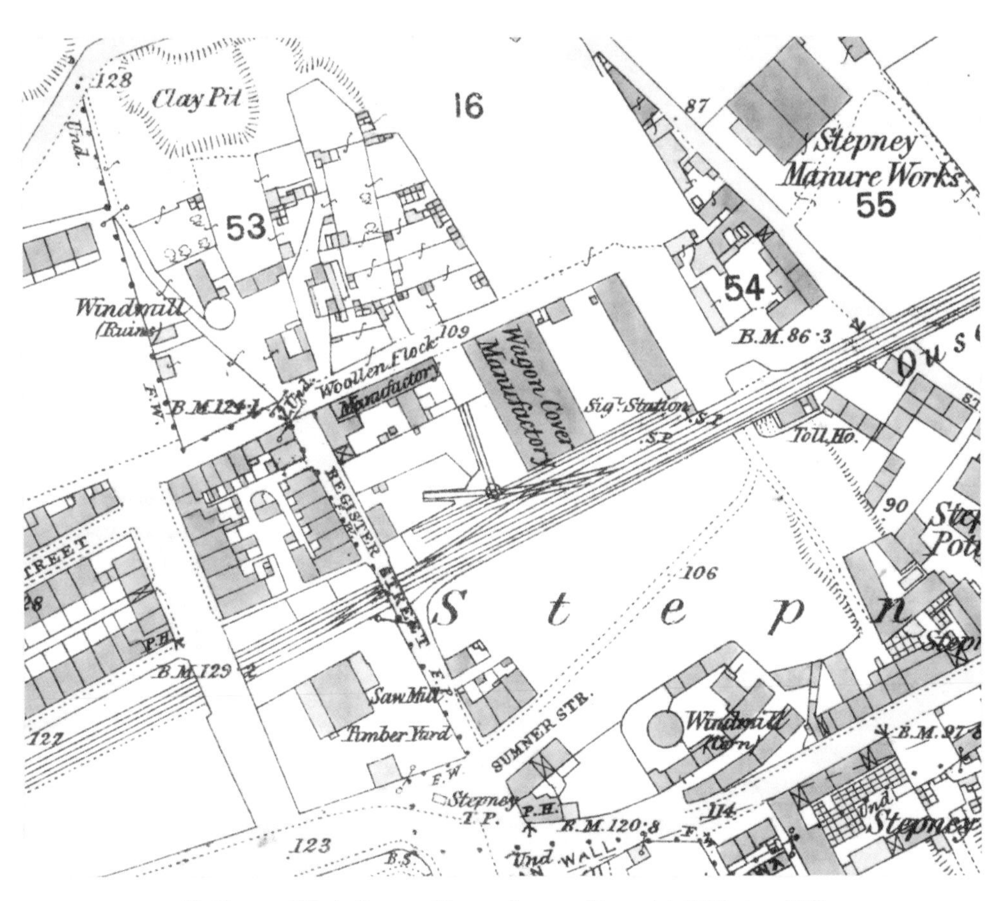

17. Stepney Windmills near Newcastle upon Tyne at 1:2500 *circa* 1860.

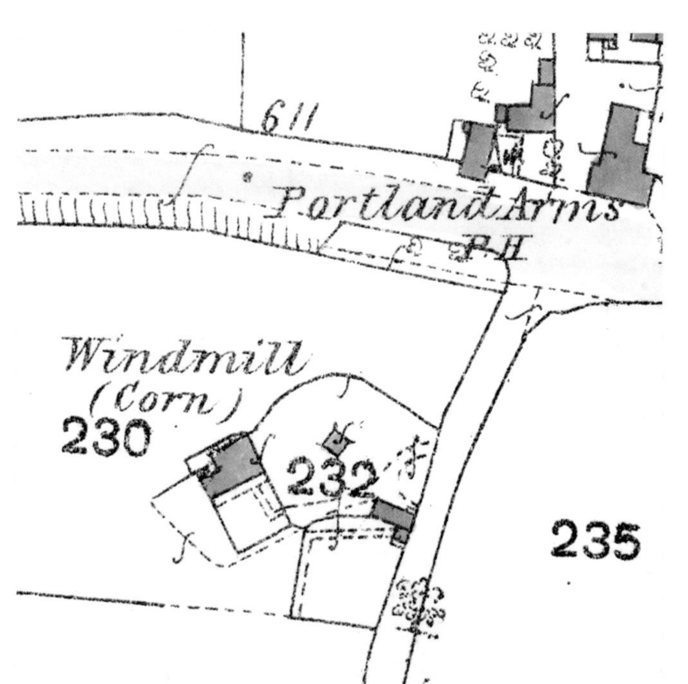

18. *(top left)* Post mill at Wetwang, East Riding of Yorkshire *circa* 1905.

19. *(left)* Hucknall under Huthwaite Post Mill, Nottinghamshire at 1:2500 *circa* 1879.

20. *(top right)* A view of Club Mill at Hucknall Torkard, Nottinghamshire, which shows the external design of the trestle's support piers.

21. *(top)* Windmill at Herne Bay, Kent in 1876.

22. *(left)* Woodham Mortimer Mill, Essex in 1891.

23. *(below)* Two smock mills near Margate, Kent at 1:2500 *circa* 1874.

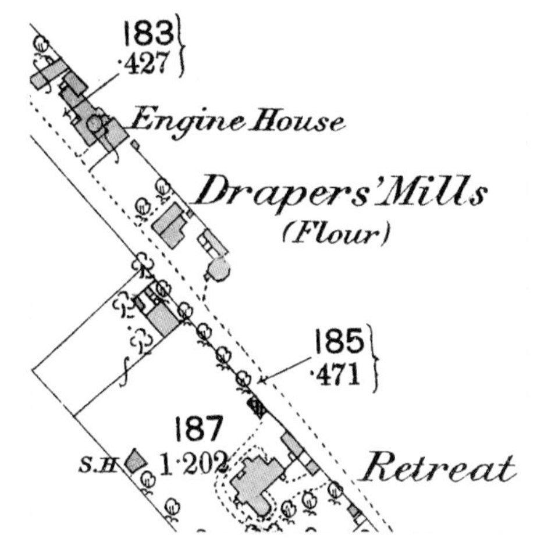

24. Part of Chatham in Kent as depicted by the second edition 1:10,560 (six-inch) *circa* 1898. Many users considered this a medium scale and its ability to show urban detail is limited, as can be seen. Colour has been applied here post-production to indicate the major roads and a pair of cuttings on the London, Chatham and Dover Railway. Only one windmill is recorded by ground plan and annotation, leaving at least one other to go unrecorded.

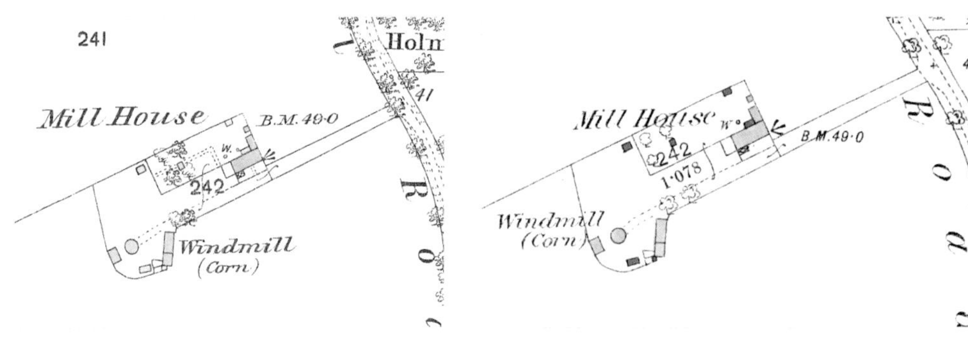

25. Windmill at Lound in Suffolk as seen on an early state 1st edition 1:2500 *circa* 1884.

26. Windmill at Lound in Suffolk as seen on a late state 1st edition 1:2500 *circa* 1896.

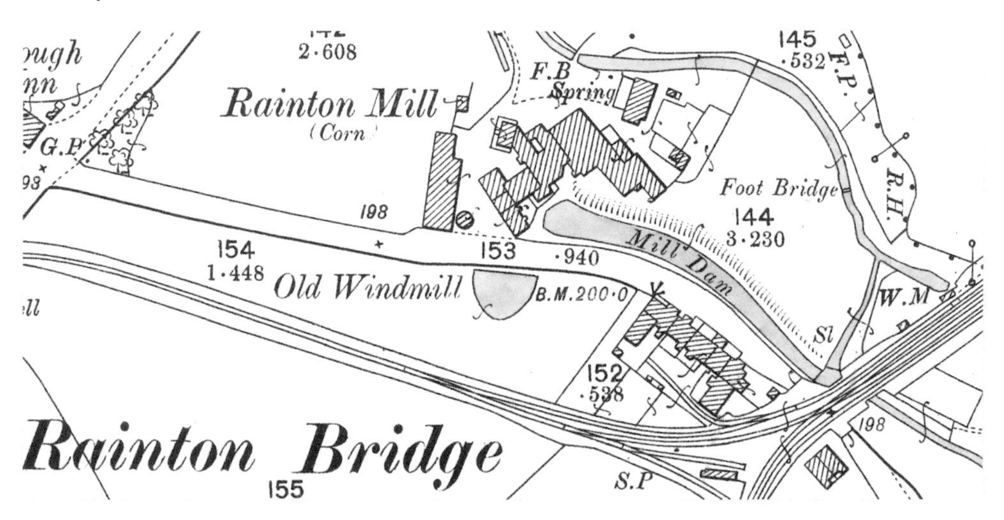

27. Rainton Windmill, County Durham at 1:2500 *circa* 1896.

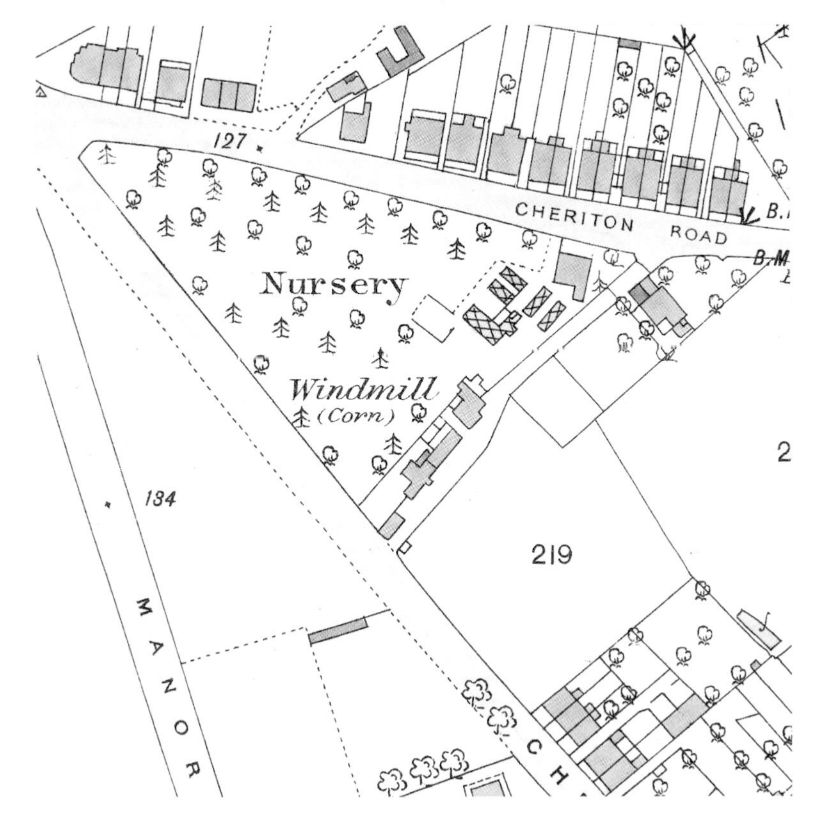

28. Part of Folkestone, Kent at 1:2500 *circa* 1873.

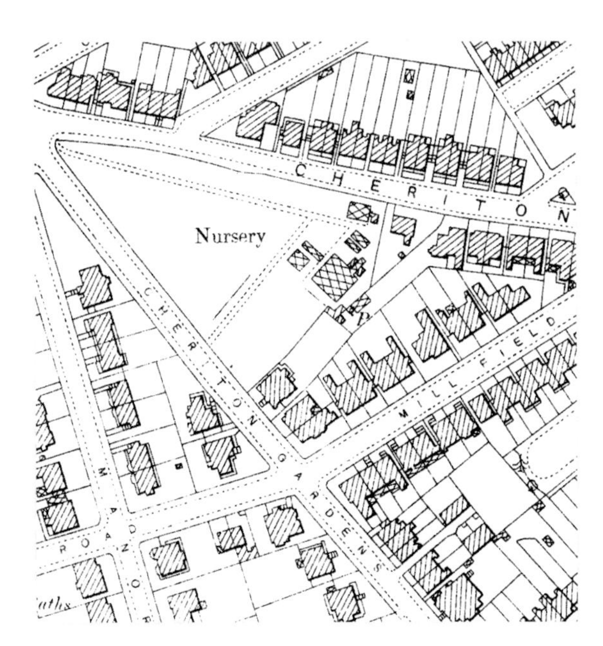

29. Part of Folkestone, Kent at 1:2500 *circa* 1898.

30. Darn Crook Mill in the centre of Newcastle upon Tyne *circa* 1890.

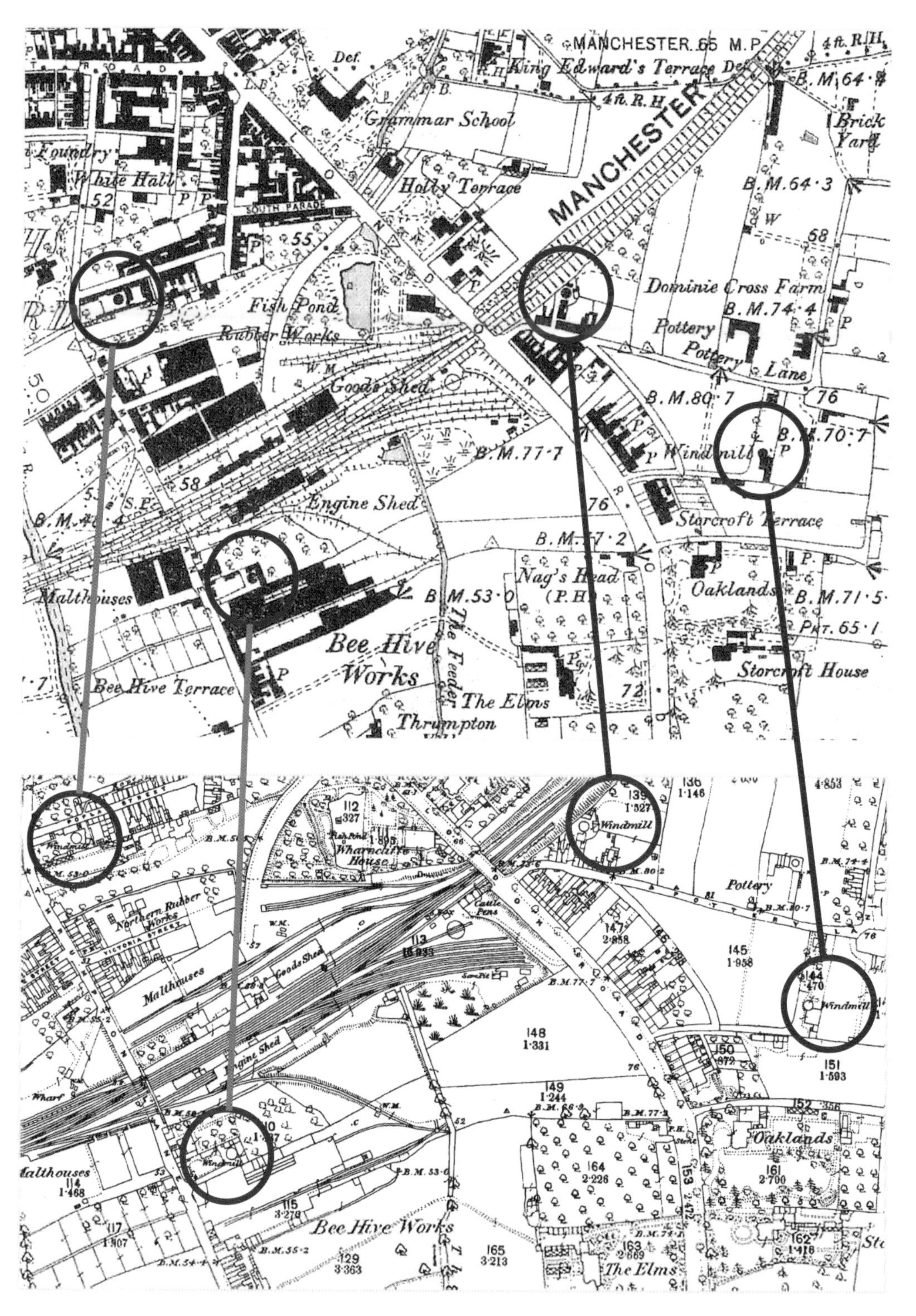

31. Part of South Retford in Nottinghamshire shown at: *(top)* the scale of six inches to one mile (1:10,560), and *(above)* the scale of 1:2500. The four windmills are shown clearly at 1:2500 but only Storcroft Mill is identifiable at six-inch – due to its annotation.

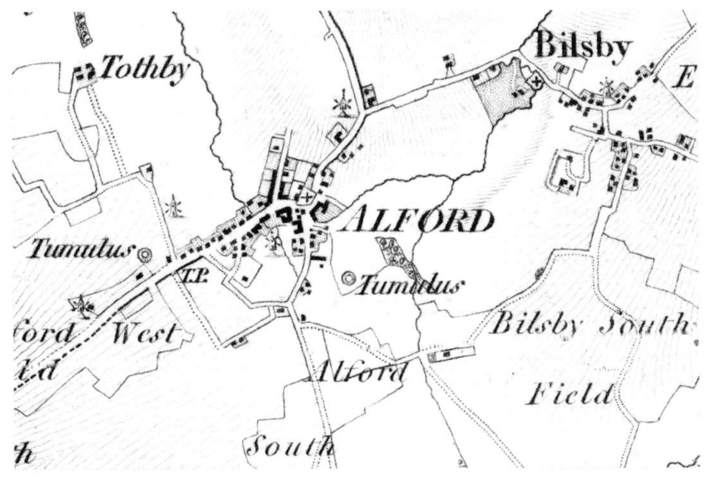

32. The area around Alford in Lincolnshire as seen on one-inch Old Series sheet 84 *circa* 1827.

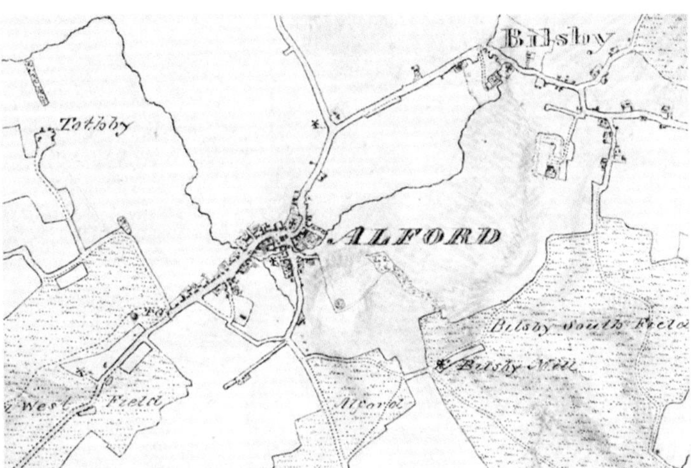

33. The area shown in Plate 32 as it had appeared earlier on OSD 280 *circa* 1819. Notice the difference in mill sites. *(Courtesy of the British Library)*

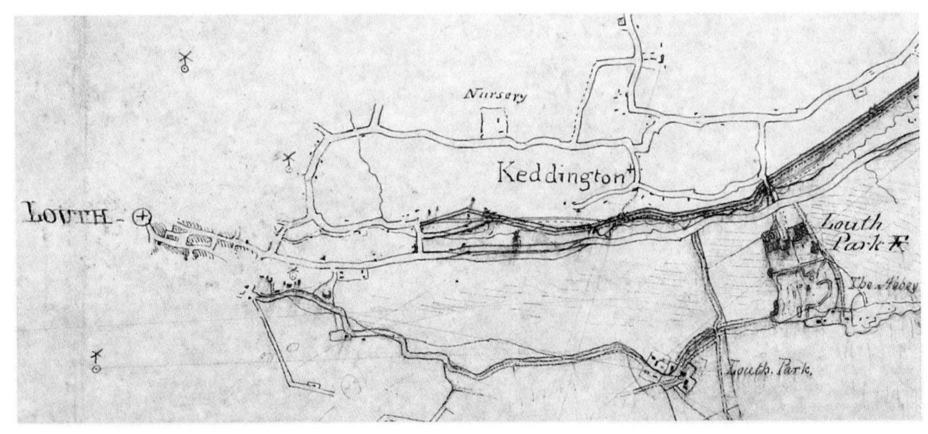

34. A Revision Drawing made in the field *circa* 1823 for the area immediately east of Louth in Lincolnshire. The crude nature of this manuscript is apparent. The placing of three of the cross-and-circle symbols corresponds exactly with that of windmill symbols on the later Old Series sheet 84. The fourth location differs between the two by 275 yards, so implying that later revision took place. Note also the 'normal' windmill symbol on this drawing (below the *'e'* of *Keddington*); this too corresponds exactly with a later mill symbol on sheet 84. *(Courtesy of the British Library)*

35. *(left)* The smock mill on Grange Road in Ramsgate, Kent *circa* 1908.

36. *(above)* An early view of the 'barn-top' mill at West Ashling in Sussex *circa* 1864.

37. Detail from 'The Tyne from Windmill Hills' by T.M.Richardson (snr) 1818. *(Courtesy Laing Art Gallery)*

38. The extremely derelict post mill at Madingley, Cambridgeshire *circa* 1910.

39. Remains of a post mill at Burton on the Wirral, Cheshire *circa* 1912.

40. Post mill at Clacton, Essex *circa* 1909.

41. Mill at North Hykeham, Lincolnshire in 1907.

42. *(top left)* The derelict shell of a mill at Hambledon in Hampshire *circa* 1908.

43. *(top right)* A converted tower mill, complete still with its cap, at Aldeburgh in Suffolk *circa* 1920.

44. *(above left)* A drainage pump on the River Bure in Norfolk *circa* 1920.

45. *(below left)* Windpower in use on a farm in Cambridgeshire *circa* 1910.

46. A large contrast in circumstances for the two mills at Shrewley in Warwickshire *circa* 1908. The post mill is the focus of attention despite the fact that it is the mundane tower mill behind it which is probably still working, albeit by a secondary power source.

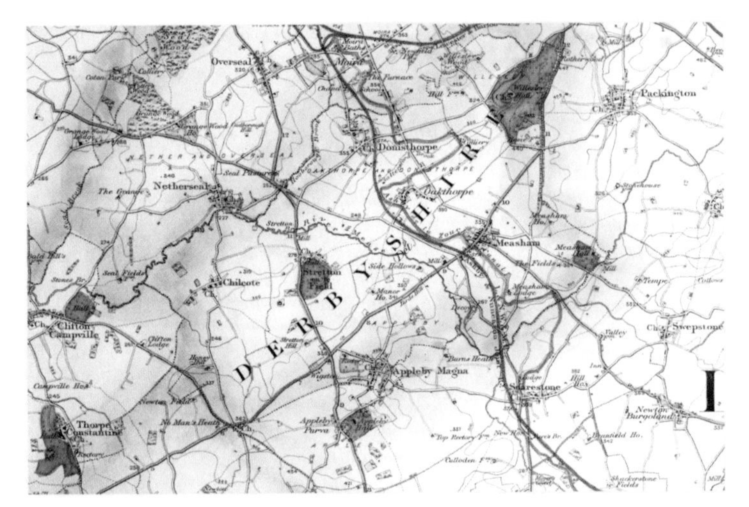

47. An extract from one-inch New Series sheet 155 *Atherstone* of 1892. It shows a detachment of Derbyshire almost totally enclosed within another county – Leicestershire (tinted in lilac), but joined also with Staffordshire (in yellow) and Warwickshire (in green).

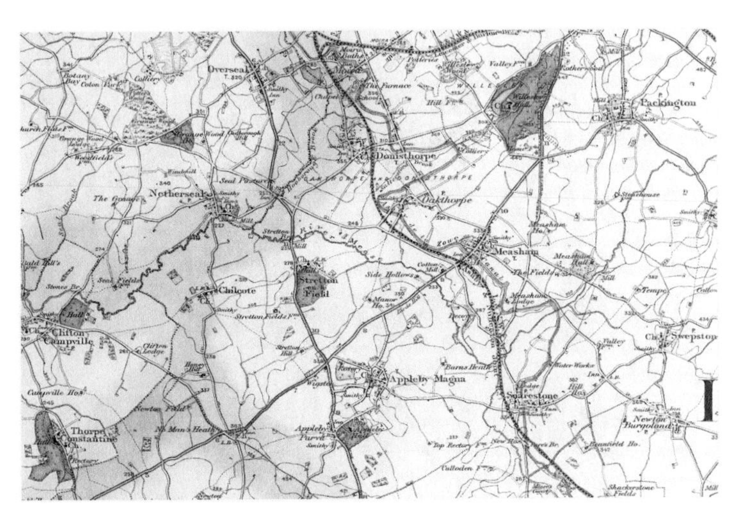

48. The same area seen in Plate 47 as it was later shown on Revised New Series sheet 155 of 1899 with county borders now much altered. In addition, it may be thought that the charm of hand-coloured maps such as these is demonstrated particularly well here.

49. Loppington Mill, Shropshire *circa* 1902.

50. Thorpeness Mill, Suffolk after its conversion to drainage in 1923.

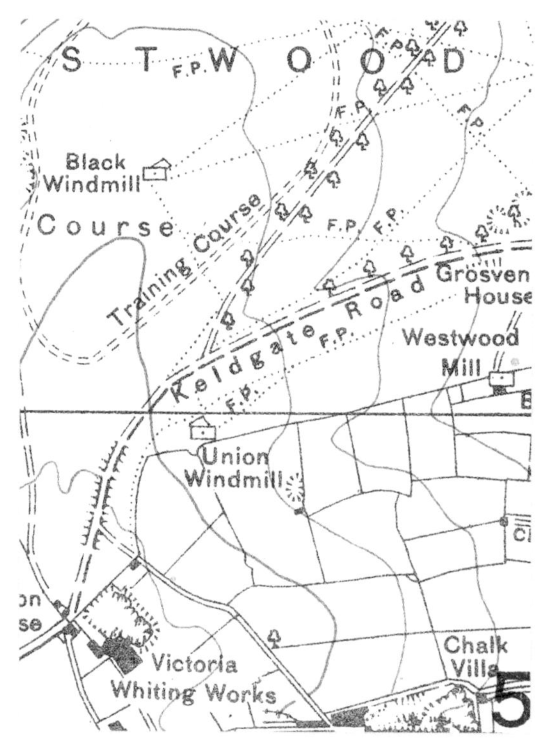

51. *(left)* The group of three tower mills that still remained on Beverley Westwood, Yorkshire in 1936, shown at 1:25,000.

52. *(below)* Scarborough Windmill, Yorkshire at 1:25,000 (1968).

Name of Mill	1:10,560 (six-inch) map depiction				Mill Demise
Windmills not recorded by the New Series	Use of ground plan and/or annotation	Sheet	Dates of sheet survey and publication		Apling's date for cessation of work by wind
Aylsham: Mill Road	✓	28SW	1885	1890	1895
Banham: Mill Road	×	95SE	1882	1890	*circa* 1915
Barton Turf	✓	40SE	1881-5	1889	1896 - 1904
Bixley	*Windmill (Disused)*	75NE	1881	1889	1865
Briningham: Belle Vue	×	17SE	1885	1891	*ante* 1780
Broome	✓	98SE	1882-4	1891	*circa* 1890
Bunwell	×	86SW	1881-2	1891	*post* 1888
Burgh St Peter	*Old Windmill*	99NE	1883-4	1889	*circa* 1933
Diss: Shelfanger Road	✓	110NW	1885	1890	*post* 1905
Diss: Victoria Road	✓	110NW	1885	1890	1929
East Dereham	✓	49SW	1881-2	1887	*circa* 1922
Fakenham	*Old Windmill*	25NE	1885	1889	1882 - 1910
Great Walsingham	✓	16NE	1885	1891	1895
Harpley	*Harpley Flour Mill*	24SW	1885	1891	1921
Holt	×	18NW	1885	1891	*ante* 1883
Horning	*Draining Pump (Disused)*	53SW	1880-4	1889	1863 (as a drainage mill)
Little Cressingham	✓	71SE	1882-3	1891	1916
Mattishall: Mill Road	✓	61NE	1882	1891	*circa* 1890
Norwich: Lakenham	✓	63SE	1880-3	1887	*circa* 1912
Paston	×	20SE	1885	1891	1930
Reedham	*Windmill Pumping*	78NW	1883	1889	*post* 1920
Smallburgh	*Draining Pump*	40NE	1884	1889	*post* 1913
Topcroft	✓	97NE	1883	1889	*post* 1916
West Walton: Highway	*Dobb's Mill (Corn)*	44SW	1881-2	1887	1908

Table 4.4. Recognition at the 10,560 (six-inch) scale of Norfolk sample windmills absent from the New Series.

Key: ✓ use is made of a recognisable ground plan and an annotation of *Windmill (Corn)*.

× use is made of a solid unannotated circle that, in size and setting, varies from being very suggestive to distinctly ambiguous in any intention it may have of signifying the circularity of a post mill roundhouse or a tower mill, or the near-circularity of the base of a smock mill.

At this yet larger scale, two of the five windmills have the uncertainty surrounding them resolved with the use of *Windmill (Corn)* annotations. This leaves mills at Briningham (Belle Vue), Bunwell and Holt still to be settled – their unclear six-inch depictions simply being a reflection of unclear 1:2500 depictions. The mill at Briningham had stopped work during the previous century, and that its brick tower was still in existence in the 1890s (and would remain so, prompting Apling to record it much later) should not trouble this analysis any further. The post mill at Holt had ceased working by the time of survey though only just, so it is perhaps reasonable to attribute merely a modicum of error to the Survey on account of this mill. The third of this trio of windmills, however, poses more of a problem. That the unannotated and ambiguous six-inch depiction of a smock mill at Bunwell is no different to the depiction at 1:2500 sounds warning bells. Given that the treatment accorded to this mill by the Survey seems to be unique among our large sample of windmills, its position needs clarifying. This mill was only briefly dealt with by Apling who, in admitting that no picture of it had come to light, offered a reading of the historical record that suggested its working life may have ended by 1888. This is despite an observation recorded as late as 1931, and which Apling cites, that 'at Bunwell Street is a finely proportioned tower *[sic]* with a typical Norfolk boat-top'.[27] Apling adds this comment to his own understanding that there were still significant remains on site as late as 1971. All things considered, the windmill at Bunwell remains, for the time being, something of an enigma.

Name of Mill	1:2500 map depiction				Mill Demise
Windmills recorded neither by the New Series nor the six-inch	Extra annotation not shown at the six-inch	Sheet	Dates of sheet survey & publication		Apling's date for cessation of work by wind
Banham: Mill Road	*Windmill (Corn)*	95.16	1882	1883	*circa* 1915
Briningham: Belle Vue	(same as six-inch)	17.15	1885	1886	*ante* 1780
Bunwell	(same as six-inch)	86.14	1882	1884	*post* 1888
Holt	(same as six-inch)	18.1	1885	1886	*ante* 1883
Paston	*Windmill (Corn)*	20.11	1885	1886	1930

Table 4.5. Recognition at the 1:2500 scale of Norfolk sample mills absent from the New Series and the 1:10,560 (six-inch) scale.

So, despite the considerable virtue of the 1:2500 in accurately depicting windmills, and the only slightly reduced value of the six-inch in doing the same, the one-inch New Series is significantly less good in this respect. Looking for a common factor that embraces the 24 absentee mills, to try and isolate a reason for their non-recognition by the New Series, but not a factor contradicted by the great majority of windmills that are recognised, can only be speculative. Preclusion of space at the one-inch is a reason that could account for Norwich (Lakenham) Mill and possibly two others, but no more. That most of these 24 windmills were still at work when the surveyors visited them is established since Apling gives specific dates for when many of them ceased to work.[28] Apart from the understandable omission

[27] R. Thurston Hopkins and S. Freese, *In search of English windmills,* London: Cecil Palmer, 1931. It was customary during the nineteenth century to talk of a tower mill when, in actual fact, it was a smock mill being discussed, but by 1931 this idiom had gone out of fashion. Hence the suggestion that a mis-statement has been made here. This all adds to the impression that the set of clues for the continuing existence of this windmill up to recent times is a bit dishevelled.

[28] It will be noticed that two windmills (at Aylsham (Mill Road) and Great Walsingham) ceased working in the

from the New Series of Briningham Mill and Holt Mill, along with Norwich (Lakenham) Mill and maybe Fakenham Mill as precluded town mills, twenty mills are unrecognised by the New Series for no apparent reason.[29] The only conclusion that can be drawn is that the New Series representation of windmills for Norfolk is significantly flawed.

Given the large number of Norfolk windmill survivals, and the even greater number of windmills that once worked there, but which have not survived, the overall number of mills that the New Series needed to depict in Norfolk would have been high compared with lots of other counties. But Suffolk and Essex, which did have comparable numbers of mills, had already been mapped by the New Series, and the Survey was maybe aware of the lamentably incomplete windmill recognition that they divulged. The continued use of annotated ground plans to depict windmills would have led to greater difficulties of legibility, or loss of map integrity, had the Survey not reintroduced the map symbol for windmills when it did. Aware of this imminent problem and conscious that partial mill depiction, as had been provided earlier, was undesirable, the Survey may have been further pushed into invoking the space-saving utility of the map symbol, and to do this just prior to publication of the New Series sheets for Norfolk. At a time when the representation of landscapes by topographic maps was inclining towards the judicious use of symbols, the very visible yet mutable windmill may have been more responsible than most landscape features for this new direction that the one-inch mapping was taking.

The late 1880s were a time of conflicting social pressures within the country, and senior staff at the Ordnance Survey would not have been blind to this. There had been a change of Director-General in 1886 when Colonel Sir Charles Wilson had assumed command. Wilson arrived as something of a fêted individual due to his involvement with the recent campaign in the Sudan and, at this time of high-Victorian confidence, his was a popular appointment guaranteed to promote the interests of the Survey.[30] But alongside the elation surrounding

---

same year – 1895. In fact, both windmills lost their sails on the same day in the great storm of Sunday, 24 March that, as reported in *The Miller* and many local newspapers, notably the *East Anglia Daily Times*, accounted for the loss of a vast swathe of windmills across the eastern counties. This was at a point in time when the loss of sails, for example, would probably put a windmill beyond economic repair and that two mills from the sample of 24 should come to the end of their working lives on the same day is not so much a coincidence as illustrative of the fact that the 1890s marked a watershed for the working windmill population in England.

[29] No date is provided by Apling for the post mill at Holt having had its body removed but this had probably happened soon after the mill ceased working in 1883, as it was in the curtilage of Heath House. The inference to be drawn from both the map and its subsequent occupancy by the County Court Registrar is that this fine house had become a residence of some distinction. While having a decaying post mill in the garden might be this writer's idea of heaven it is not an arrangement, apparently, that would suit everyone and the presence of a disused and perhaps unsafe post mill so close to the house was probably not appreciated.

[30] Charles Wilson had spent much of his military service involved in mapping all over the world and it had been his reporting of the deficiencies in French mapping at the time of the Franco-Prussian War that had, in part, prompted authorisation for wholesale revision to be made to the one-inch scale, so leading to the New Series. His participation in the Sudan Campaign arose from being in Cairo when it became necessary to send a relief expedition up the River Nile to support General Gordon at Khartoum. At this point in his army career, Wilson, already a Fellow of the Royal Society, was attached to the Intelligence Department and it was as one of its heads that he travelled with the expedition. However, his seniority was such that when the commander of the column heading for Khartoum was seriously wounded, it was Wilson who replaced him. Subsequently, he acquitted himself exceptionally well in his new unsought-for role, but he was too late to save Gordon. The commander of the expedition – General Wolseley – sought to scapegoat Wilson for what was fast coming to be seen as a national fiasco, but the political fall-out struck elsewhere. Eventually Charles Wilson would retire as a Major-General. He is less well-known today than other high-ranking Sapper officers of the time – chiefly

the Golden Jubilee of 1887 lay the spectre of political radicalism, fomenting social unrest that led to serious outbreaks of rioting in London later that year. Against this backdrop, the need to be progressive as a government agency was important. Treasury approval for a rolling revision of the now nearly completed 1:2500 county surveys had been given in 1886, but conditions were attached which *inter alia* unfortunately led to incaution when it came to considering the last two counties to be surveyed. Lancashire and Yorkshire had, of course, already been subject to large-scale survey much earlier. But, instead of now resurveying these counties so that their 1:2500 sheets could exhibit the same level of accuracy as those of other counties, the earlier survey information was simply re-plotted at 1:2500. This single shard of OS history, where the accuracy of the largest scales for Lancashire and Yorkshire was compromised for more than half a century, serves to show that concessions were made to the demands of this time that prejudiced the integrity of map production. Certainly the pressure on the Survey to come up with this kind of money-saving measure was as severe during this period in its history as any other.[31]

The New Series continued to evolve. The map symbol now reintroduced for depicting windmills was well established by 1890 and it was also uniform, unlike its counterparts on the Old Series that could look distinctly individual in their nature. From this point on, and for decades to come, any departure from the use of a standard symbol at the one-inch scale is very rarely found, though this reflects, as much as anything else, the fact that the engraver was now using a punch rather than, as before, having to engrave symbols individually.[32] By the early 1890s, the New Series coverage for England and Wales was nearing completion. By this time also, features of its sheets were changing. Within sheet borders, the names of counties were added. Then, starting in August 1894, the style of contouring changed from streams of dots to sequences of dashes and dots: this alone could appreciably alter the 'look' of a sheet. While this change only affected the last eighteen New Series sheets for England, it then remained a feature for the Revised New Series and the Third Edition.[33] Since the derived New Series maps had followed closely on the heels of larger-scale surveys – which were organised by county or groups of counties – it was possible for a situation to develop where not all of the survey data needed for a one-inch sheet that straddled a county border was available when needed. If it happened that adjacent counties were surveyed at different times, such as Hampshire whose large-scale survey had been very early, and Dorset where the county survey for all areas – apart from Portland – was not completed until 1888, the Survey had to decide whether to delay publication of Hampshire's New Series sheets, which in the south of the county were needed to replace very early Old Series sheets, or to issue some sheets with areas of Dorset left as blank spaces, in the same way that early large-scale

---

Gordon and Kitchener – but, for all his many contributions in life, he richly deserves to be remembered.

[31] For further details and a rueful comment by a later Director-General on this episode see Brigadier H.St.J.L. Winterbotham, 'The replotted counties', supplement to *The National Plans (The Ten-foot, Five-foot, Twenty-five-inch and Six-inch Scales),* Ordnance Survey Professional Papers: New Series 16, Southampton: Ordnance Survey, 1934, in which this course of action taken in the late 1880s was described as having been 'a thoroughly unsound thing'.

[32] Apart from anything else, the new symbol was consistently bold and upright. But the symbol used on sheet 267 *Hungerford* to depict a tower mill at Aldbourne, for example, owes more in appearance to the quirky traditions of the Old Series. The depictions of Borstal Hill Mill and Minster Mill, both in Kent, had been other exceptions.

[33] These eighteen sheets are 153-4, 167-8, 180-3, 197, 214, 253, 293-5, 309-11 and 325.

sheets had only shown single parishes. Taking the latter course of action in this case led, for example, to the early states of sheet 329 *Christchurch* being published with a large blank area where the coverage of Dorset should have been, though this was remedied for the last state (by which time the name of this sheet had been changed to *Bournemouth*). More noticeably, with the surveys of Lancashire and Yorkshire yet to be done, the parts of these counties that lay below the Preston-Hull line do not feature on New Series sheets, and those sheets that showed only these counties below that line were not published in New Series form.[34]

The practice of periodically revising a sheet, thus leading to the issue of a sequence of states, had been applied across the Old Series. Indeed, it was still happening in the cases of sheets yet to be replaced by their New Series equivalents, but it was also now being applied to the New Series where sheets had been provided. While this process of revision had led to uneven results across the Old Series in respect of its windmill-related information, at least the recognition of some windmills *was* updated, which was undoubtedly appropriate, one might think, considering the long span of years that this process stretched across. Yet little evidence has yet come to light for the depictions of windmills changing between successive states of New Series sheets, other than for a handful of mills on the *Tenterden* sheet and just one solitary mill on the *Reigate* sheet. Inevitably, it could only be a matter of time before a wholesale revision of the New Series was needed. Accordingly, the Revised New Series was established and then, some few years later, this was replaced by the Third Edition. Gaining approval for this process of revision was not without its difficulties owing to obvious cost implications, and the genesis of both editions needs to be considered. But, importantly for this study, the reintroduction of the mill symbol part-way through the lifetime of the New Series was to remain the practice for showing mills on these, and later, one-inch editions.

Summing up the New Series representation of the windmill can best be expressed by emphasising its variability. For many areas such as Norfolk, where sheets of the New Series were issued no earlier than *circa* 1890, and so not long before the Revised New Series sheets became available, the number of states for a New Series sheet becomes minimal – and this is especially the case when changes to the railway network were negligible. Yet in such areas, the recognition of windmills was still an improvement on what had gone before, even if not very reasonable in percentage terms. For a large swathe of the country north of the Preston-Hull line, and for an extensive area of southern England where some sheets were issued in a significant number of states, the recognition of mills had effectively been non-existent. Any desire to engage in a broad appraisal for the coherency of windmill recognition on the New Series is severely handicapped by this variability, so explaining why a researcher like Farries, who had looked into poorly-served areas, was not slow to express his dissatisfaction with it, as well as with the Old Series that had preceded it.[35]

---

[34] That said, a later state was issued for sheet 101 *East Retford* that retained its initial publication date of 1890 – when the part of the sheet covering Yorkshire had been left blank – but that also bore the additional annotation *Yorkshire portion added 1895*. On this complete state the new style of contouring is used, replacing the older style of contouring previously used for the non-Yorkshire parts, though other characteristics of New Series sheets are kept such as the representation of churches. It had been necessary to issue a later New Series state for this sheet, together with one for sheet 90 *Great Grimsby* since, unlike other sheets below the Preston-Hull line either partly or totally showing parts of Lancashire or Yorkshire, their replacement Revised New Series sheets were not ready for publication when the large-scale survey data for these two counties was adapted for small-scale use.

[35] Farries, *Essex windmills millers & millwrights,* 1, 45.

The Revised New Series edition

Consequently, our reflections on windmill representation by the New Series needs to move on to what can in hindsight be seen as a *de facto* second edition for the family of New Series editions but which, at the time of its preparation in the mid-1890s, was described as the first national revision of the New Series. The push for the Revised New Series came from many directions. Inevitably, what chiefly came under scrutiny was the obsolescence of the earliest sheets of the New Series. The northern sheets that derived from mid-century six-inch maps were looking very dated by 1890. The oldest of the southern sheets – those considered to have military significance – derived from surveys made before 1870 and were unacceptably old by 1890. The governance of the Ordnance Survey changed in 1890. It now came under the newly-constituted Board of Agriculture and its first secretary was Sir George Leach, a retired Colonel of Royal Engineers who, having served with the Survey, was, usefully, quite conversant with its customs. Meanwhile, dissatisfaction with what the Survey was achieving at this time was finding its voice in the personality of Henry T. Crook, a civil engineer from Manchester, who had written to *The Times* at some length on the failings, as he saw them, of the Survey and its maps, and who had gathered further support for his views by publishing a polemic entitled *Ordnance Survey Maps - As they are and as they ought to be.*[36] To both Director-General of the OS (DGOS) – Wilson – and Secretary to the Board of Agriculture – Leach – it must have seemed propitious when propelled into having to engage in a debate sparked by the forceful Crook. This arose when the Board of Agriculture was pressed to instigate a Departmental Committee charged with investigating the general condition and remits of the Ordnance Survey.[37] There were issues other than the problem of sheet obsolescence that were concerning the Survey and Wilson may well have been pleased of the opportunity to bring some of his resource-based problems out into the open, though whether he colluded with Crook to bring this about, as has been hinted at, is questionable, the more so as the fuller account of this episode by Oliver makes no mention of this as likely.[38] Whatever the background to the establishment of this Committee, it was about to conduct a wide-ranging appraisal of OS practices and the ideas that underpinned them.[39] It was tasked to undertake this through terms of reference which were to consider:

> 'What steps should be taken to expedite the completion and publication of the new or revised 1-inch map (with or without hill-shading) of the British Isles?
>
> What permanent arrangements should be made for the continuous revision and speedy publication of the maps (1:500, 25-inch, 6-inch, and 1-inch scales)?
>
> Whether the maps as at present issued satisfy the reasonable requirements of the public in regard to style of execution, form, information conveyed, and price; and

---

[36] See, for example, *The Times,* 15 April 1892, p.4, c.5.

[37] The pressure to establish this Departmental Committee derived from a Commons Motion. See Hellyer and Oliver, *One-inch engraved maps of the Ordnance Survey from 1847,* 55.

[38] See Seymour, *A history of the Ordnance Survey,* 188; Owen and Pilbeam, *Ordnance Survey: map makers to Britain since 1791,* 72; Hellyer and Oliver, *One-inch engraved maps of the Ordnance Survey from 1847,* 54-5.

[39] In due course presented as the *Report of the Departmental Committee appointed by the Board of Agriculture to inquire into the present condition of the Ordnance Survey,* British Parliamentary Papers (House of Commons series) 1893-94 [C.6895], LXXII, 305 (hereafter abbreviated as *Dorington*).

whether any improvement can be made in the catalogue and indexes?' [40]

The Committee sat under the chairmanship of Sir John E. Dorington Bart MP, and it took evidence principally through interviewing people concerned with map production and map sales, together with end-users such as land agents, surveyors, civil engineers and geologists. As a matter of protocol, Colonel Wilson as DGOS was the first witness to be called. He was followed immediately by Henry Crook. The questioning of all witnesses over eighteen days extended from concerns for an effective map-selling apparatus to the legibility of the indexes for all scales; from procedures of notification of new building work in towns to issues of contours and the continuance of vertical hachuring; from the quality of the paper used to the orthography, or accuracy of names (particularly Welsh ones). As far as the one-inch sheets were concerned, discussion largely centred around road classification, and in all that time the nearest that the depiction of windmills came to be mentioned was when the map agent and map producer James Wyld argued for the better naming of farms, houses and (water)mills on a specimen sheet of the Bath area.[41] The most illuminating remarks for the nature of the one-inch were probably those of Wilson himself. As first witness, it is clear that he had set his mind on a revision of the one-inch, but that this would be distinct from the imminent major scheme of work – a first revision of the 1:2500, sanctioned in 1886.[42] He was sufficiently insouciant in his remarks, playing down the costs and the manpower needs that any programme of work for revising the one-inch would involve, alleging that it would be possible to revise the whole country within four years with hardly any effort. Even Crook, who was canvassing elsewhere for this scheme of work to be undertaken, was taken aback when he heard of Wilson's low estimate for the costs involved.[43]

In among all the issues raised and the suggestions explored at the time of the hearings, the proposal to revise the one-inch independently was upheld by the Committee when it completed its report on 31 December 1892. The Summary of Recommendations included one (the sixth) which considered that 'special arrangements [should] be made to revise the 1-inch map within the next four years independently of the maps [at] the larger scales, and that subsequently this map [should] be constantly revised within periods of fifteen years'.[44] To understand more fully the origins of the Revised New Series, and the changes made to the later states of the New Series, requires an exploration of the thinking that circumscribed the one-inch before the Dorington Committee was appointed. For one insight into how the Old Series, and later the first of the New Series editions, had been viewed by the military command structure of the Survey in the years leading up to this time, an excerpt from the written memoranda of submissions made by Wilson and others after the main hearings, but

---

[40] *Dorington,* minute appointing the committee, p.iii.

[41] *Dorington,* evidence, appendices and index, q.2056.

[42] *Dorington,* evidence, appendices and index, qq 239-40 for Wilson outlining his plan for revision of the one-inch. Also q.5650 when Wilson was recalled towards the end of the hearings, and his remarks clearly indicate that he had considered the resource implications for this plan and found them not to be taxing.

[43] *Dorington,* evidence, appendices and index, qq 800, 1226 and 768.

[44] *Dorington,* committee report, p.x: 'The evidence of Sir Charles Wilson satisfies us that there need be no difficulty in bringing this new series map up to date at a reasonable cost. He estimates that, if the revision of the 1-inch map is carried out independently of that of the larger scales, it can be brought up to date in four years time at an expenditure of £7,000 or £5,500 per annum (according as [to whether] the Highlands of Scotland are or are not included), in addition to present expenditure, which amounts to about £2,000 per annum'. See also p.xxxviii.

forming part of the evidence considered by the Committee, reads as follows:

> 'The old 1-inch map of England was surveyed on the 1-inch and 2-inch scales as a military map; nothing was introduced that could not be easily shown on that scale; features of military importance were exaggerated and details not of military importance were omitted. The adoption of the Irish System and reduction from a large to a small scale led to the insertion of names and details that should not have been shown on a map on the 1-inch scale that were not required by the general public and that, to a certain extent, lessened the value of the map for military purposes. The military character of the map was, in fact, made to give way to the real or presumed civil requirements of the State.' [45]

Whether when writing this in October 1892, Wilson was concerned to emphasise that for any recommendation the Committee might make, there was a strong need not to shut the door on redressing this undesirable state of affairs, as he saw it, or simply just stating an historical fact, is not clear. In the event, he was largely given a free hand in how to depict topographical detail on the revised one-inch map. Moreover, it was further recommended by Dorington that a variant of the one-inch, more appropriate to military requirements, was to be provided, and that it should incorporate colour. This may have come as no surprise to Wilson. The Dorington Committee had already noted and acquiesced to recommendations from an earlier War Office Committee, which, though not as comprehensive in its enquiries as the Dorington Committee, had examined OS maps from a military perspective just a few weeks beforehand.[46] That the one-inch had been a military map in the minds of successive Director-Generals, though never having had to be tested in that context, is axiomatic. But this was a precept that by the end of the nineteenth century was increasingly coming under threat from the requirements of a civilian population more mobile and more aware of the value of small-scale maps. Something of a tension was growing between military protocol and civilian desires. The Dorington Committee may have clarified points of contention and resolved issues but many of these were administrative in nature; so that it was probably the War Office Committee preceding Dorington – underpinned by a military perspective on what constituted a usable map – that did more to set the tone for the content and style of the Revised New Series.[47]

For the depiction of windmills on the Revised New Series, this was arguably very true. Although the War Office (Baker) Committee merely consisted of three members – two generals and Colonel Wilson – and it devoted only two days to the questioning of just eight, exclusively military, witnesses, from the Minutes of Evidence two of the eight were strident

---

[45] *Dorington,* evidence, appendices and index; written submission by Colonel Wilson 'Memorandum on the 1-inch map and on various processes used in the reproduction of maps', 15 October 1892, p.224.

[46] *Report of Committee on a military map of the United Kingdom,* printed at the War Office, 1892 [A.237]. This was unpublished but a copy is lodged in TNA PRO WO 33/52, p.639 (hereafter abbreviated as *Baker*).

[47] A simple, even amusing, example of the military and civilian mindsets having to embrace each others' point of view occurred at one stage in the Dorington Committee hearings when the discussion turned to the topic of the contours in the area of Newhaven Fort, and whether or not they should be shown. Those military officers present argued that to show them would provide information to a potential enemy. Others who were there reasoned with some conviction that anyone wanting to extrapolate their own contours on a blank map could do so, probably in a matter of minutes. On this occasion military prowess did not rule the day. *Dorington,* evidence, appendices and index, q.756.

in their support for having windmills more clearly marked on the one-inch map. The more senior of the two, Lieutenant-General W. H. Goodenough, stated that he thought 'a careful arrangement should be made by which the names of all conspicuous windmills should be kept in' and, in responding to a later question on the use of conventional signs, 'we already show churches and I think we might show what windmills there are'.[48] The less senior of the two witnesses was a Major Verner who, when questioned whether more conventional signs should be applied at the small scale, answered:

> 'Yes, military conventional signs for churches, bridges, cuttings and embankments would be most useful; I would also add post offices, telegraph offices and windmills, and also railway stations, and they should be well defined, because [they are most] important; map reading requires good landmarks and the marks on the map should be sufficiently clear and conspicuous so that you would be able to read the map, if necessary, upside down in the field... Windmills are of course going out of date, but there are a good many about still, and they are most valuable landmarks in a strange country.' [49]

Importantly, one should add to these testimonies the submission of Lieutenant-Colonel J. Farquharson R.E. who was one of 21 – again, exclusively military – addressees whom the Committee considered could advise them in writing, and who, when asked for his opinions as to what conventional signs he would add to those already in use, responded 'the only one I should like to add is that for windmills; they help one to identify his position better than anything else I know'.[50] This comment is particularly significant given that Farquharson, commissioned into the Royal Engineers in 1859, had spent half his service career with the Survey at the time it was written. He had first been posted to the Survey in 1872 and was destined to become Director-General in 1894 on Wilson's retirement. Farquharson's tenure as DGOS lasted for five years until 1899 and so it was he who oversaw the implementation of the Dorington Committee recommendations. The significance of his view on the worth of the windmill, and therefore presumably its depiction on the one-inch, derives from the position of responsibility as Executive Officer (second-in-command) of the Survey that he had held since 1887, when quite probably it had been his decision to reinstate the windmill symbol on the New Series.[51]

The revision of the New Series to create a nationally consistent one-inch map by virtue of a compact period of preparation, once recommended by the Dorington Committee, was

---

[48] *Baker,* minutes of evidence, qq 13, 21. It might be considered from the way that Goodenough is expressing himself here that he was unaware that windmills had become much more visible on the one-inch since 1888 and that, maybe in keeping with other senior army officers, his knowledge of New Series one-inch maps was limited to those that covered the garrisons and training areas of south-eastern and southern-central England. These sheets were of course largely silent in respect of windmills including, for example, sheet 285 *Aldershot.* This fragment of conjecture is somewhat offset by realising that Goodenough had, as a Major-General, been GOC North Western Command and so based in Chester from July 1889 onwards. But then again, sheet 109 *Chester* had been issued as early as 1887, and so too recorded no windmills. General Goodenough can perhaps be let off this particular hook!

[49] *Baker,* minutes of evidence, q.90. Verner could certainly be relied upon to support Wilson. They had both been on the Sudan campaign where Verner had been deputy to Wilson in the latter's initial role as Head of Intelligence.

[50] *Baker,* minutes of evidence, appendix A, reply by Farquharson, 16 April 1892.

[51] Seymour, *A history of the Ordnance Survey,* 195.

placed in the hands of an OS officer named Grant (later to be DGOS from 1908 to 1911).[52] The options open to Grant when deciding how windmills should be shown were: maybe to remove their representation altogether – and the diminishing utility of the windmill by 1892 might have persuaded him to do this – or, as mills were becoming increasingly significant as pastoral icons, to have retained some minimalist representation, possibly by continuing to use unannotated symbols in an unobtrusive way. It is doubtful that much military deference would have been paid to any notion of pastoral sentimentality in deciding whether or not to continue depicting windmills, but the orthodoxy of military thinking already articulated does suggest a reason for retaining recognition of windmills and why Grant not only did this, but went further in enhancing their representation with a third option of the near-universal use of annotated symbols. What was clearly important to the military mind was the notion that windmills are potentially useful as locational devices in landscapes that may otherwise be featureless.[53] The naming of 'conspicuous windmills' where this was done, or even of public houses that occupied strategic locations at crossroads, would have needed no justification other than this, and it can strongly be argued that the greater visibility given to windmills on the Revised New Series principally derives from this criterion.

Major Verner was quite correct to declare that numbers of windmills were declining in 1892. The attrition rate for windmills in the 1890s was higher than for any decade before or since. By then, if a windmill suffered damage it was more likely to be considered beyond economic repair than it would have been only a few years before. With the last 'authentic' windmill to be built *de novo* dating from 1892, by the late 1890s the line between windmills constituting a contemporary technology and an obsolescent one had certainly been crossed. This assessment of the mill had definitely been true for urban areas well before the turn of the century, and it was also now equally becoming the case for rural areas. Quite literally capturing this position graphically, Farries was able to demonstrate how the windmills of Essex had gone into decline from the 1840s onwards by plotting a decay curve to show the rate of that decline. This rate is shown to have been a constant one in the years leading up to 1890, and then since 1900 a lesser, but still constant, rate seems to have applied. So there is an obvious discontinuity in the graph around the 1890s that Farries felt unable to explain, but which is certainly indicative of this decade representing a watershed in the fortunes of the windmill. Fortuitously, the revision of the New Series one-inch sheets was accomplished in a timescale that allowed it to capture the windmill population on the edge of this cusp. Since Wilson proved to be as good as his word in ensuring that the revision took no longer than the four or five years he had assured the Dorington Committee it would take, this very short period of revision (but not re-survey) together with just as short a publication period, provides a snapshot view of windmills. This is unlike the New Series edition that has mills shown as contemporaries, but which in reality were never all standing at the same time. On

---

[52] Seymour, *A history of the Ordnance Survey*, 195-6.

[53] This is a statement well borne out by the events of history. A catalogue of episodes in military history can be cited of times when windmills have been appropriated by the militia or suffered at their hands. Starting maybe with battles from the English Civil War, Charles I is reputed to have witnessed Edgehill from the convenience of a post mill and parliamentarian forces used another as a rallying point at Naseby. In examples from more recent European history, windmills have served as useful observation posts for locating enemy positions as well as being useful reference points for the direction of artillery fire. For these reasons armies in both defensive and offensive situations have sought their destruction as, for instance, in the struggle for the Seelow Heights immediately west of the River Oder in early 1945. As a general rule, windmills and wars do not go well together.

the whole, the Revised New Series is very appropriate for use as a benchmark for windmill depiction and, coming after years of inadequate representation for windmills at the one-inch scale, it merits a central role in this study.

When sheets of the Revised New Series first appeared, they did so in outline version (NS-2-O by the Oliver notation). Many were then issued in a hills version (NS-2-H) where the colour of the hachuring could be either black or brown.[54] The adding of further colour later provided a third basic variant (NS-2-C) and the popularity of this type of map ensured that colour would, within a few years, become *de rigueur* on all OS one-inch mapping. At this point, the public was becoming increasingly aware of the benefits that one-inch maps could bring: sales were escalating and this was leading to the need for higher levels of production. Quite separate from this, the move to have colour on the one-inch was calling into question the continuing use of the time-consuming process of engraving, and so forcing a rethink of how maps could more easily be made. As the chosen way forward for the small scales, early efforts at colour-printed mapping were not problem-free – but this is rather moving ahead. In the mid-1890s, it is the raft of enhanced conventional signs brought to the Revised New Series that interests us most.[55] Churches, as landmarks, are differentiated not by dedications as on earlier New Series sheets but by whether each has a tower or spire.[56] The classification systems for roads and railways are better and there are, of course, the improved depictions for windmills. Whereas the treatment of windmills on the New Series was variable both in mode and consistency, the representation of windmills on the Revised New Series offers a nationwide consistency whereby a symbol, usually annotated as *Windmill* or *Old Windmill* (or sometimes by the diminutive forms *W'Mill* or *Old W'Mill*), is given for each mill recorded.[57] Every so often, the annotation *Windmills* suffices for two or more symbols. This saves on space and avoids repetition, but with the great majority of depictions using an individual symbol and *Windmill* annotation, and from all that has been said, it will be appreciated that the representation of windmills by the Revised New Series was a huge improvement over anything seen before at this scale. The result is a coherent statement for the mill population where part of that coherency unquestionably stems from the condensed periods of revision and publication. Engraving of the new sheets had in fact only fully got under way after the New Series outline sheets were completed at the end of 1895. It then only took until 1898 to complete, so ensuring a publication programme that lasted barely four years – from 1895 to 1899.[58] The date of a sheet's revision was usually only a few months before publication

---

[54] Although all Revised New Series sheets were issued in an outline version, not all were as individual sheets. A few coastal ones that principally showed sea were published only in combination with neighbouring sheets.

[55] Hellyer and Oliver, *One-inch engraved maps of the Ordnance Survey from 1847*, 79.

[56] Just as the manner in which contour lines were shown had been changed, a rethinking of the way that churches should be depicted on the New Series had made its way onto sheets published after 1887. Use of the simple New Series style cross for depicting churches was retained for all sheets of this edition, but dedications of the sort that had been provided on earlier sheets were omitted after this date.

[57] See the explanation in Chapter 2 concerning these styles of depiction. As mentioned there, use is sometimes also made of the annotation *Disused* to describe a windmill's condition, though this is quite rare. More common are instances where a hill is named as *Windmill Hill* but the recognition of a windmill that stood on it is limited to a symbol so as not to have to repeat the word *Windmill* on the map. This occurs at Chalton in Hampshire and at East Knoyle in Wiltshire. Conversely, other annotation separating *Windmill* from *Hill* can be confusing for a casual observer who is left looking in vain for a windmill symbol. Such an arrangement can be seen at Cliffe Pypard, also in Wiltshire.

[58] Colonel Sir John Farquharson, 'Twelve years' work of the Ordnance Survey, 1887 to 1899', *The Geographical*

and the dates of both are noted in the bottom margin. All sixteen sheets that cover Norfolk, for example, happen to have been revised the same year, in 1897. So, if we now return to our sample of Norfolk windmills – all those considered by Apling to have survived, in one form or another, until 1984 – we can carry out a second assessment for it, this time looking at each windmill's depiction on the Revised New Series. All 123 mills predated the 1890s so, barring the few mills reasonably omitted from the New Series as well as any others that are known to have disappeared before 1897, we can expect the rest to appear on the Revised New Series or at least for the new maps to offer a pattern of recognition that represents an improvement on the 80 per cent success rate achieved by the New Series.

How the sample of 123 windmills fares on the Revised New Series can be expressed as raw data – as was done in Table 4.3 for the New Series. At a glance, the Revised New Series is seen to offer a better account for windmills than the earlier edition had done (Table 4.6). Of the four windmills annotated as *Mill(s)* on the New Series, two – at Barnham Broom and Burnham Overy (Union Mills) – retain that style on the revised edition. Apart from these two mills, all but twelve others are recognised conventionally either by annotated symbol or, in just six cases, by the use of an unannotated symbol. The twelve exceptions include mills already accepted as having reasonably been omitted from the New Series – at Briningham (Belle Vue), Fakenham, Holt and Norwich (Lakenham) – this on account either of having ceased to work and no longer to look like windmills, or of being town mills where lack of space prevented their being recorded. The other eight mills are those at Bixley, Bunwell, Erpingham, Foxley, Honingham, Mattishall (Mill Road), Smallburgh and Thrigby. Apling offers evidence for four of these eight windmills having largely been dismantled or severely damaged by *circa* 1893. Moreover, he also offers no evidence for three of the remaining four to be working any later than 1883, 1888 and *circa* 1892. This leaves just one mill. A précis of all the evidence assembled by Apling for these eight mills and how each may have appeared to the OS field reviser in 1897 is provided in Table 4.7. Depending on how unambiguous this evidence is judged to be, in the case of mills that had indeed disappeared by 1897 it is entirely reasonable for the Revised New Series not to have recorded them. But even if the evidence provided by Apling is disregarded, and the number of inexplicable errors made by this edition for the sample of mills was to remain as high as eight, this would still represent a far better record of achievement than the New Series had managed.

One of these eight mills – Bixley Mill – got very little more than a mention when the omissions attributable to the New Series were being discussed earlier. Given that this mill is known to have stopped working as early as 1865, and that in 1881 it was recognised by the large-scale maps as disused, this lack of a mention can perhaps be seen as understandable. Given the extra evidence that this structure was certainly out of commission as a windmill by 1872, albeit still a seven-storey tower, then the New Series was justified in not recording

Journal 15 (1900), 581. As noted already, the detail provided on a Revised New Series sheet includes retrospective information for the survey that had led to the corresponding New Series sheet. For the parts of Lancashire and Yorkshire not catered for by the New Series, the information given on the Revised New Series escalates. So, on sheet 85 *Manchester* – catering for small parts of Cheshire and Yorkshire, but which is very largely dominated by Lancashire – the notes read as *Surveyed in 1871-72 and Published...1886. Revised in 1895 and Published...1896. Lancashire and Yorkshire portions added in 1896.* On the other hand, sheet 86 *Glossop* – more equally covering parts of the same three counties, but also showing part of Derbyshire – provides the same information in respect of Cheshire and Derbyshire as the New Series had done, together with the revision information, but this time they are coupled with *Yorkshire and Lancashire portions added in 1895* (note the reversal of county names).

it as a mill. But if adhering strictly to the idea that whatever is recognised at 1:2500 should be cascaded down to the New Series, then this edition had been remiss. Maybe, since nearly a decade elapsed between the large-scale survey and the publication of the derivative New Series sheet covering Bixley, the depiction of this mill, or more precisely of a structure that had increasingly lost its status as a one-time windmill, got the benefit of an enhanced system of field-checking that was introduced *circa* 1887-8, or of some last-minute field revision that revealed the 'mill' as not worth recording on the New Series. But whatever the truth behind the New Series depiction for this mill, and the large-scale annotation of *Windmill (Disused)* in 1881 can be considered rather generous, the failure of the Revised New Series to recognise it a decade later was surely correct despite the structure having 'survived' until recent times.

Type of Mill	Mills depicted by the Revised New Series using an annotated symbol	Mills depicted by the Revised New Series using an unannotated symbol	Mills not recognised by the Revised New Series – including two cases of *Mill(s)*
Post Mill	18	0	2
Tower Mill	81	6	10
Smock Mill	2	0	2
Composite Mill	2	0	0

Table 4.6. Summary of recognition of Norfolk sample mills by the Revised New Series (circa 1897).

The harshest interpretation that the evidence gathered by Apling sensibly allows us to put to the recording of these eight Norfolk windmills by the Revised New Series is that it was remiss in failing to recognise mills at Bunwell, Erpingham, Honingham, Smallburgh and perhaps Mattishall (Mill Road). Being left with just five misrecorded mills represents a huge improvement over the New Series, and arguably this number is still an overestimate given that, with the exception of the mill structure at Smallburgh, all that any of these five mills later manages on the large-scale editions – revised 1903-4 – is an unannotated ground plan. More realistically, these shortcomings can be narrowed to a failure to recognise the three mills at Bunwell, Honingham and Smallburgh. Of these, Smallburgh Mill was one of those that from its location and appearance belied any suggestion that it might be anything other than a drainage mill. Nevertheless, under the Revised New Series regime it should at least have been given an unannotated symbol. Bunwell Mill again goes unrecorded; this lack of recognition further supports the claim that this mill stopped work as early as *circa* 1888, since by the mid-1890s there would be much less justification for recording what may have been left of this mill than when the New Series was being prepared for publication. If we really are to believe that ground evidence for a windmill at Bunwell survived for very nearly another century, then only the alleged small size of this mill offers any shred of a reason for its cartographic humiliation.[59] The small mill at Honingham worked in conjunction with a watermill and under conditions that offer no obvious reason for its non-recognition by the Revised New Series. Whatever the circumstances of these three particular mills, the record of windmill recognition by the Revised New Series surpasses that of the New Series by an extremely wide margin, with the number of errors attributable to the latter being six times as many as those attributable to the former.

Name of Mill	Date from Apling for mill ceasing to work by wind	Date from Apling for mill being dismantled, suffering severe damage, or ceasing generally to have landmark status as a windmill
Bixley	1865	1872: the tower was stripped of machinery, reduced in height and roofed over; this one-time windmill was still seven storeys high at the time of large-scale survey.
Bunwell	*circa* 1888	1971: the brick walls of a small smock mill demolished.
Erpingham	*post* 1883	(not offered)
Foxley	*post* 1883	*circa* 1890: the tower dismantled and reduced in height.
Honingham	*circa* 1892	1975: the tower stripped of remaining machinery.
Mattishall: Mill Road	*circa* 1890	*circa* 1893: the cap was blown off; some machinery was reused in Mill Street Mill; the rest dismantled *circa* 1900.
Smallburgh	*post* 1913	1979: house converted.
Thrigby	1889	1892: body of post mill taken down.

Table 4.7. Evidence from Apling for eight windmills absent from the Revised New Series (circa 1897).

This narrative for the depiction of windmills by the Revised New Series reinforces what was revealed earlier when discussing Owslebury Mill, and is compatible with other remarks already made about this edition. There were two types of depiction for windmills: annotated symbols and unannotated symbols. Remaining with our sample of Norfolk windmills, our knowledge of Smallburgh (Drainage) Mill suggests that an unannotated symbol should have been used here – if we accept that this practice was, indeed, reserved for mill structures that were visually similar to grinding windmills, yet acted as pumps. As features in the landscape, drainage mills have already been mentioned, and for the Survey to have considered them as very different to grinding mills would have been surprising given that the two types of mill are visually so alike and, as our sample shows, a mill can even switch between roles or, in the cases of mills at Catfield (Swim Coots), Oby and Smallburgh, combine both. Recalling the importance attached to landmark status by the military authorities, these two types of structure would have been of equal worth in this regard, but other military imperatives were

[59] To add to the earlier unease about the evidence for this windmill, Apling further notes that the octagonal foundations measured less than thirteen feet between opposite faces, making this an extremely small and thus probably ungainly-looking mill. It would certainly not have made it a windmill that was in keeping with being described as 'finely proportioned'. So, on the evidence of the literature, it would probably be unsafe to judge the Survey as being in error over this mill, the more so if it does indeed turn out to be the only 'error' made in respect of these mills at the 1:2500 scale.

perhaps playing a part here. To a defending army needing to live off the land or wanting to flood low-lying areas, being able to distinguish between corn mills and drainage mills makes sense. Whatever the exact thinking here, the correlation between the occurrence of drainage mills and the use of unannotated symbols is very high, as perusal of any Revised New Series sheet showing the fens of Cambridgeshire or Norfolk will confirm.

The six unannotated symbols used to record windmills from the Norfolk sample relate to structures found by the Survey at Catfield (Swim Coots), Great Walsingham, Horning, Oby, Reedham and Thurlton. Of these, two were essentially drainage mills at the time of revision – Catfield (Swim Coots) and Oby – but each also had an ancillary role – operating a small set of stones and a sawmill respectively. Reedham Mill was a drainage mill at this time. Horning Mill, though a drainage mill before losing its cap and sails, was now celebrated as a picturesque, if not a necessarily prominent, landmark. Thurlton Mill had stopped working by wind *circa* 1896 and then worked as a steam-driven flour mill; it certainly never had any association with drainage. Lastly, Great Walsingham Mill was one of those that suffered so badly in the storm of March 1895 that it was redundant by 1897, though its tower remained at full-height, and would continue as such until *circa* 1945. This mill had been overlooked by the New Series and the revision team may have felt something of a need to compensate for this oversight. If so, this could demonstrate that an unannotated symbol was used not only for the recognition of drainage mills, but also as a way of acknowledging other types of mill, including cereal mills, that were not quite up to the mark in some way or deserving of a full annotated-symbol depiction. The three mills not actively working as drainage mills could each claim to satisfy this criterion: full recognition would have been inappropriate in each case but some sort of depiction was possibly merited. If the steam-driven mill at Thurlton was viewed as a deserving case for this brand of treatment, then the recognition of Great Walsingham Mill becomes a little more logical. The unannotated symbol used to record this disused windmill is possibly there to imply the existence of a milling capacity, not so much of a normal windmill, but of the roller mill that had been built on this site to work alongside the windmill while the latter was still working. This then begs the question, of course, that if acknowledgement of a milling capacity is important, then why are watermills so completely excluded from the one-inch scale? The truth is that consistency with the pattern established by nearly all other mills should have led to Great Walsingham Mill being accorded an *Old Windmill* annotation in the readily available space. But a sample of six unannotated symbols is too small to assess the range of reasons for some marginal windmills being recorded in this way and, in any case, these mills are far outweighed in number by unannotated symbols that are used to recognise conventional drainage mills. Further consideration of the use of unannotated symbols on the Revised New Series is needed, and this will be provided at the end of this chapter when the use of solitary symbols to recognise urban mills – for which space precludes the use of any annotation – comes under discussion.

The second basic class of differentiation between types of windmill depiction first seen on the Revised New Series relates to whether annotations incorporate the descriptor *Old*, or are simply left as *Windmill*. The great majority of windmills are recognised with the use of an annotated symbol, but of these only a small minority have the annotation *Old Windmill* (or a diminutive version of it) accompanying the symbol. Clearly, no purpose would be served by the public being informed which windmills had their origins further back in time than their neighbours, so the descriptor *Old* is being used to define some attribute, undoubtedly visual, of a windmill's condition at the time of revision. The foremost candidate for this attribute is surely whether or not the mill still had sails and was working by wind. The sails of a mill are particularly prone to bad weather when not turning regularly and, under such conditions, it would soon appear, even to a casual observer, that a mill was disused. Moreover, a good set of sails could be cannibalised for use on another mill, and this was often done to help defray any expenses associated with cessation of trading, so that the conditions of having sails and still be working could be thought to go hand in hand. Apling offers no evidence to suggest that any of the eleven sample mills accorded the descriptor *Old* on the Revised New Series worked beyond 1897, other than stating that four mills – those at Aylsham (Cawston Road), Ludham (How Hill), Sedgeford and Stiffkey – ceased working at some point between 1896 and 1900. Five other mills – those at Scole, Broome, North Creake, Booton and Ringstead – are, in that order, described as stopping work in 1883, the next two *circa* 1890 and the last two in 1892. No evidence is offered for the post mill at Briningham (Mill Lane) after 1879 but, helpfully, precise details are provided for Wighton Mill, which worked up to the gale of 24 March 1895 but sustained enough damage that day to cause its remaining machinery to be taken out and sold in July that year. It is entirely conceivable that by 1897 – the year of revision in Norfolk – each of these eleven mills had stopped working and each was falling into dereliction.

All the remaining mills in our sample – those straightforwardly recorded by the Revised New Series with a symbol and a *Windmill* annotation – accordingly add up to 92 in number and they clearly form a significant majority of the sample. The description of *Old* applied to a much smaller number of mills has just been noted as credible looking at the circumstances of each of the eleven mills so described, but to conclude that the application of *Old* was not fallible would be premature. Some of the 92 mills accorded what may be described as the standard depiction of the revisers would have been more accurately described had they too, like Great Walsingham Mill, been given the annotation *Old*, though there are perhaps only as few as three windmills from the 92 that fit this category. The mill at Aylsham (Mill Road) was in a similar position by 1897 to Great Walsingham Mill in that both had lost their sails in the same storm two years before, never to work again by wind. The fortunes of this mill at Aylsham had been linked to a granary and bakery located on the same site that had then continued for a while to make use of the adjacent Cawston Road Mill. This latter mill was owned by the same miller *cum* baker and, as one of the eleven mills described as *Old*, had stopped work when both mills were sold to pay off business debts in March 1896. Hence the Mill Road site still retained its milling presence in a sense and, as at Great Walsingham, the circumstances of this mill suggest that an *Old Windmill* annotation would have been a better way of expressing the situation here. The other two candidates that call into question the accuracy with which the word *Old* was added to the standard annotation are tower mills at Ludham (High Mill) and Methwold. Apling is somewhat opaque in his biographies of these two mills. The last miller at Ludham Mill only arrived in 1890 and is last recorded as a miller in 1892, and thereafter only as a corn merchant. Apling is silent on the later history of this windmill except that his illustration of it derives from a postcard of *circa* 1900 showing it still to be in full-working order. The array of evidence cited by Apling for Methwold Mill is similarly vague. This mill had been worked by wind until it was put out of action during a gale. Subsequently, the miller carried on using only an oil engine until 1904 but then, finding this to be uneconomic, he relinquished his trade. Quite when this windmill had been ruined through storm damage is unknown though a date of 1895 is speculatively offered.

This, admittedly fairly minimal, evidence is enough to suggest that the inclusion of *Old* in an annotation was reasonably considered. Furthermore, this claim is greatly reinforced by considering the case of Stiffkey Mill. The evidence provided by Apling for this tower mill – whose annotation did include *Old* – puts it in the category as one of those windmills whose date of ceasing work can be fixed only as some point between 1896 and 1900. The revisers were clearly anxious in 1897 to ensure that the annotation for this mill did include the word *Old* as, within the very limited space surrounding this mill, the annotation *Old W.M.* is used. The annotative use of *W.M.* as an abbreviation for *Windmill* is particularly rare and, since the descriptor *Old* could have been sacrificed and the depiction then not been abnormal, quite clearly the enthusiasm for applying *Old* was obviously of especial concern on this occasion. As photographs of this windmill have survived showing the tower and cap to be intact well into the following century, the degree of dereliction in 1897 would have been minimal, even if damage had already befallen the sails. The degree of exactitude displayed by the reviser for Stiffkey Mill is striking and could be explained by the coastal position of this windmill. Of all such mills, stretching from the Wash around to Great Yarmouth, this would have been the only one possibly to have lost its sails. The interests of the Admiralty, acknowledged by Farquharson in 1900 as second only to those of the War Office in importance, may well have been a guiding influence here.[60]

Declaring windmills to be working or disused, drainage mills or corn-grinding mills or whatever, were refinements that were part and parcel of the remodelling required for the Revised New Series. The sample of Norfolk windmills certainly confirms the presumption that field revision for sheets took place here since sheets of the 1:2500 were up to eighteen years old in 1897, and it would be another three years before revision for the larger scales began. The discretion available to the one-inch revision teams for making the distinctions now required of them was no match, however, for the variety of windmill structures that they would have come across. Inevitably, compromises had to be made. But that is in part the task of maps – to translate the infinite variety of the landscape into a practical number of categories. The issue of variety in the circumstances of the windmill was touched on in Chapter Two and there will be more observations to make on this subject, but the lucid representation of windmills seen on the Revised New Series is still to be credited with being an improvement on what had gone before. The same process of incremental revision that had been applied to the Old Series and to the New Series was now taken to sheets of the Revised New Series, but no cases have yet come to light for alterations being made to any windmill depiction between early and late states. This is a finding that gets support, if not total endorsement, from the recent cartobibliography (2009) for successive sheet states of the three earliest New Series editions, since Hellyer's work makes no reference at all to any change in individual windmill depictions across states of Revised New Series sheets.[61]

Inevitably, one can find cases where mills are not shown as clearly as they might have been, or where their recognition does not match the norms of depiction. In one instance, a

---

[60] Farquharson, 'Twelve years' work of the Ordnance Survey, 1887 to 1899', 591.

[61] Hellyer and Oliver, *One-inch engraved maps of the Ordnance Survey from 1847*. The caveat noted before when using this text still applies – lack of a mention of windmills, or any other sort of feature, does not imply their collective immutability between states, merely that they have not been chosen for any sheet to serve as a diagnostic tool to distinguish between states (see Note 15).

group of four windmills at Dunkirk and Boughton, just north of the London to Canterbury road, can be seen on sheet 289 *Canterbury*. Their depictions reveal a three-fold shortcoming, any element of which would be sufficient to discredit a depiction. First, two of the symbols for these mills are overlain by heavy writing that almost obliterates them, in this case by the name of the settlement – *Boughton Street*. This is a hazard that several mills had to contend with – one of the mills at Normanton Common on sheet 112 *Chesterfield* is another example of a windmill that is easy to overlook because of this difficulty. Second, although the great majority of windmill symbols overlie blank parts of a map and as a consequence are not in competition with other elements of map detail, the symbols are 'see-through' and if they are lain over other detail, then they become less visible. This occurs for the other two symbols of the group of four at Dunkirk and Boughton, where the mill at Dunkirk is supposedly in woodland. Last of all, not one of these four mills has an annotation of any kind to alert the map-user to their presence. Fig. 4.1.1 shows three of these mills, and Fig. 4.1.2 two of them. This third reason, which has contributed to the depictions at Boughton being undermined here, occurs primarily because there is simply insufficient space for an annotation and this is a problem commonly associated with urban mills.

Figure 4.1.1. Windmills at Boughton, Kent: one-inch Revised New Series sheet 289 (1895).

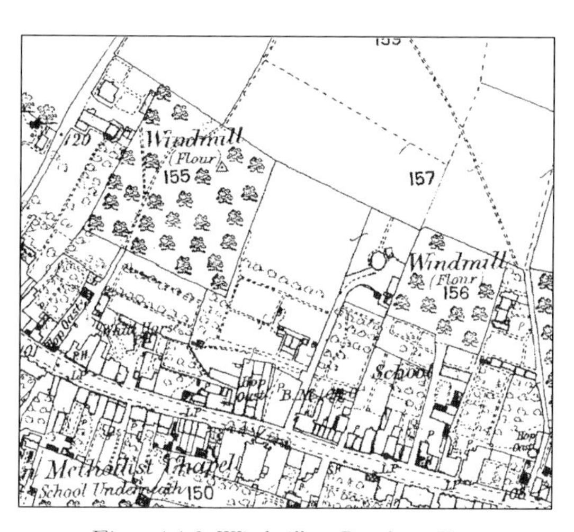

Figure 4.1.2. Windmills at Boughton, Kent: 1:2500 sheet 34.15 (circa 1872).

Since the Third Edition was to supersede the Revised New Series within a decade, the lifetime of many Revised New Series sheets was short compared to their earlier New Series counterparts, and this meant that not nearly as many states were published for each as had happened with the earlier edition. Where later states were issued for a Revised New Series sheet the usual motivation was, predictably, a change to the railway network. Even though most sheets could only boast a couple of states or so, a few were issued in a markedly larger number of states. Sheet 270 *South London* was certainly atypical in being regularly updated over as many as five identifiable states. Changes to the railways on later states could be subtle: one state for sheet 169 *Coventry* inconspicuously shows two of the three short road realignments and overbridges intended for what the next state would identify as just one and a half miles of the Great Central Railway crossing the bottom right-hand corner of the sheet. Changes to Revised New Series sheets of a more generic nature that took advantage of later railway-enhanced states included the use of the capital letter 'T' against a settlement name to indicate the presence there of a *Post & Telegraph Office* or, alternatively, a letter 'P' to indicate just a *Post Office*. Other common additions to Revised New Series later states were boundary revisions. Sheet 321 *Denge Ness*, which had a very small amount of land compared to sea, was revised in 1893 and published in 1895. Later, a state appeared with the statement *Boundaries revised to December 1898* placed in the margin, though it certainly takes a very keen eye to detect any changes made on the face of this sheet.

More importantly, in keeping with the assurances given to, and the recommendations made by, the Dorington Committee, a variant of the Revised New Series that incorporated colour was produced between 1897 and 1904 after an initial hiatus over funding. Fashioned as a military map by virtue of elements of its design that were particularly suited to military needs, such as borders that enabled quick and succinct reference to any point on the map, this variant was made available to the public and would even appear to have been reliant on sales for its completion. The colours on these sheets: hachures in brown, contours in red, blue for water, burnt sienna for main roads, the outline in black and, for sheets north of the Preston-Hull line, the woods in green, made for attractive maps that in the case of some sheets went to a number of printings.[62] The topographic content of each NS-2-C sheet was the same as the engraved outline sheet: certainly, the windmill representation of the Revised New Series outline sheets is mirrored on the colour sheets with no discrepancies between the two versions (as yet) being encountered for any individual mill depiction. Before moving on to discuss the next replacement one-inch edition, one last observation for the Revised New Series can be made. Selecting sheet 272 *Chatham* – although virtually any sheet taken at random that shows windmills would suffice to make the same point – Hoo Common Mill, north of the River Medway, is depicted by a symbol placed on the south-eastern side of a paved road, while at the larger scales the ground plan of this mill is shown only a few yards away but on the north-western side of the same road. This error on the one-inch has crept in despite ample space for showing the mill in its correct position. The precision (or lack of it) with which symbols were plotted raised by this example – there were many, many others – is another issue over which the one-inch scale could perhaps be berated but the extent to which symbols are noticeably misplaced, in offering completely the wrong juxtaposition to adjacent features, such as the road in this case, is fairly minimal. This characteristic of the Revised New Series and later editions can safely be left to the more pedantic among us!

The Third Edition

Inevitably, the cyclical revision programme authorised in 1893 for the one-inch New Series led to a second wholesale revision. The resulting replacement edition to the NS-2 Revised New Series followed hard on the heels of the NS-2-C 'military' sheets, but distinguishing the new sheets from previous ones is now easy. The label *Third Edition* explicitly appears in the top left corner of each sheet, rendering NS-3 sheets instantly distinguishable from those of any other edition – providing the margins have not been cropped! Other, more subtle,

---

[62] For a list of the publication dates and any subsequent printing dates for these sheets see Nicholson, *The birth of the modern Ordnance Survey small-scale map*. Although later printings may postdate the corresponding Third Edition sheets, they retain the Revised New Series style of depictions including that for windmills. After a while, this map lost its connotations of being a military map and became generally referred to as the 'one-inch map in colour'.

changes were incorporated as well. Whereas the depictions of some landscape features had altered during the lifetime of the Revised New Series sheets – the depiction of railways, for example, had become bolder after about the middle of 1896 – other changes were new to the Third Edition. Importantly, a fresh symbol was introduced to depict windpumps. This was undoubtedly done to accommodate the new skeletal metal structures then appearing on farms and estates for use in the raising of water.[63] The wider picture for the one-inch family of OS maps becomes ever more convoluted by the closing years of the nineteenth century. It will be remembered that a 'hills' version of the New Series was part of the output of that edition, but by 1900 the plates on which the hachures for the hills sheets of the New Series were being engraved (separated of late from the plates bearing the topographic detail) had been appropriated for the successor Revised New Series sheets with hills. In the fullness of time this would also happen for the Third Edition with hills printings (NS-3-H), without necessarily either NS-1-H or NS-2-H ever being completed.[64] Alongside such complicating factors as these in the story of the family of New Series editions, the Third Edition outline sheets were the outcome of a revision programme that lasted from 1901 until 1912. The standard-sized sheets were issued in the usual formats but, in an indication of the path that future mapping would take, the coloured version of the Third Edition, once it had become available, was quickly reconfigured at the instigation of the military into a Large Sheet Series (LSS) where the notional 360-sheet layout gave way to a new 152-sheet layout for England and Wales (Fig. 4.2).[65] This successor to NS-2-C – and also the shortlived NS-3-C that had kept to the same smaller-sheet format, but which was soon relegated, ending up as a *de facto* small sheet series – was published from 1906 onwards and became relatively popular. Sales of this series were such that most of its large coloured sheets went through many printings – we may refer to them as NS-3-LC – where, once again, the user could expect the issue of consecutive sheet states to have revisions largely centred around alterations to the railway network.[66]

---

[63] In some cases these metal windpumps were impressively large structures, as Plate 13 suggests. As a feature in the landscape, they had begun to appear from as early as the mid-1870s and could partly trace their origins in the few annular sails that had been experimentally used on standard windmills, or windmill-like structures such as the drainage mill at Owslebury. But by the turn of the century their numbers were increasing dramatically and they were sufficiently large enough and numerous enough to presumably be considered equal in value to windmills as landmarks, both to the army and to a newly mobile public wanting to negotiate its way around the countryside. Yet inappropriate annotations relating to these new structures can be found. A very large and impressive annular-sailed structure was erected on the skyline at Odd Down near Bath to provide power for a factory that processed Fullers Earth. Yet the Revised New Series still managed to describe it as an *Old Windmill*.

[64] The tangled story of the 'hills versions' of the successive New Series editions up to *circa* 1905 is told in Roger Hellyer, 'One-inch engraved maps with hills: some notes on double printing', *Sheetlines* 44 (1995), 11-20. This story of a single topic relating to the turn-of-the-century production of the one-inch OS map family demonstrates the degree of complexity left for the historian to unpick; complexities owing much to changing use of technology, shifting remits and a growing variety of available map formats. For further explanation surrounding these points see Hellyer and Oliver, *One-inch engraved maps of the Ordnance Survey from 1847*.

[65] The 360-sheet layout had long since been whittled down to a smaller number of discrete sheets that could be purchased. This was due to combining some coastal sheets, which showed only vast areas of open water, with their neighbours. An example of this was the sensible combining of sheet 82 *Kilnsea Warren* (that shows only a few hundred square yards of land) with sheet 81 *Patrington* to become sheets 81.82 *Patrington*. In any case, District Maps were becoming an increasingly available alternative to buying the standard sheets for the area in which one was interested.

[66] Although rather dated, one account of the cartobibliography of this Large Sheet Series is Guy Messenger, *The Ordnance Survey one-inch map of England and Wales: Third Edition (Large Sheet Series): a map study monograph*, London:

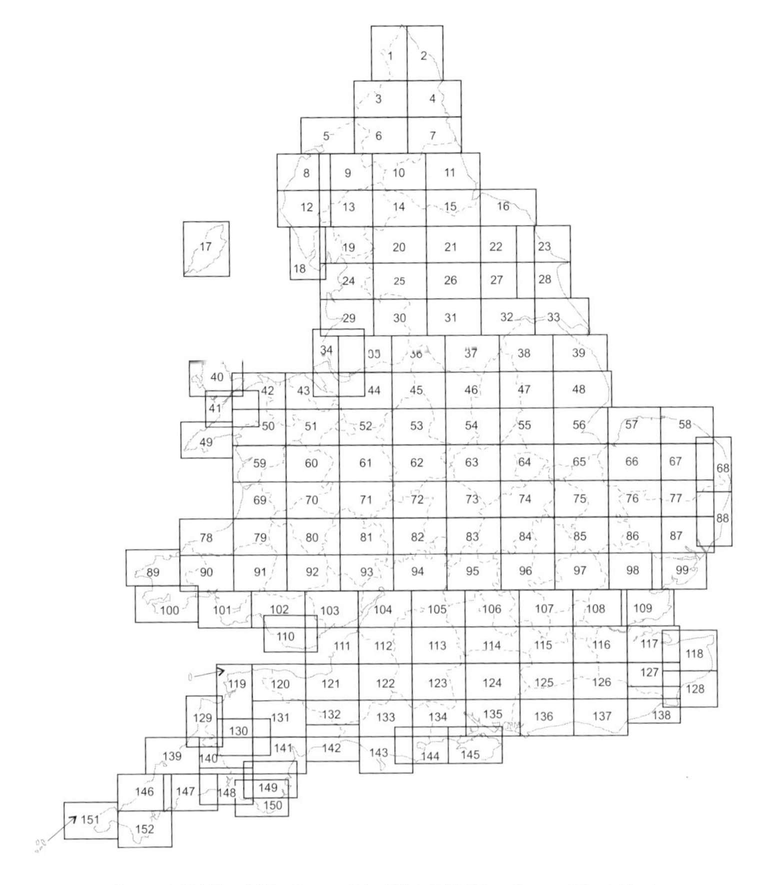

Figure 4.2. The 152 sheets of the Third Edition Large Sheet Series.

Any legacy of being a military map that the Large Sheet Series coloured sheets and the Third Edition in general may have inherited from the Revised New Series was outweighed over its lifetime by public enthusiasm for it, and in all probability by an improving sense of national security that followed the *Entente Cordiale* of 1904. In addition, military needs were now also being met by derivative half-inch scale mapping. As the general tone of the Third Edition topographic depiction stayed akin to that of the Revised New Series, the protocols developed for the representation of windmills on the earlier edition were carried forward to the next. But there was now the complication of the newly-introduced windpump symbol. This meant that the Survey was given an opportunity to reconsider a stance it had taken earlier. How it had once viewed grinding and industrial windmills compared to how it had viewed the visually similar structures whose function was to pump water, could be updated.

Charles Close Society, 1988. Importantly, this sheet-by-sheet account records no instances of windmill alterations between Third Edition states. A more comprehensive approach is offered in Roger Hellyer and Richard Oliver, *A guide to the Ordnance Survey one-inch Third Edition maps, in colour*, London: Charles Close Society, 2004. See also Hellyer and Oliver, *One-inch engraved maps of the Ordnance Survey from 1847* where not only is no case cited for the depiction of a windmill altering between states of Revised New Series sheets (see Note 61), but that this is also the case for states of all Third Edition sheets. But then, this is not proof positive that changes to windmill depictions between succeeding states of these two editions did not occur, as Hellyer's cartobibliography makes no claim to have noted *all* changes between states (see Note 15).

Reviewing any Third Edition sheet of an area where drainage windmills proliferated shows that the Survey continued, however, to recognise these mills using the old and arguably now inappropriate mill symbol (still unannotated). However some windmills depicted in this way by the Revised New Series did now find themselves upgraded (or relegated depending on your outlook) and depicted on the Third Edition with the new windpump symbol. In short, windpump symbols and unannotated windmill symbols co-exist on a map, where often the two types of structure being recorded are visually and functionally much the same. A point of departure from the Revised New Series for the representation of windmills on the Third Edition, though, is the greater reliance sometimes put on unannotated symbols to record corn-grinding mills, along with drainage mills. This seemingly retrograde step complicates the picture on offer and obscures any assessment that might be attempted for the traditional milling capacity of an area. Using annotated as well as unannotated symbols for windmills that appear to be comparable will, presumably, have been done for a reason but, as with so much else, the records of the Survey give little away in respect of this. That said, what the record does provide for this period – more so than for any earlier period – is a précis of the instructions given to the revision teams informing them how landscape features should be recognised on the Third Edition.[67] These collated instructions included:

> 'All windmills are to be shown and the purpose for which they are used should be noted. If the windmill be permanently disused the words *Old Windmill* are [to be] written.
>
> Windfan Ontario and other such pumps in isolated positions will be shown by 1-inch field revisers on the 6-inch plots and will be engraved on the 2nd revision of the 1-inch map.
>
> They should not be shown in or close to towns or large villages where there is not room to show them properly. Should it occur exceptionally that these pumps are so numerous in a country district that all cannot be shown on the 1-inch map, the reviser will select the most important ones to show. These pumps will be indicated on the revised maps by the symbol [x on inverted T]. The reviser besides showing them on the 6-inch plots should indicate them by this symbol in position on the reviser's fair sheet.'[68]

Despite a concern for the depiction of windpumps in places where it might have seemed they were threatening to be an overwhelmingly plentiful landscape feature, confirmation is at least provided for the idea that, in the ordinary run of things, all 'normal' windmills could expect to be recognised. But how the conventions for depicting mills on the Third Edition actually panned out, though, depends on a scrutiny of the maps themselves. So, returning to the sample of 123 Norfolk windmills, and again with the help of Apling, the

---

[67] Earlier instructions had certainly been issued; some from 1896, for example, that would have been available during the preparation of the Revised New Series, have survived. However the insights they offer are limited. In the case of mills, they merely state that 'all windmills are to be shown, and the purpose for which they are used should be noted' – neither of which happened, at least as far as the published maps were concerned! See 'The one-inch revision instructions of 1896', *Sheetlines* 66 (2003).

[68] Colonel D.A. Johnston R.E., *Instructions to one-inch field revisers,* 1901, pghs 87-9. Unpublished: copy in TNA PRO OS 45/9.

evidence for these mills displayed by the Third Edition can be studied. A summary of this evidence is again provided for easy reference (Table 4.8). Additionally, and retrospectively supporting the analyses done for the New Series and the Revised New Series, full details of just exactly how each of the 123 windmills was depicted by the first three of the New Series editions is also offered (Table 4.9).

Type of Mill	Mills depicted by the Third Edition using an annotated symbol	Mills depicted by the Third Edition using an unannotated symbol	Mills not recognised by the Third Edition – including two cases of *Mill* and three pumps
Post Mill	9	2	9
Tower Mill	39	32	26
Smock Mill	1	1	2
Composite Mill	1	0	1

Table 4.8. Summary of recognition of Norfolk sample mills by the Third Edition (circa 1908).

Comparing the Third Edition representation for the sample of 123 Norfolk windmills with that of the Revised New Series supports the claim repeatedly made that the windmill population was in serious decline at this time: more specifically, the decreasing worth of the mill may have persuaded the revisers to be less rigorous in their recognition of mills for the Third Edition revision of 1907-8 than they had been for the Revised New Series ten years earlier. The figure of twelve complete non-recognitions of a mill on the Revised New Series increases to 33 for the Third Edition. Unquestionably, many of the 21 mills recognised by the earlier, but not the later edition, will have undergone substantial change in the interval between the two, either by virtually disappearing altogether, falling into disrepair, or being partially dismantled, even though all these mills still provided evidence on the ground of their existence almost a century later. The evidence offered by Apling varies from broadly supporting the map evidence – the tower mill at Roughton burnt down on 17 September 1906 and the post mill at Dersingham was demolished in February 1907 – to cases where support is not forthcoming. Aylsham (Cawston Road) Mill, though disused, retained its sails until *circa* 1920 and the post mill at East Ruston (High Mill) seems to have worked partly by sail until 1916. As well as these two mills, each credibly able to refute the evidence of the Third Edition, there is ready affirmation of the idea that substantial mill remains existed on many other sites *circa* 1907. Hindringham Mill last worked *circa* 1908 and other tower mills such as Stiffkey Mill still had their caps at this time. This all suggests that either the concern to record every windmill as a landmark feature was on the wane by the new century, or that the criterion for what constituted a notifiable landmark was perhaps becoming very flexible.

Name of Mill & type		New Series editions		
		NS-1	NS-2	NS-3
Aslacton	T	✓	✓	✓
Aylsham:				
Cawston Road	T	✓	✓ *Old*	×
Mill Road	T	×	✓	×
Banham	C	✓	✓	✓
Banham:				
Mill Road	P	×	✓	✓
Barnham Broom	T	*Mill*	*Mill*	*Mill*
Barton Turf	P	×	✓	×
Billingford	T	✓	✓	✓
Bixley	T	×	×	×
Blakeney	T	✓	✓	sym
Booton	P	✓	✓ *Old*	×
Briningham:				
Belle Vue [1]	S	×	×	×
Mill Lane	P	✓	✓ *Old*	×
Broome	T	×	✓ *Old*	×
Bunwell	S	×	×	×
Burgh St Peter	T	×	✓	✓
Burnham Overy	T	✓	✓	✓
Burnham Overy:				
Union Mills	T	*Mills*	*Mills*	*Mill*
Carbrooke	T	✓	✓	✓
Caston	T	✓	✓	✓
Catfield	P	✓	✓	✓
Catfield:				
Swim Coots	T	✓	sym	wp
Cawston	T	✓	✓	sym
Cley-next-the-Sea	T	✓	✓	sym
Croxton	T	✓	✓	✓
Denver	T	*Mill*	✓	✓
Dersingham	P	✓	✓	×
Dilham Staithe	T	✓	✓	sym
Diss:				
Shelfanger Road	P	×	✓	×
Victoria Road	T	×	✓	✓
East Dereham	T	×	✓	sym
East Harling	T	✓	✓	✓
East Runton	T	✓	✓	sym
East Ruston:				
High Mill	P	✓	✓	×
New Mill	T	✓	✓	✓
East Wretham	T	✓	✓	✓ *Old*
Erpingham	T	✓	×	×
Fakenham [2]	T	×	×	×
Feltwell	T	✓	✓	✓ *Old*
Foulsham	T	✓	✓	sym
Foxley	T	✓	×	×
Frettenham	T	✓	✓	sym
Garboldisham	P	✓	✓	✓
Gayton	T	✓	✓	✓
Gooderstone	T	✓	✓	✓
Great Bircham	T	✓	✓	✓
Great Dunham	T	✓	✓	✓
Great Ellingham	T	✓	✓	sym
Great Snoring	P	✓	✓	✓
Great Walsingham	T	×	sym	sym
Hales	T	✓	✓	sym
Halvergate	T	✓	✓	✓
Happisburgh	P	✓	✓	✓
Harpley	T	×	✓	✓
Hempnall	T	✓	✓	✓
Hickling	T	✓	✓	sym
Hindolveston	T	✓	✓	✓
Hindringham	T	✓	✓	×
Hingham	T	✓	✓	sym
Holt	P	×	×	×
Honingham	T	✓	×	sym
Horning [3]	T	×	sym	×
Horsford	T	✓	✓	sym
Ingham	T	✓	✓	✓
Kelling	P	✓	✓	sym
Little Cressingham	T	×	✓	sym
Little Melton	T	✓	✓	×
Ludham:				
High Mill	T	✓	✓	sym
How Hill Mill	T	✓	✓ *Old*	sym
Marham	S	✓	✓	sym
Mattishall:				
Mill Road	T	×	×	×
Mill Street	T	✓	✓	sym
Methwold	T	✓	✓	✓
Mileham	T	✓	✓	✓
Mulbarton	T	✓	✓	sym
Neatishead	T	✓	✓	sym
Necton	T	✓	✓	sym
North Creake	T	✓	✓ *Old*	×
North Tuddenham	P	✓	✓	sym
Norwich:				
Lakenham	T	×	×	×
Oby [4]	T	✓	sym	wp
Old Buckenham	T	✓	✓	✓
Palling	T	✓	✓	✓
Paston	T	×	✓	sym
Potter Heigham	T	✓	✓	✓
Reedham	T	×	sym	sym

Name of Mill & type		New Series editions			Name of Mill & type		New Series editions		
		NS-1	NS-2	NS-3			NS-1	NS-2	NS-3
Ringstead	T	✓	✓ *Old*	×	Thrigby	P	✓	×	×
Rockland					Thurlton	T	✓	sym	×
St Peter	T	✓	✓	✓	Tittleshall	P	✓	✓	✓
Roughton	T	✓	✓	×	Topcroft	P	×	✓	✓
Saham Toney	T	✓	✓	×	Tottenhill	P	✓	✓	✓
Scole	T	✓	✓ *Old*	✓ *Old*	Upwell	T	✓	✓	✓
Sedgeford	T	✓	✓ *Old*	×	Walpole St Peter	T	*Mill*	✓	✓
Shouldham					West Walton:				
Thorpe	T	✓	✓	✓ *Old*	Walton Highway	T	×	✓	✓
Smallburgh [5]	T	×	×	wp	Ingleborough	T	✓	✓	✓
Stalham Staithe	T	✓	✓	sym	West Winch	T	✓	✓	✓
Stiffkey	T	✓	✓ *Old*	×	Weybourne	T	✓	✓	✓
Stoke Ferry	T	✓	✓	sym	Wicklewood:				
Stokesby	T	✓	✓	sym	Hackford Road	T	✓	✓	×
Stratton					High Street	T	✓	✓	sym
St Michael	T	✓	✓	✓	Wighton	T	✓	✓ *Old*	×
Sutton	T	✓	✓	sym	Winfarthing	S	✓	✓	✓
Swanton Abbott	T	✓	✓	✓	Witton	P	✓	✓	✓
Swanton Morley	P	✓	✓	×	Worstead	T	✓	✓	✓
Terrington					Wymondham	T	✓	✓	sym
St Clement	T	✓	✓	sym	Yaxham	T	✓	✓	sym
Thornham	C	✓	✓	×					

Table 4.9. Recognition of Norfolk sample mills by the first three editions of the New Series.

Most windmills are listed here using less than the full names used by Apling who, anticipating a time when he would want also to write on mills that had disappeared, and so need to distinguish between a number of mills for each place, added details – either the name of miller or owner, or a more exact location. Only in those cases where multiple sets of remains have survived at the same settlement has a simpler version of this procedure been used here, since it is only surviving mills that are of concern to us. The type of mill has been abbreviated to a capital letter, hence C for Composite Mill, and so on.

Key to NS-1 abbreviations (the New Series)

✓	indicates use of an unannotated symbol to record this mill
Mill(s)	indicates use of this annotation (and plan) to record this mill
×	indicates that this windmill is not recognised by NS-1

Key to NS-2 abbreviations (the Revised New Series)

✓	indicates use of *Windmill(s)* and a symbol to record this mill
✓ *Old*	indicates use of *Old Windmill* and a symbol to record this mill
sym	indicates use of an unannotated symbol to record this mill
Mill(s)	indicates use of this annotation (and plan) to record this mill
×	indicates that this windmill is not recognised by NS-2

Key to NS-3 abbreviations (the Third Edition)

✓	indicates use of *Windmill(s)* and a symbol to record this mill
✓ *Old*	indicates use of *Old Windmill* and a symbol to record this mill
sym	indicates use of an unannotated symbol to record this mill
Mill	indicates use of this annotation (and plan) to record this mill
wp	indicates use of a windpump symbol to record this structure
×	indicates that this windmill is not recognised by NS-3

1. This windmill was built as a smock mill in 1721 but within sixty years the wooden structure was dismantled; the brick substructure was heightened considerably and the whole edifice then became a residential folly, which is how it remains.

2. This tower mill had been built onto the base of an earlier smock mill sometime just before the middle of the nineteenth century. This in turn became a residence early in the twentieth century.

3. This structure ended its working days as a drainage mill but previously it had been an oil mill and, as the prominent feature incorporated into the gateway of the one-time St Benet's Abbey, it has often found itself the subject of attention from artists. An image of this windmill showing it in its pre-1861 condition survives as one of our earliest mill photographs.

4. This was primarily a drainage mill but combined this function with also operating a saw mill. This function was apparently reserved, though, for estate work rather than for commercial saw-mill work.

5. This again was primarily a drainage mill before its conversion into a holiday home in 1979. Its inclusion in this sample of corn and industrial windmills derives from it having worked (unusually for its size and type of structure) a pair of millstones in addition to its scoop wheel. A similar account is claimed for the mill at Catfield (Swim Coots).

The frugal use made by the Third Edition of the windpump symbol is understandable for our particular sample of mills; the minimal use of *Old* as part of an annotation less so, but it is very hard to understand the increased use of unannotated symbols for depicting mills that were working, especially as there is only one instance where lack of space might have prevented the insertion of an annotation. As an example of a windmill recognised by just a symbol, Sutton Mill is known to have worked by sail until 1940. It continues to stand in an isolated position and remains one of the most substantial tower mills in the county. Conversely, some mills that were disused were fully recognised: a post mill at Tittleshall that had been disused since *circa* 1904 and another at Great Snoring that had operated with a single pair of sails since *circa* 1900 are each recorded by an annotated symbol. Some of the mills recorded only by a symbol are however on sites where an annotation, while possible, might have given the map a rather congested look. Maybe in these cases the unannotated symbols represent a compromise between congestion on the map and something else that, though short of a full and accurate depiction, is at least in keeping with the treatment of windpumps as outlined in the instructions to the revisers. While the Revised New Series is accepted as having been an improvement over the earlier New Series in the depiction of mills, any assumption that the clarity of this depiction would progressively improve on the Third Edition seems threatened by this particular characteristic. There is, after all, no getting away from the fact that on the Revised New Series only six of the tower mills, out of a total of 87 from the sample of Norfolk windmills that were recorded by the use of a symbol, did not have an accompanying annotation – and these were by and large drainage mills. This compares with 32 from 71 on the Third Edition, or nearly half the tower mills recognised by a windmill symbol. This trait of the Third Edition, it might be thought, goes some way to explain the types of anomaly seen in the sequences of evidence offered by the three New Series editions for Farndon in Nottinghamshire, or Brigg in Lincolnshire. It is possible to find other instances too where the depictions of a windmill by the New Series and later by the Revised New Series seem logical and compatible with one another, but then the Third Edition comes along and undermines that coherency. The tentative suggestion that these three editions, NS-1, NS-2 and NS-3, exhibited increasing accuracy and transparency in the way they recognised windmills as each edition was rolled out – as happened at Owslebury – should now perhaps be viewed as not terribly persuasive for all windmills, and not, it would seem, in respect of those of Norfolk.

The narrative offered so far for the New Series has been almost exclusively dependent on the cartographic fate of one sample of mills, albeit a fairly large one. So the reader might already have been thinking that it is somewhat being taken for granted that the treatment of Norfolk's windmills by the Revised New Series and Third Edition signifies an approach to the mapping of mills, for these editions, that was followed in all counties. This would be a shrewd observation to make and any assumption that the approaches adopted for Norfolk were applied elsewhere does need verifying.[69] To see if Norfolk is typical of other counties in this context, a small sample of twelve sheets – in four groups of three – can be examined for their Revised New Series (NS-2) and Third Edition (NS-3) depictions. The changing style of mill depiction across the lifetime of the New Series edition (NS-1) is a complication that makes it not nearly as useful for this analysis and in any case the issue, for Norfolk at least, appears to be the lack of coherence between what NS-2 depicts and then what NS-3 shows a few years later. This, then, is a simple first test for coherence of windmill depiction between these two later editions across an area greater than just a single county. Hopefully, it will demonstrate that the evidence offered for a windmill's history from the Revised New Series and then from the Third Edition can, in the great majority of cases, be as clear-cut as we have seen it to be for the tower mill that stood at Owslebury in Hampshire. So the four groups of sheets, each of three adjoining sheets and centred on areas of Essex, Lincolnshire, Sussex and Warwickshire, need examining to find the extent of their usage of each coupling of a standard Revised New Series depiction for a windmill with a standard Third Edition depiction for the same mill. The proportional usage of each pair of depictions can then be compared with how the sample of 123 Norfolk mills fares in the same respect. Studying Table 4.9 for the many different pairs of depictions used by the Revised New Series and then the Third Edition; thirteen present themselves. These can be grouped neatly into five categories. Category One embraces the only three ways in which a grinding mill remaining *in situ* throughout the lifetime of both editions could be recognised 'conventionally'. These are

---

[69] The appraisal of the Third Edition just offered was particularly critical of the use of unannotated windmill symbols for recognising mills that were probably still working. To appreciate how misleading any assessment for the mills of Norfolk is likely to be when using this edition, sheet 161 *Norwich* appears to offer no symbols annotated with *Windmill*. Instead, this sheet recognises twenty windmills – many of which would have been working, corn-grinding mills – with unannotated symbols. The position with the remaining fifteen sheets that cover Norfolk is a lot less severe in this respect, leading to the numbers of windmills in each of the categories shown in Table 4.8 being as they are. Why would sheet 161 have treated windmills any differently to other sheets? Just as with other map series and editions, the depiction of detail on the Third Edition was subject to ongoing change. Sheet 161 was one of the last sheets in this edition to be issued (in 1910), not just of those covering Norfolk but of those for other areas of the country that supported windmills. It will be shown later that the importance attached to mill representation at the one-inch scale continued to mirror the declining significance of the windmill through the first half of the twentieth century. It is quite possible that the loss of annotations on NS-3-O:161 *Norwich* signals a first stage in this course of events and, though fieldwork may have continued to follow normal practice, it was the outcome of an 'office' decision made shortly before sheet 161 was issued. This explanation linking the changes seen on sheet 161 *Norwich* to later developments in mill representation, convenient as it might be, is somewhat countered, though, by evidence from the small number of Third Edition sheets first issued after 1910, particularly those covering parts of Lancashire. Annotations of *Old Windmill* accompanying symbols can be found, for example, on sheets first issued as late as 1912, such as sheet 83 *Formby* and sheet 84 *Wigan*. But, looking at the broader picture, the depiction of other detail on Third Edition sheets was also changing and the way in which mills were shown on sheet 161 *Norwich* was probably just one of a raft of changes, still only experimental in 1910, that would take another three or four years to be fully implemented.

the use of a symbol with a *Windmill* annotation on both editions – suggestive of a windmill continuing to work – or for the Third Edition annotation to have the word *Old* included or, lastly, for both annotations to have done so. These three individual pairs of depictions can be abbreviated (adopting the legends used in Table 4.9) to:

✓ *(on NS-2 and, on NS-3)* ✓, ✓ *(on NS-2 and, on NS-3)* ✓*Old* , ✓*Old (on NS-2 and, on NS-3)* ✓*Old*

The second category simply includes the only two ways of recognising a grinding windmill on the Revised New series – either as *Windmill* or as *Old Windmill* – but where the mill then disappears by the time of the Third Edition. The third category embraces the five ways of depicting windmills that are supposedly drainage mills. Three derive from the appearance of an unannotated symbol on the Revised New Series with the mill's depiction on the Third Edition then staying the same, being replaced by a windpump symbol, or even disappearing altogether. The remaining two possibilities derive from this drainage mill being a brand new windpump making its first appearance on the Third Edition and being recorded either with a windpump symbol or with an unannotated mill symbol. These three categories all consist of coupled pairs of depictions that are non-anomalous in that, viewed together, they present an understandable progression of status for a cereal-grinding or drainage-related windmill. The fourth category, on the other hand, encompasses the two sets of paired depictions that prompted concern over the competence of the Third Edition at handling the big sample of Norfolk mills. A symbol on the Revised New Series – annotated either with *Windmill* or *Old Windmill* – which is then superseded by an unannotated symbol on the Third Edition is hard to fathom and was the main reason for extending this part of the review of the New Series beyond Norfolk. The remaining category is a catch-all for any other pair of depictions that might occur, and includes from the sample of Norfolk mills the recognition of windmills at Barnham Broom and Burnham Overy (Union Mills), where the annotation *Mill(s)* was used for both editions.

All these pairs of depictions from the Revised New Series and the Third Edition can be abbreviated, in the same style used above for the paired depictions of Category One, by also making use of the legends in Table 4.9. These can then be seen in Table 4.10, which groups by category and by area the number of times each pair of depictions appears on the twelve-sheet sample. In trying to establish if the stance taken by the Third Edition over its use of unannotated symbols in Norfolk was a normal one, the choice of sample sheets has been such that, for three of the four areas chosen, a large mill population is involved. The fourth area, mainly of Warwickshire, has been chosen to see if somewhere rather more moderate in its mill numbers was handled any differently. Even though the thoroughness brought by the Revised New Series to the handling of windmill depiction is evident across the country, the proportions of each of the different pairs of depictions can sensibly be expected to alter from one region to another, simply due to their reflecting diverse sets of mill circumstances. The numbers of drainage mills mixed in with cereal-grinding mills feature more heavily in the localised Broads area of Norfolk and for the Fens areas of Cambridgeshire, for example, than they do elsewhere.[70] In wanting to bring as many one-time windmill structures under

---

[70] The use of the windpump symbol was naturally muted for our Norfolk sample of corn-grinding windmills, though the sample does include five drainage mills, but only because three were doubling-up as grinding mills and the other two had been converted from earlier industrial-grinding roles. A very brief look at the handling by the Third Edition of the wider population of one-time drainage mills in Norfolk can compensate for this

the gaze of this study as possible – so increasing the chances of including those mills whose depictions were non-standard – the aim, as always, is to search for insights into the opaque working practices of the Ordnance Survey. Moving beyond Norfolk in our appraisal of the Revised New Series and the Third Edition helps with this, but choosing only twelve sheets may appear a little paltry. This would be a valid misgiving were it not for the second part of this book that consists of eighteen detailed descriptions of the recognition of windmills at all the major OS scales, each one linked to the area covered by a New Series sheet.

Sample of Norfolk windmills	Sheets 102 *Market Rasen* 114 *Lincoln* 127 *Grantham*	Sheets 169 *Coventry* 184 *Warwick* 201 *Banbury*	Sheets 222 *Great Dunmow* 240 *Epping* 257 *Romford*	Sheets 318 *Brighton* 319 *Lewes* 320 *Hastings*
113 mills *	92 mills	57 mills	91 mills	101 mills
Category One (✓ and ✓ ; ✓ and ✓*Old* ; ✓*Old* and ✓*Old*)				
50	54	28	49	53
Category Two (✓ and × ; ✓*Old* and ×)				
21	7	2	21	20
Category Three (sym and sym ; sym and wp ; sym and × ; × and wp ; × and sym)				
8	28	26	14	18
Category Four (✓ and sym ; ✓*Old* and sym)				
32	0	0	2	6
Category Five (eg *Mill(s)* and *Mill(s)* ; × and ✓*Old* ; ✓ and wp ; × and ✓)				
2	3	1	5	4

Table 4.10. Numbers of windmills by category of paired depictions for grouped New Series sheets.

* The number of windmills considered here is reduced from 123 to 113 due to the loss of ten mills

shortcoming of our chosen sample. With its emphasis centred on the area of the Broads, a review by Smith of Norfolk drainage windmills revealed 62 of tower-mill construction as still standing in the mid-1970s. Drainage windmills can also be of smock-mill construction as well as being of a much less substantial nature: to avoid unnecessary complications here, only these 62 sturdier tower mill examples of a drainage mill are considered. All will have been built and in a condition fit to be recorded by 1908. Leaving the five drainage mills included in the large sample of Norfolk corn windmills out of this quick analysis, the Third Edition recognised 23 of the remaining 57 mills with the new windpump symbol, 25 with unannotated windmill symbols, and missed nine altogether. The demarcation line between the use of almost equal numbers of unannotated mill symbols and windpump symbols is very pronounced: the mill symbol is used along the River Bure between Acle and Great Yarmouth and for drainage mills further south; the windpump symbol for the area lying to the north of the Bure. A.C.Smith, *Drainage windmills of the Norfolk marshes*, second edition, Stevenage Museums Publications, 1990.

that are not shown by the Revised New Series (NS-2) or the Third Edition (NS-3). This is irrespective of whether or not these mills had appeared on the New Series (NS-1).

Key to NS-2 and NS-3 abbreviations in each paired depiction		
	✓	indicates use of *Windmill(s)* and a symbol to record this mill
	✓*Old*	indicates use of *Old Windmill* and a symbol to record this mill
	sym	indicates use of an unannotated symbol to record this mill
	Mill(s)	indicates use of this annotation (and plan) to record this mill
	wp	indicates use of a windpump symbol to record this structure
	×	indicates that this windmill is not recognised by this edition

What do the figures in Table 4.10 tell us about the mapping of Norfolk mills compared with those of elsewhere? In any area, the number of windmills that fall within Category One – those whose depictions by the Revised New Series and the Third Edition conform to one of the three standard pairs of depictions used for active and not-so-active mills – can usually be expected to form a substantial element of a mill population, irrespective of wherever and however large that area might be. For each of the four new sample areas, the proportions of windmills in this category are much the same – something in the order of a half. The value for the sample of Norfolk mills is a little less, but not by so much that it is notably different to the other four areas. Looking at the windmills that fall within Category Two – those that appear on the Revised New Series but not later on the Third Edition – the proportions for three areas, including Norfolk, are similar, though not as close as their Category One values are. The proportions for the other two areas – Lincolnshire and Warwickshire – are a lot less, maybe reflecting, in the case of Lincolnshire, the relative durability of a large tower mill population. In the case of Warwickshire, though, the low figure is harder to understand, but could well have its roots in the possibility that the revisers of the 1890s were disinclined, in an area where windmills were not as numerous as they were in some more easterly counties, to record mills that were already severely ruinous.[71]

The numbers of windmills falling into Category Three need some explaining. The great majority of the paired depictions represented by the figures in this category, other than that for Norfolk, are in respect of windpump symbols on the Third Edition where nothing had been recorded by the Revised New Series: cases therefore of '× and wp'. In each of these four new areas, therefore, the Category Three figures are a reflection of the adoption of the new metal windpump. In Essex and Sussex this take-up seems to have been comparatively modest, with less than a fifth in each case of total numbers for the area being attributable to them. Warwickshire would appear to score much more heavily in this respect. Indeed, if the new windpumps found here across the three sheets are subtracted from the total figure of 57, the number of mills in this area falls to 34. Of course, any comparison of the four new areas with Norfolk is misleading in the case of this category, as the sample from that county was one of corn-grinding windmills, and the figure of eight appearing in Table 4.10 is made up from the other four paired depictions. Jumping ahead to the figures that appear in Table 4.10 under Category Five, these reveal, both by their paucity and near-parity, that the paired depictions they represent are indicators of anomalies. These include, for example, windmills that were missed by the Revised New Series and then correctly added to the Third Edition.

[71] This suggestion is obviously one that threatens the idea of consistency between counties, and so our ability to compare like with like – in our case the windmills of one county with those of another. This is an idea that will be considered more thoroughly in the next chapter.

The figures for Category Four tell a different story. The use of a mill symbol on the Third Edition to signify what on the Revised New Series had been conventionally recorded as a grinding mill (*Old* or otherwise) is a feature rarely seen in the four new areas. The contrast between these areas and Norfolk is noticeable to say the least. This is proof positive that the treatment of Norfolk had definitely been abnormal in this respect, with much of the 'error' being attributable to one sheet – sheet NS-3-O:161 *Norwich*. This peculiarity of the Norfolk sample also rather overshadows another apparent feature of the Third Edition. One might think that the number of windmills working at the time of the Revised New Series, and so recognised with symbols and *Windmill* annotations, but that subsequently fell into disuse, justifying a change of annotation to *Old Windmill* on the Third Edition, would be common. But, as one of the Category One paired depictions, only three instances of this occur among the 50 Category One windmills from Norfolk. Moreover, the other four areas also conform to the same pattern of illiberally providing *Old Windmill* annotations for the first time on the Third Edition. Indeed there are just 38 from the 184 Category One windmills for these four new areas. Thinking back to the discussion of Owslebury Mill, one facet of depiction then aired was that the addition of *Old* to the annotation on the Third Edition was assisted by the fact that there was plenty of space in which to do this, but that if there had not been, this extra annotation may have been put in jeopardy. The idea, that any declaration of a windmill to be non-working by using the annotation *Old* may depend on the availability of space on the map in which to insert the extra annotation, may seem an attack on the integrity of the map and aberrant, but appears valid. With very few exceptions, where *Old* has been added on the Third Edition to a *Windmill* annotation that had been on the Revised New Series, the root word *Windmill* has not been moved. Also, where *Old* has been applied, it can often look cramped, making the overall depiction of symbol and full annotation on the Third Edition appear awkward. The suspicion has to be that the use of *Old* as part of an annotation on the Third Edition would have occurred far more often had it not been for a lack of willingness to go to the trouble of moving the root word *Windmill* from its old position on the Revised New Series. One last observation for the windmill depictions on the Norfolk sheets and the other twelve sheets can be offered. In a large sample of Norfolk corn-grinding mills, it is unsurprising to find only two mills suspected of being drainage mills – by virtue of being recorded on the Revised New Series with unannotated symbols – which are then correctly given a windpump symbol on the Third Edition. But a disinclination to make better use of this new means of recording drainage-mill structures on the Third Edition is evident across the new sample sheets. As one of the five paired depictions of Category Three, the use of an unannotated symbol on the Revised New Series followed by a windpump symbol on the Third Edition occurs only a further two times out of a total of 86 examples in this category across the twelve sheets. Conversely, it is not unheard of for windmills to be suitably treated by the Revised New Series as grinding mills only to be wrongly recorded later on the Third Edition by the use of windpump symbols. Appearing on one of the Sussex sample sheets, Race Hill Mill, sited on the high ground to the north-east of Brighton, was handled in this way.

Yet, compared to what had gone before, the representation of many landscape features on the Revised New Series and the Third Edition – including that of windmills – achieved a uniformity across the country that must have satisfied at least some map users. This was also now a time far less permissive of individual draughtsmen applying their own local rules.

The anomalous circumstances of Norwich aside, it is apparent that the use of unannotated symbols for the depiction of windmills on the Third Edition was not a normal convention, and where this did happen, it was often because of congestion on the map. Any assessment of early New Series sheets has to conclude that they are marred by internal anomalies and inconsistencies such as those seen at Farndon, but far from acutely so. That the New Series edition is derivative is also clear; it provides no evidence of windmills from Norfolk that is not seen at the larger scales, and this is a reality repeated elsewhere, but the extent to which New Series sheets for Norfolk are fallible is repeated elsewhere.[72] For further evaluation of the fullness of windmill representation on the Revised New Series and Third Edition sheets, more evidence is needed from the margins of those activities whereby mills got recorded in the first place and then later got included on a map. In particular, given that the landmark status of a windmill has been noted as significant, the appearance of a windmill – especially a hybrid one – to an observer is an issue that needs to be explored further. But before that, the last section of this chapter looks at the particular situation of the urban mill, to see how problems of space on the map could compromise recognition.

Windmills in towns

A concern that lack of space on the one-inch sheets for built-up areas might cause urban windmills to go unrecognised has been expressed both in the first chapter, and subsequently when examining the mills of Retford in detail. The study of Retford revealed that the Survey was quite prepared for its small-scale New Series sheets to be grossly lacking in their depiction of windmills in built-up areas, and for this to happen, not only when space really was at an absolute premium and the insertion of a windmill symbol all but impossible, but also when adequate space *was* available for the insertion of a mill symbol, as was the case for Storcroft Mill. Elsewhere, mill symbols can be found, for example not far from Retford at Gainsborough, where someone has clearly been determined to incorporate one onto the map despite its near-obliteration by surrounding detail. Even for a city like Lincoln, where the scale of windmill-related activity had been drastically reduced in the second half of the nineteenth century, sheet 114 *Lincoln* of the Revised New Series (revised in 1898) manages to show the last three survivors from the line of nine windmills that had once occupied the western edge of the city ridge (Fig. 4.3). Yet, despite the depiction of the most southerly of these three mills simply being by unannotated symbol – where the annotation for the mill immediately to the north of it could easily have been made to double up for both mills – this compares not unfavourably with the depiction of the second edition 1:2500 (revised 1904-5) where a very muted unannotated circle is all that is offered for the central windmill. The diligence displayed here by the Revised New Series in recording the two mills from this trio, sited in congested surroundings, may be admirable but it is not one repeated uniformly across all major cities and towns, or even extended to the southern district of Lincoln. An

---

[72] The Arthur Smith volume dealing with the windmill remains of Shropshire, for example, implies that the New Series achieved the same tolerable – though not definitive – degree of competency in recording the few mills here as it had done for the few mills in Derbyshire. Using the Dolman guide for Lincolnshire's windmill remains, the degree of competency achieved by the New Series for a very different county, where there were far more mills, is commensurate with the findings for Norfolk. See Wilfred A. Seaby and Arthur C. Smith, *Windmills in Shropshire, Hereford and Worcester*, Stevenage Museums Publications, 1984; Dolman, *Lincolnshire windmills: a contemporary survey*.

extremely tall tower-mill shell survives here, a remnant of one of those mills that managed to elude the one-inch scale and its successor 1:50,000 scale for well over a century.

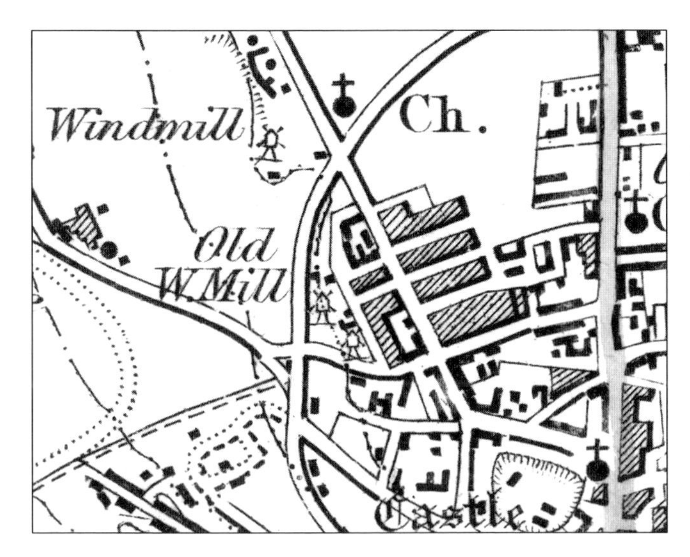

Figure 4.3. Windmills in Lincoln: one-inch Revised New Series sheet 114 (1899).

Of the mills from the Norfolk sample that were omitted from the New Series, there is only one obvious case from the total of 24 mills where space restriction can be offered as a likely reason for omission – Lakenham Mill in Norwich. But this was a sample of windmills that have survived to the present day in one form or another, and as urban mills are acutely prone to being swept away for the redevelopment of their sites, then this was a mill sample likely to be biased against urban windmills. This now suggests that the fallibility of the New Series in its recording of Norfolk windmills may have been understated because of the mill sample used. As a stark indication of the truth of this assertion, the three first edition 1:2500 sheets of the north-eastern quadrant of the City of Norwich (Norfolk sheets 63.7, 63.11 and 63.12, all published 1884-8) recognise twelve windmills by annotated ground plan (Plates 15 and 16 show three of them). Yet only three of these mills were recorded by the New Series though space was available on sheet 161 *Norwich* to have easily inserted symbols for another six; the corresponding figures for the Revised New Series (1899) were three (though not the same three on the New Series) and four. Inevitably, the circumstances of mills considered to be part and parcel of the fabric of a major city or town would have varied. Some certainly would have been town mills in that, at the end of the nineteenth century, they were part of a well-established urban infrastructure. Others would have only fairly recently been urbanised by encroaching development, or were still at the point of being so; others yet again were still on the fringes of built-up areas, but having to acknowledge that their commerce was tied in with the urban economy. Those windmills still within sight of open fields may have offered an engraver space in which to add an annotation. As a result, Hellesdon, Mousehold Heath and Sprowston Mills from the twelve Norwich mills seen on the 1:2500 sheets were suitably treated by the Revised New Series. Otherwise, it was simply a case of a symbol being taken to the one-inch if the windmill was fortunate, but for many mills embedded within an urban sprawl it was physically impossible for them to have any recognition at this scale.

If this is seen as a bleak situation for the one-inch representation of windmills to have found itself in, there is some evidence to suggest that the numbers of urban windmills that struggled on, gazing out over an expanse of tile and slate, may have been quite limited. This was certainly so for mills that kept their sails for any length of time under urban conditions. Looking at the significant point in time for windmills near the end of the nineteenth century that coincided with the preparation of the Revised New Series, an idea of what this edition should have been recording in towns can be gained from viewing the second edition 1:2500. As noted in Chapter Two, the sometimes extravagant depictions of mills on the first edition 1:2500 often gave rise to more muted six-inch depictions, and these were often followed by even more uninformative depictions on the second edition 1:2500. Windmills, if they were not totally obliterated in the interval between the two editions, are frequently shown only as unannotated circles on the later edition, so it is only by knowing exactly where to look on a map that such depictions can be understood for what they are. The proportion of windmills that fade from the scene between the first and the second editions of the 1:2500 is striking, irrespective of where one looks in the country, and establishing the number of urban mills that continued to work, not by sail but by an alternative source of power, is understandably not easy to ascertain. For anyone interested in windmills, closely examining 1:2500 sheets for these less than obvious depictions can be engaging, but our principal concern here is with the nature of the one-inch representation of windmills. The Survey may not have been too concerned with the plight of the urban mill at this smaller scale on two accounts: first, field experience may have been suggesting that most were fast losing their essence as mills in that they were resorting to the use of alternative power sources, particularly if the ready flow of wind was now affected by adjacent buildings. This would lessen any commitment to having to provide a full recognition of a windmill amid difficult circumstances. Second, the need to have windmills – especially those that had lost their sails – to fulfil the requirement for notable landmarks would have been less acute in an urban setting. The risk of confusing a mill known to be nearby with tall chimneys and suchlike would have been ever-present, and for anyone wishing to negotiate their way around an urban area, there would have been easier clues to one's whereabouts, such as churches. So, looking realistically at the usefulness of the Revised New Series and Third Edition, whatever the nuances of depiction given at the 1:2500 and six-inch scales, such subtleties were not appropriate at the smaller scale. If a lack of clarity over urban windmills was sometimes deemed to be acceptable at the larger scales, then why should it have been incumbent on the smaller scale to compensate for that shortfall, or even just to be as informative? Realistically, we can expect only urban windmills that were still working by sail to have stood any chance of being recorded, and then by just a symbol. Such windmills by 1900 were a small fraction of those that the first edition 1:2500 had chosen to record.

Looking at a range of cities and towns and the the mapped evidence for their windmills will corroborate what has just been said. On completion in the 1890s of the first editions of the 1:2500 for Lancashire and Yorkshire – the last counties to be surveyed – the main tasks of the Survey became the revision of counties surveyed much earlier at this scale, together with preparation for the one-inch Revised New Series. The findings from both emphatically endorse the suspicion that by this time not much remained of the windmill populations of key industrial towns and cities in the north. Kingston upon Hull, Liverpool and Newcastle upon Tyne had, mid-nineteenth century, each supported numbers of windmills that reached well into double figures. Fifty years later, it was very different. The great majority of these mills had been encroached upon and relegated, or obliterated altogether – so reducing their numbers to very few, or even none at all. Liverpool, earlier in the century, could boast of having supported more than a dozen windmills within a mile of its centre, but these had all gone by 1890 or been subsumed within new enterprises and, at the turn of the century, one had to travel out as far as Crosby or Wavertree to see a windmill, and these two would not then have been thought of as urban mills. In Newcastle, of the many windmills recognised by a relatively early first edition 1:2500 (including those seen in Plate 17), only two disused tower mills, as well as a smock mill still able to work by sail – a rarity on two accounts for this far north and a likely explanation for its poor recognition by the second edition 1:2500 – were left by 1900.[73] Of the eleven windmills once lining the Holderness Road in Hull (rather than the far larger number of mills within easy reach of it) that were recorded by the mid-century six-inch edition, only two survived long enough to appear on the 1:2500 sheets of the early 1890s. Even then, one appears simply as an unannotated circle while the other – Eyre's Mill – was the only windmill from these three important urban areas recorded by the Revised New Series. But possibly it was the only mill, other than Chimney Mill – the smock mill in Newcastle – that deserved to be acknowledged by it. So the record of urban windmill depiction by the Revised New Series for these three urban areas cannot be considered to be bad.[74] This story is repeated for smaller industrial towns and for cities and towns further south. With a similar industrial profile to its larger neighbours, Preston could only boast the remnants of three tower mills in 1900. All had lost their sails by then but, in the cramped space of the 1:2500, two of these mills had acquired circular ground plans overwritten with *Corn Mill* annotations (see Fig. 4.4.1). This style of depiction may seem atypical to anyone familiar with the norms of windmill depiction but, in the context of this large-scale sheet for Preston, it is in keeping with the numerous other examples of large rectangular hemmed-in buildings overlain with annotations declaring them to be foundries, cotton mills or any one of a host of other types of manufactory. The third windmill tower – ironically, today's sole survivor – had no such annotation and, understandably, none of them was recognised by the Revised New Series. Moving south, Nottingham also had a moderate number of mills much earlier in the century: a few were moved sometime around the middle of the century and they continued to work in their new rural locations; others had gone by 1900 with only Green's Mill at Sneinton surviving in a poor state as a truly urban mill. On the other hand,

---

[73] There were other disused tower mills nearby such as those at Matthew Bank and Cowgate whose sites are well within today's city limits, but in 1900 these mills were still surrounded by open fields.

[74] The realisation that urban windmills had had their day and were rapidly being replaced by steam-driven roller mills, especially in dockside towns, and that this was happening amid general programmes of civic improvement, is made very obvious by Burnett. In his biography of Joseph Rank, Burnett relates how Rank started as a miller at one of the Holderness Road windmills in east Hull and later went on to achieve fame and fortune in the roller-mill industry. The decline of the windmill in Hull is placed in the context of progress that was both necessary and desirable, especially in the 1890s, when 'it [Hull] was altered almost out of recognition during the last decade of the nineteenth century... main roads were widened and in place of narrow, crooked, crowded thoroughfares and congested areas, spacious avenues and stately streets appeared. A new pure-water supply was provided. Transport facilities were improved... vast schemes were initiated for the modernising of the docks, with the building of new and better bridges to remove causes of delay, inconvenience, and loss of time, tides and trains'. What Burnett fails to add is that many erstwhile tower-mill structures found alternative uses and did not finally disappear until well into the next century. See R.G. Burnett, *Through the mill,* London: Epworth Press, 1945, 64. For further discussion and explanation of urban growth in the nineteenth century, see H.J. Dyos and Michael Wolff (eds), *The Victorian city,* volumes 1 & 2, London: Routledge & Kegan Paul, 1973.

many fairly large towns such as Stafford also turn out to have been one-windmill towns by 1900, not because they had lost many of their mills, but rather that they had never really had any in the first place. Careful scrutiny of the first two editions of the relevant 1:2500 sheet is needed to uncover the single windmill here; again, the Revised New Series fails to record it.

Moving further south and east to parts of the country seen by many as having been the traditional areas for windmills, far greater numbers of them had survived here and several were still thriving in 1900, particularly those well away from urban centres. Here, in contrast to further north, there were many small towns sustaining two or three windmills. Alford, Lewes and Reigate are just three examples that are well-known to windmill historians and that can speak for the many other similar places. Each of these three towns had a variety of windmills still working in 1900; Alford's particularly noteworthy claim to fame is that it then had three tower mills, one carrying four sails, another five sails and the third six sails! Urban encroachment was happening on a far less severe scale in these smaller towns and windmills often got recorded on the one-inch, if only by a symbol. The style of recognition followed in these places often mirrors what occurred at Lincoln. Sometimes, the Revised New Series was even capable of outperforming the 1:2500 in its windmill depiction. For example, when it recorded four mills at Newark just by symbol and no individual or collective annotation – which could easily have been added – the Revised New Series still managed to outclass the second edition 1:2500, which just recorded two unannotated shaded circles and two ground plans each annotated by *Windmill (Disused)*. But for all this, windmills *were* disappearing in the smaller towns of central and southern England. Market Harborough, as one example, had supported two windmills for much of the nineteenth century but on the first edition 1:2500, published in 1886, only one is shown. This mill is also recorded on the New Series and then on the Revised New Series, but not the Third edition. The appearance of this mill at 1:2500 is understandable but it is interesting that the Revised New Series should also show it since NS-2-O:170 *Market Harborough* carries a revision date of 1897, and this mill was allegedly taken down in 1895 according to one reliable source.[75] This is not as inappropriate as it may seem when it is realised that a steam mill, later driven by a gas engine, had been built next to the windmill in 1891 and continued to work until the 1930s. Such was the layout and size of this small town in the 1890s that the unannotated windmill symbol is shown just yards from the centre of the town in sufficient space for an annotation to have easily been inserted. We are left to wonder if, with the windmill only just gone in 1897, the use of an unannotated symbol here on the Revised New Series is a further example of the acknowledgement of a continuation of milling by alternative means.

Even in the larger towns of the windmill-rich areas of the south-east, the urban mill was moving surely into decline by 1900 though examples could still be found, just as they could in Lincoln. Ramsgate was certainly one of the larger urban centres in Kent yet it appears to have only had one windmill at this time. Grange Road Mill (Plate 35) is of interest to us for a number of reasons, not the least of which is its footprint that the Survey duly recorded as the shaded ground plan on the second edition 1:2500 (Fig. 4.4.2). The urban 'clutter' at the base of this mill is clearly apparent in the photograph of it and no doubt contributed to a ground plan that offers no inkling of the smock mill's actual polygonal shape. This issue of surrounding buildings has already been raised in Chapter Two but is brought up again here

[75] See Moon, *The windmills of Leicestershire and Rutland.*

to reinforce the idea that it may well have been a more common feature of windmill ground plans than is realised. The activity and bustle immediately surrounding the base of this mill may well have been unrelated to its operation and was obviously something of a distraction to the OS surveyor. The Revised New Series and Third Edition both chose not to record this windmill – it survived until at least 1905 – but the availability of space could have been an issue here. These circumstances of Grange Road Mill as a working yet unrecorded smock mill on the Revised New Series should concern us, but on the other hand it is unlikely that the number of mills in similar circumstances would have been significant. Kent probably had as many urban mills as any other county since the practice of milling by wind lingered here for as long as it did anywhere. Yet examples of working windmills in, say, Folkestone, Margate, Ramsgate or the Medway towns were unusual by 1900 and even more so by 1910. Delce Mill in Rochester was a rare example of an urban windmill that continued to operate by sail well into the 1930s. Gradually hemmed in by more and more housing, this was a mill that the first edition 1:2500 had shown in an open rural setting. Increasingly this altered; so much so that by the fourth edition of the 1930s this mill was totally surrounded by housing and had truly become an urban mill. In 1896 there had still been space for the Revised New Series to record this mill: this did not happen.

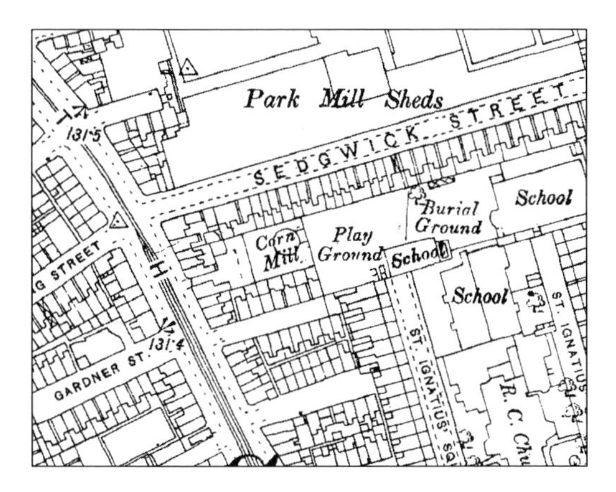

Figure 4.4.1. Windmill at Preston, Lancashire: 1:2500 sheet 61.10 (1893).

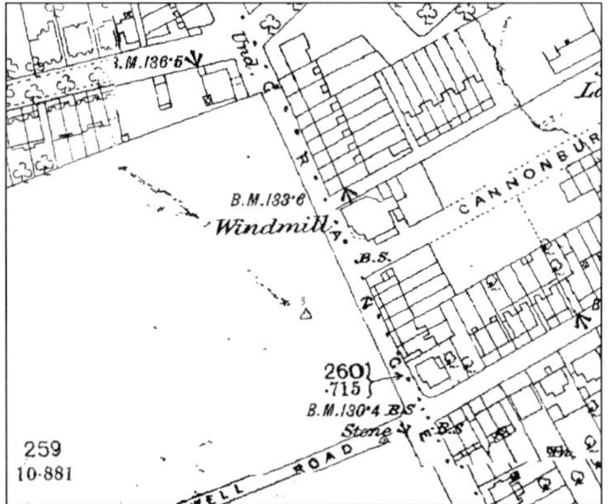

Figure 4.4.2. Windmill at Ramsgate, Kent: 1:2500 sheet 38.1(circa 1874).

To finish this account of urban mills, and indeed the chapter, on a more positive note: Brighton is a town with a record of having supported many windmills during the nineteenth century. One account of these mills relates that in the years leading up to 1900, half-a-dozen were still at work. The topography of Brighton is such that any mill trying to catch the wind had to be sited on the surrounding high ground and, for those built early in the nineteenth century, this meant that by 1900 they found themselves engulfed by a still-expanding town. The Revised New Series manages to provide a good account of these mills to the point of recognising those that were relatively decrepit and two that, at least by one person's reading of the available evidence, had already vanished. Yet the story for Brighton on the earliest editions of the New Series is not flawless. The New Series itself had not recorded any mills here, but that was the arrangement in place for that edition at that time, rather than the fault of the surveyor – or of Brighton! On the Revised New Series, and later the Third Edition, the presence of urban windmills was better recognised, but their prominence or visibility to all was still no guarantee that the status of each would be correctly recorded (Table 4.11). A chronology of evidence supplied by the New Series, the Revised New Series and the Third Edition that is non-conflicting and that offers increasing clarity for the status of a windmill – hopefully, akin to the evidence for Owslebury, rather than that for semi-urban Brigg – is

Name of Mill	Revised New Series (NS-2)	Notes for the last known date of working and the disappearance of the mill
Brighton:		
Bear Mill	✓	Still at work in 1903
Cutress' Mill	×	Worked after 1900; demolished 1913
Hanover Mill	✓	Demolished in 1890s when out of use
Port Hall Mill	✓	Worked until demolition *circa* 1882
Preston Mill	✓	Demolished *circa* 1890
Race Hill Mill	✓	Worked until *circa* 1905; collapsed 1913
Kingston upon Hull:		
Anti-Mill Mill	×	Worked until 1897
Bell's Mill	×	Worked (as a seed crusher) until 1890s
Blockhouse Mill	×	Worked (as a seed crusher) until 1904
Dale's Mill	×	Demolished *circa* 1930s
Eyre's Mill	✓	Worked after 1900; tower remains
Preston:		
Craggs Row Mill	×	Sails removed *circa* 1882; tower remains
Kirkham's Mill	×	Sails wrecked 1863; tower in place 1914
Parrs Croft Mill	×	Demolished *circa* 1907

Table 4.11. Recognition of urban windmills in three towns by the Revised New Series (circa 1895). (Brighton, Kingston upon Hull and Preston)

The windmills included are all those within the limits of urban development (as defined at the time of the Third Edition) which either were recorded by the Revised New Series (using either an annotated or an unannotated symbol) or which should have been, based on the sources offered below.

Key to NS-2 abbreviations:
✓ indicates use of a symbol to record this windmill
× indicates that this windmill is not recognised by NS-2

Sources:

H.T. Dawes, The windmills and millers of Brighton, *Journal of the Sussex Industrial Archaeology Society* 18 (1988); Mary Fowler, *Holderness Road: through the heart of East Hull*, Beverley: Highgate Publications, 1990, 10-20; T. Harrison Myres, *Story of the windmill: past and present historically treated*, Preston: Guardian Printing Works, 1914.

not forthcoming for Race Hill Mill at Brighton. This perfectly ordinary post mill complete with roundhouse enjoyed something of a landmark status close to the town boundary and the racecourse. Overlooked by the New Series, it was conventionally shown by the Revised

New Series but then mistakenly recorded as a windpump on the Third Edition (as noted a few pages ago), even though recognition of this mill by the first three editions of the 1:2500 has it correctly recorded at times when it was working, after it had ceased to work *circa* 1905, and after its eventual collapse in 1913. Nonetheless, in contrast to at least two urban centres already discussed – Hull and Preston – the evidence for Brighton suggests that its windmills were well-served by the Revised New Series. Moreover, even though the treatment by this edition of windmills in all three towns was demonstrably unequal, the merit of the Revised New Series for us lies in it being the one-inch edition that best captured the culmination of the windmill as a contemporary landscape feature, and this remains the case despite any inconsistency of urban depiction. Across large parts of the country, the rural windmill was still much in evidence in 1900 but numbers were relentlessly falling and this trend was far more advanced in urban areas. All of this only serves to strengthen the idea that the Revised New Series really should be seen as having embraced the high point of small-scale windmill depiction by the Ordnance Survey.

5

Deferring to the larger scales

The last two chapters, reviewing in turn the one-inch Old Series and New Series, took for granted that the effectiveness of windmill representation is sensibly best predicated on the idea that every individual mill should be depicted. This accordingly leads to a quantitative account of ground-truth accuracy, with the non-recognition of any single mill revealing the map account to be flawed. The idea that this approach may not necessarily lead to the most 'correct' appreciation of how effectively windmills were documented will be touched upon later, but in the meantime this chapter is happy to extend the critique for an assessment of windmill representation that is essentially quantitative. It does so by extending the simple yet revealing ploy of comparing the Revised New Series with a suitable (*ie* somewhat later) larger-scale edition.[1] The improved accuracy of the larger scales compared with the small one-inch has already been established, and is hardly surprising. The reasoning that a mill should appear on the Revised New Series if it appears on a later large-scale edition is easy to accept and comparisons, such as made already in the case of the mills at Brill, offer a simple yet thorough assessment of the small-scale mapping with its intrinsic use of symbols. After all the issues of depiction discussed in earlier chapters, this chapter concerns itself with just two: the completeness of representation for an area – easy as a concept to understand – and, more contentiously, just what constituted a notifiable windmill in the first place. So, the emphasis now moves away from consideration of individual windmills to an assessment of how they were recognised *en masse*. Unless the whole country is to come under review, some geographical parameters for this kind of grouped consideration are necessary. Sensibly, mills can be selected as those from a single sheet or from a group of sheets or even from a whole county. Different types of outcome will also result from whether the areas chosen are those where mills played a marginal part in the life of an area or where, by dint of large numbers, they made a different type of contribution. For the latter type of area, it is not uncommon to find individual Revised New Series sheets offering anything up to 30 windmill sites. Yet, recalling that large tracts of England, and even more so of Wales, supported no windmills, a

---

[1] As already achieved for the pair of windmills at Brill, the mills of Retford and, in a sense, all those of Norfolk. Comparison of the New Series with the larger scales as a way of assessing the one-inch featured somewhat in the last chapter where, theoretically of course, the New Series was a derivative of the larger scales. The choice of the Revised New Series as the one-inch edition worth focusing upon for further inquiry stems from all that has been discussed: its supposedly non-derivative origins and its contention for being the most timely of the New Series editions for depicting windmills, as well as its sheets being easier to verify than those of the Old Series, belonging, as they do, to a later time period.

majority of sheets offer no sites at all, or perhaps a couple at most. The eighteen New Series sheets selected for closer examination in Part Two are sensibly all of areas that supported reasonable numbers of windmills, or had done so. The same is true of the main part of this chapter where the comparison exercise is applied to Essex, a county noted for its windmills and for an excellent windmill biography by Farries. But, as an introduction to this, and in an attempt to be even-handed when building on the findings of Brill and Retford, some sheets from areas of low-windmill density will be examined first.

The extreme north-western part of England has received scant mention so far in this study and for good reason. Cumberland – more specifically, the area around Carlisle – was never home to many windmills and the few that had been there were mostly gone by 1900. However, early sheets of this area are still of interest. Inspecting sheet NS-2-O:17 *Carlisle*, a search for windmill symbols together with their annotations reveals nothing. Yet when this sheet was published in 1897 there were three mill towers eligible for some sort of recording. Two, at Monkhill and Cardewlees, were totally disregarded, which is ironic since their full-height shells both still stand, lately converted into houses. The other windmill was recorded by the Revised New Series but, in an oversight, its depiction had not been changed from the earlier New Series sheet, which, this being Cumberland, was one of those adopted from the Old Series and renumbered. As such, it dates from before 1888 when the windmill symbol was reinstated on the New Series. So *Holmelow Windmill* with its diminutive ground plan – a perfectly normal type of depiction before 1888 – appears on the Revised New Series where it looks distinctly out of place. Another windmill had been similarly recorded by the New Series, this time annotated as *Laythes Windmill,* but presumably this mill had gone before the time of revision as the later Revised New Series fails to record it. In its recognition of just two windmills, the New Series had not only neglected to show the mills at Monkhill and Cardewlees but also a fifth mill at Bowness not dismantled until *circa* 1887. The recognition of these five windmills by the larger scales was predictably better but not faultless. That of Laythes Mill upholds what we can infer from the New Series and the Revised New Series. This windmill is registered by the first editions of both six-inch and 1:2500 scales as *Disused* before its subsequent removal from each of the second editions. Holmelow Mill is accorded a circular ground plan of sorts (overlain by the sign for a triangulation station) on the first editions. This is then deleted on later editions though the annotation remains. A modified annotation of *The Windmill* survives for this site on current OS mapping, but care should be exercised since this is only an echo of a past presence; the remains of this mill disappeared long ago. The larger scales were also fully aware of Monkhill Mill: a circular ground plan was accompanied at first by *Windmill* but this was changed to *Old Windmill* for all later editions, continuing up to the present. Cardewlees Mill, on the other hand, obviously made less of an impression on those preparing the first large-scale editions since no annotation for this mill then appeared, or has done so since to accompany the circular ground plan which again has survived on maps to the present day. There is not a lot of evidence for when this mill gave up working. Certainly the building has found other uses in the past century and it may even be the case that the New Series and the Revised New Series were justified in ignoring this mill structure. But where the large scales are less than faultless and exhibit their own internal contradiction is in the case of Bowness Mill. Here the first edition 1:2500 clearly recognises a windmill, which the derivative six-inch ignores. Given that this mill was dismantled *circa* 1887 that then becomes the end of the matter. So the situation we are left in for this north-western part of the country is one where the large scales give plausibly definitive evidence for remaining windmills with the single exception of one mill where the six-inch does not repeat the information of the 1:2500. The New Series and Revised New Series sheets of this area are both flawed, either because they each defy their conventions for showing windmills or they get the mill count wrong, or both. How (or if) these five windmills were recorded by the Survey can be demonstrated using the annotated facsimile of NS-2-O:17 *Carlisle* shown as Fig. 5.1. Alternatively, this same information could be made the basis of a table, although for this sheet it would be a very concise one. Both ways of recounting how windmills were depicted by different OS map series are used for the second part of this book.

In contrast to the minimal number of windmills around Carlisle, the long coastal strip stretching from Berwick at the tip of Northumberland down to Whitby in Yorkshire had at one time supported a significant number of windmills, as already noted when considering Durham. But by the second half of the nineteenth century these mills had largely vanished. Certainly those which survived as late as the 1890s had fallen on hard times with only a tiny number struggling to continue working. Heading south, the Yorkshire Moors offer a natural break before coming to the East Riding of Yorkshire where a different survival pattern for mills is found. For County Durham, the first revision of the 1:2500 and the derivative six-inch sheets took place between 1894 and 1897. With the initial survey for the 1:2500 only reaching northern Yorkshire around 1890, this resulted in six-inch reductions for the area around Middlesbrough carrying survey dates of 1891-3. So, contemporary or slightly earlier six-inch sheets are available for comparison with the Revised New Series sheets dated 1895-6 for much of this coastal strip. With the discrepancy between dates of revision on the six-inch and the one-inch sheets being minimal, one could therefore expect some parity of mill representation between them. Looking at the Revised New Series sheet covering the estuary of the River Tees and therefore the boundary between the North Riding of Yorkshire and Durham, together with the sheet immediately to the north of it – sheets NS-2-O:27 *Durham* and NS-2-O:33 *Stockton on Tees* – the number of windmills they record seems quite modest. The six-inch sheets depicting the area covered by NS-2-O:33 *Stockton on Tees* are obviously either from the Durham set or the Yorkshire set and their sheet-line arrangements can be seen below (Fig. 5.2 where the Durham sheets have solid outlines and the Yorkshire sheets have dashed outlines). A discontinuity in these sheet lines, leading to the Durham sheets not being contiguous with those of Yorkshire, is very evident. It should not be inferred that this is feature common to all county boundaries. Where it does happen, it is the outcome of the larger scales being plotted in groups of counties where the survey of each group is based on its own meridian. It would not be until well into the next century that any harmonisation of these different groupings was achieved.

By looking at the relevant sheets from the appropriate six-inch edition – for Durham, that derived from the first revision of the 1:2500 and for Yorkshire, that derived from the initial edition of the 1:2500 – a list of windmills can be compiled for the areas covered by Revised New Series sheets 27 *Durham* and 33 *Stockton on Tees*. This list turns out to contain the names of fourteen mills (Table 5.1). Each of these mills is accorded either an annotation that quite unequivocally incorporates the word *Windmill*, or one that simply incorporates the word *Mill* but where the accompanying ground plan is indisputably circular and the siting of this mill is on high ground. How these fourteen mills were recognised by the Revised New Series is included in Table 5.1. At a glance, the one-inch is shown to be much less proficient than the six-inch at recognising these windmills – all candidates for being recorded by the Revised New Series with annotated symbols. As suggested earlier, the secondary windmill literature for Durham is slight though, in its proximity to North Yorkshire, it does benefit from one good reference source.[2] But even if primary sources relating to the disappearance of these windmills were to be petitioned, so confirming that the Revised New Series is sadly remiss in this part of the country, the closeness of dates between the one-inch and six-inch editions must surely mean that recourse to these primary sources will have been pointless. Besides, this is probably quite a conservative assessment of the failing of the Revised New Series since a number of probable windmills recorded by the six-inch were not included in Table 5.1 through failure to meet the criteria set above. Egglescliffe Grange Mill, confirmed as a windmill by the contemporary 1:2500, is one of these. Another is Newton Bewley Mill where the circularity of ground plan is also only obvious at 1:2500. But this mill is annotated only by *Corn Mill* at each of the larger scales and, along with Fishburn and Easington Mills, it might have been the case that they were no longer working by sail. From past discussion it will be recalled that 1:2500 annotations were occasionally not cascaded down to the six-inch, or that they were abridged for reasons of space. Not that far north of the River Tees, at Windmill Hills in Gateshead, it was noted that shadowy mills from earlier years appeared as unannotated ground plans on the six-inch. With such mills not revealing themselves for comparison with the Revised New Series, this attribute further threatens the credibility of the one-inch representation but, on the other hand, any corrupted or degraded evidence for one-time windmills at the larger scale could hardly be expected to warrant repeating on the one-inch scale. But even with caveats such as these, the record of the Revised New Series in recognising the windmills on Teeside appears less than reliable. Of the twelve accorded an annotation of *Windmill* by a six-inch sheet, just seven are shown by the Revised New Series.

The problems do not end there. A windmill at Portrack Lane just north of Stockton went unrecorded by the second editions of both larger scales, but did appear on the Revised New Series. Yet only a mile further south, a windmill at Mount Pleasant was ignored by the one-inch. Not so far from Stockton too, a post mill at Elwick had been replaced by a tower mill built on a new site a hundred yards or so away from the earlier site. Both larger scales accurately record this change that took place between their first and second editions, but the Revised New Series stubbornly prefers to record the earlier rather than the later situation. Lastly, the failure of the Revised New Series to recognise a windmill at Yarm in Yorkshire should perhaps be seen as strangely remiss, given that special attention had been given to its six-inch depiction. The placing of the *Old Windmill* annotation for this mill and the clarity of

(Overleaf) Figure 5.1. One-inch Revised New Series sheet 17 Carlisle (1897).

All the facets of recognition discussed above for the five Cumberland mills have been condensed in this figure to one of simply differentiating between windmills that the one-inch Revised New Series recognised, and those that it did not but which had nevertheless been recognised by the first edition 1:2500. This forms the basis of the approach used in Part Two of this book for classifying windmills and chronicling their OS mapping histories.

---

[2] John K. Harrison, *Eight centuries of milling in north-east Yorkshire,* North York Moors National Park Authority, 2001. In this commendable text, concerned principally with the many watermills and the few windmills of the region appearing in the title, the author has thankfully strayed out of area to discuss also the windmills on the north bank of the Tees.

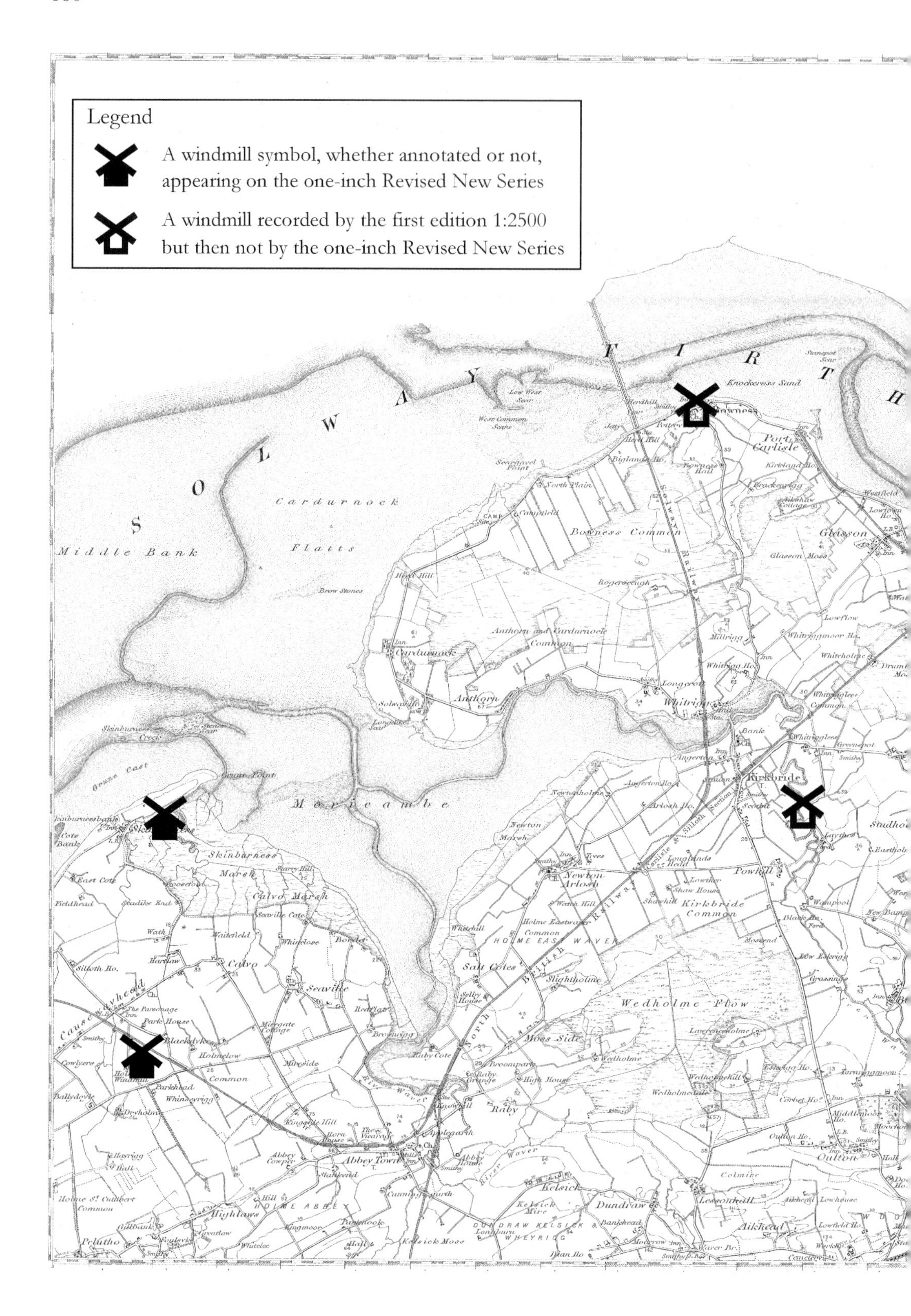
Legend
A windmill symbol, whether annotated or not, appearing on the one-inch Revised New Series
A windmill recorded by the first edition 1:2500 but then not by the one-inch Revised New Series
S O L W A Y F I R T H
Cardurnock Flatts
Middle Bank
Moricambe
Bowness Common
Anthorn and Cardurnock Common
Wedholme Flow
Kirkbride Common
Port Carlisle
Glasson
Kirkbride
Newton Arlosh
Abbey Town
Kelsick
Dundraw
Skinburness Marsh
Calvo Marsh
Causewayhead
Seaville
Raby
Oulton

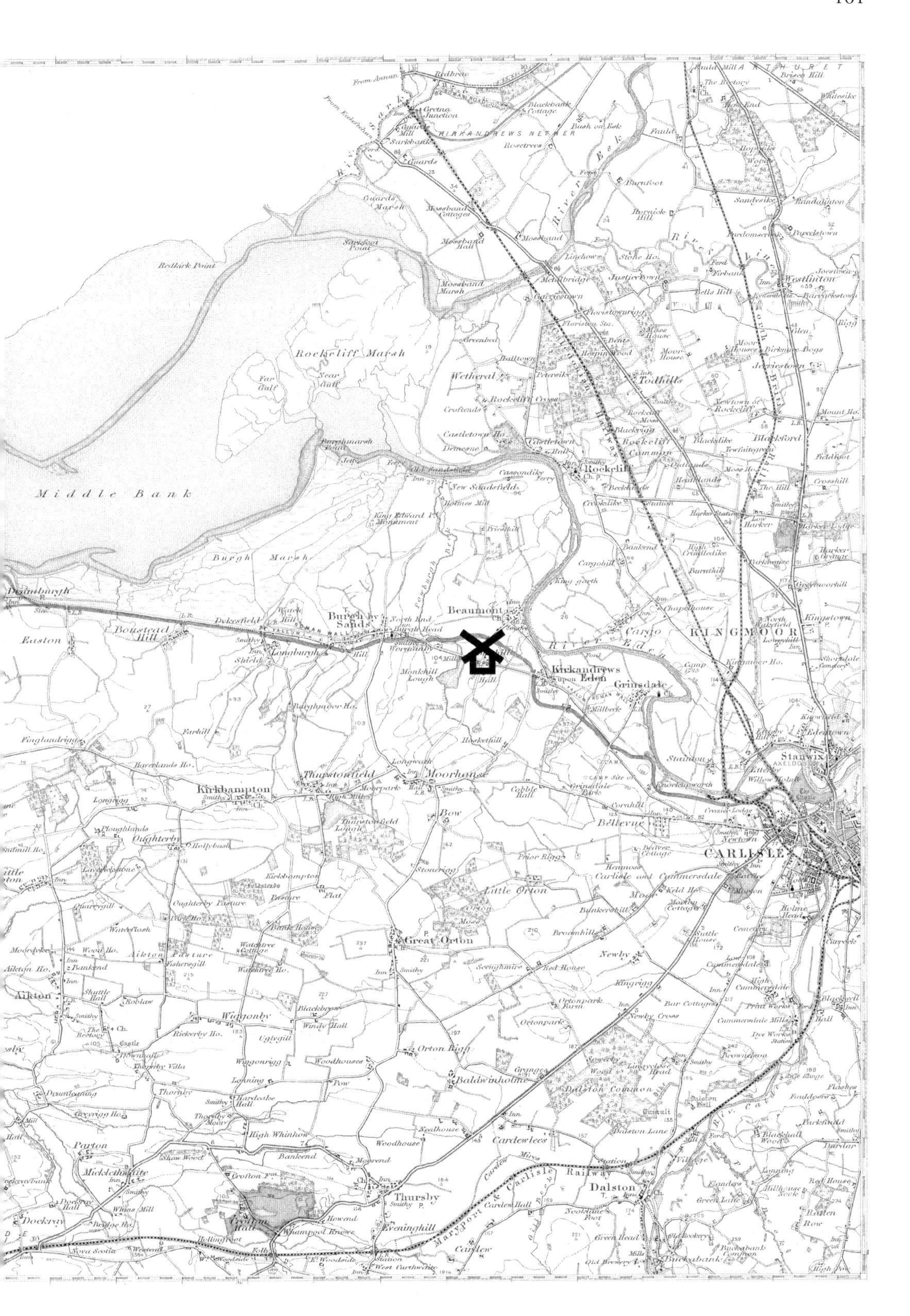

From Annan
Redbrae
Gretna Junction
Sarkbank
Guards
KIRKANDREWS NETHER
Blackbank Cottage
Bush on Esk
Rosetrees
Guards Marsh
Mossband Cottages
Mossband Hall
Mossband
River Esk
Sarkfoot Point
Redkirk Point
Mossband Marsh
Metalbridge
Garriestown
Floriston Sta.
Rockcliff Marsh
Far Gulf
Near Gulf
Greenbed
Balltown
Wetheral
Rockcliff Cross
Croftends
Castletown Ho.
Demesne
Castletown Hall
Rockcliff
Rockcliff Common
Todhills
Hespin Wood
Middle Bank
Burgh Marsh
Old Sandsfield
New Sandsfield
Holmes Mill
King Edward I. Monument
Cargo
River Eden
KINGMOOR
Drumburgh
Easton
Boustead Hill
Dykesfield
Burgh by Sands
Longburgh
Beaumont
Monkhill Lough
Kirkandrews upon Eden
Grinsdale
Burghmoor Ho.
Farhill
Finglandrigg
Kirkbampton
Thurstonfield
Moorhouse
Cobble Hall
Stanwix
CARLISLE
Oughterby
Little Orton
Great Orton
Aikton
Wiggonby
Orton Rigg
Baldwinholme
Dalston Common
Cardewlees
Thursby
Dalston
Micklethwaite
Parton
Dockray
Eveninghill
Maryport & Carlisle Railway
Newby
Prior Rigg
Bellevue
Cummersdale
Blackwell
Buckabank
Westlinton
Blackford
Kingstown
Harker

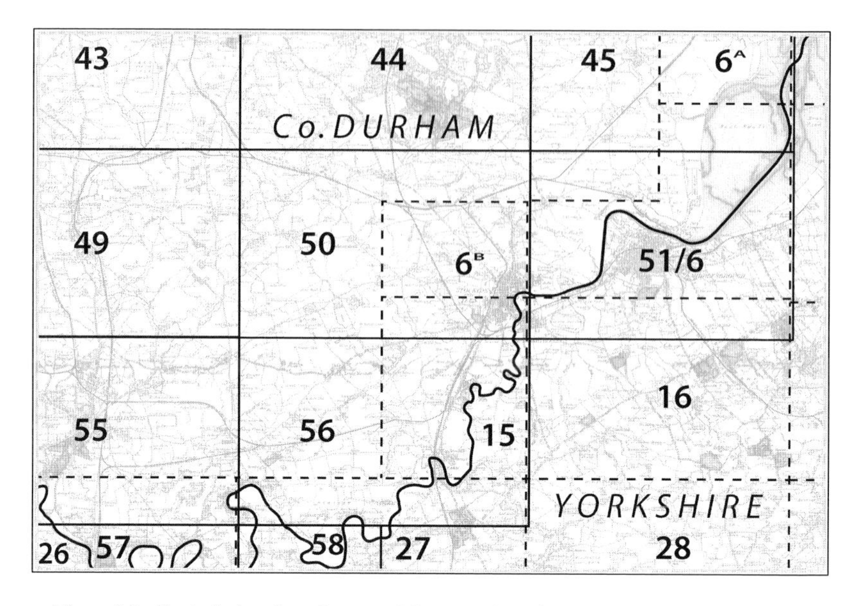

Figure 5.2. Six-inch sheet lines for part of the county boundary between Durham and Yorkshire.

ground plan both differ between Durham sheet 56SE dated 1896 and the earlier Yorkshire sheet 15SE dated 1893, despite the same (re-)survey data of those years presumably being the basis for both – this mill was only yards away from the border with Durham. That the Survey bothered to upgrade the information on the slightly later Durham sheet in respect of a windmill that by then was probably no longer working (certainly by sail) and also not even in the county, is interesting. Maybe this reflected a certain concern over the use of much earlier survey information to provide 1:2500 scale mapping for Yorkshire (as a re-plotted county), and a desire for the later Durham sheets not to be 'contaminated' in any sense by this undue process.

Moving away from northern England as an overall area of early decline in numbers of windmills, whether it be Cumberland on the west coast or Durham on the east coast, other areas where windmills were thin on the ground around 1900 are not hard to find. Scanning the southern coastal counties, for example, reveals a proliferation of windmills in Kent and Sussex, while at the other geographic extreme Cornwall and Devon could each only boast a scattering of mills, mainly sited along their coastlines. More centrally, Dorset could only ever claim a dozen or so windmills to its name, this being a county known very much more for its watermills. With Hampshire struggling to do dramatically better than Dorset in terms of numbers of windmills in the years after 1800, a change in the pattern of windmill placement can be located approximately on the border between Hampshire and Sussex. So, it is to this southern-central part of the country that our consideration of the appearance of windmills on the Revised New Series, and maybe a contrast with their appearance on the larger scales, can next briefly turn. Again selecting two suitable Revised New Series sheets: 316 *Fareham* and 331 *Portsmouth* between them show much of the boundary between these two counties and, again, the sheet lines for the two sets of large-scale sheets happen not to be contiguous. Figure 5.3 shows the overlapping sets of six-inch county sheets for Revised New Series sheet 316 where this time it is the Hampshire sheets that have solid outlines and the Sussex sheets that have dashed outlines.

Name of Mill	Depiction at 1:10,560 (six-inch scale)				Revised New Series depiction
	Sheet	Revised	Annotation root word	Ground plan	Sheets 27, 33 revised 1895
Durham mills					
Easington	21SE	1896	*Mill*	circular	no recognition
Thorpe Moor	28NE	1895-7	*Windmill*	circular	*Old Windmill*
Ferryhill	35SW	1896	*Windmill*	circular	✓ *Old Windmill*
Hutton Henry	36NE	1896	*Windmill*	circular	no recognition
Fishburn	36SW	1896	*Mill*	circular	✓ *Old Windmill*
Elwick	36SE	1896	*Windmill*	circular	no recognition
Hart	37NW	1896	*Windmill*	composite	✓ *Windmill*
Greatham	45NW	1896	*Windmill*	composite	✓ *Windmill*
Aycliffe	49NW	1896	*Windmill*	circular	no recognition
Mount Pleasant	50SE	1896-7	*Windmill*	composite	no recognition
Egglescliffe	56SE	1896	*Windmill*	circular	✓ *Windmill*
Yorkshire mills					
North Ormesby	6SE	1892-3	*Windmill*	circular	no recognition
High Leven	15SE	1893	*Windmill*	circular	✓ *Old Windmill*
Yarm	15SE	1893	*Windmill*	composite	no recognition

Table 5.1. Windmills shown at 1:10,560 (six-inch) for areas of Revised New Series sheets 27 and 33.

Circular (plan) indicates a tower mill that is either free-standing and sited completely alone or one that is dominant within a complex of buildings, but where the circularity of its ground plan is very obvious. Composite (plan) indicates a tower mill incorporated into another building where no part of the joint plan is easily identifiable as being that of a windmill. ✓ indicates the use of a windmill symbol on the one-inch Revised New Series (NS-2). All mills are listed alphabetically for each six-inch quarter-sheet in county number order.

The first revision of the New Series sheets for Hampshire occurred slightly ahead of the first revision of the larger scales, so potentially leaving itself open to criticism if it failed to recognise windmills that were later depicted on the six-inch maps. The position for the west of Sussex is comparable. Applying the same standard criteria used when searching out mills on the six-inch maps of Teeside, the windmills revealed by this next scrutiny for two sheets of the Revised New Series have also been listed (Table 5.2). The number of windmills that seem to have been present here is again only in the order of a dozen. Moreover, these few mills form the majority of windmills that were to be found within the whole of Hampshire (including the Isle of Wight) at the end of the nineteenth century. This is in huge contrast to the contemporary windmill population of neighbouring Sussex, measured at that time in three figures. But the incidence of windmills in Sussex did tend to diminish the further west one got so, even though only a very small slice of this county comes under consideration, it should not come as too much of a surprise to find that there is only one Sussex windmill on this second list.

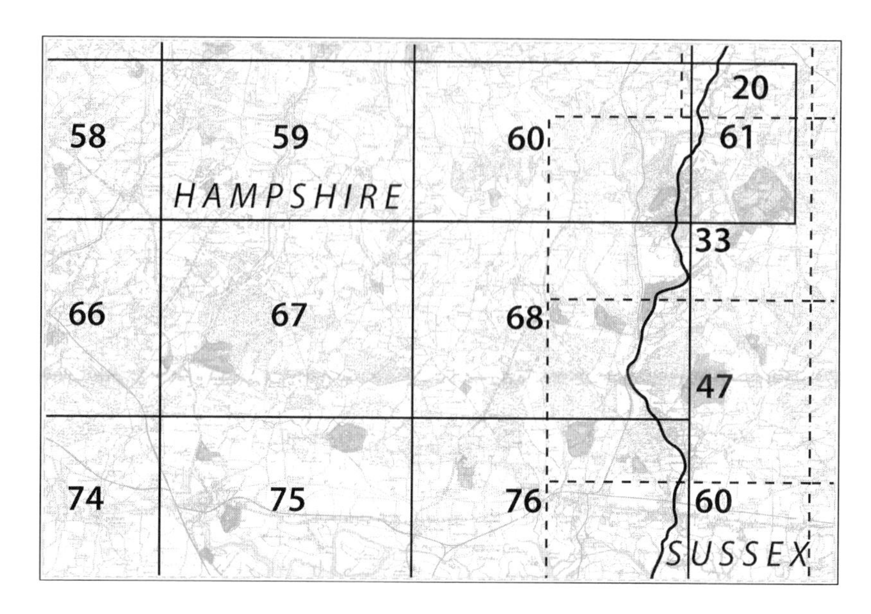

Figure 5.3. Six-inch sheet lines for part of the county boundary between Hampshire and Sussex.

With almost the same number of windmills as the northern two-sheet area, sheets 316 *Fareham* and 331 *Portsmouth* appear to offer a slight improvement in their representation of windmills over the Teeside sheets, and the six-inch depictions of the windmills in the south are also more consistent than those for the northern mills. All the new mills are recognised by *Windmill* annotations on the six-inch edition, and the record of the Revised New Series is correspondingly better with lapses in recognition not attributable to any single cause. There are anomalies but these are few and occur for a variety of reasons. Dock Mill in Southsea was surrounded by urban development towards the end of its life and was not recognised at all by the six-inch, and then only by a shaded circle coupled with the annotation *Corn Mill* at 1:2500. Due to restricted space, this mill was reasonably ignored by the Revised New Series despite being the largest, most prosperous and recently-worked windmill in Hampshire. A mill at Purbrook (later thought of as Waterlooville Mill) was annotated as *Saw Mills* but the circularity of its ground plan only becomes apparent at the 1:2500 rather than the six-inch scale; hence its exclusion also from the table.[3] The Revised New Series failed to record this mill too despite it being in open countryside but, conversely, did offer an annotated symbol for a mill at West Ashling in Sussex that is not recognised at all by either of the larger scales. This is not as surprising as might be supposed. This site sustained one of the class of wind-driven structures erected on top of industrial buildings to provide power for all manner of different purposes. Often (as here) the power would be supplementary to that already being generated by a waterwheel. Without needing to house a lot of machinery and, in effect, just be a support for the apparatus of the sails, these structures were usually slender-framed and without cladding and so not unlike the skeletal farm windpumps coming into fashion at the end of the nineteenth century. They were equipped with normal sails or an annular sail but the difference, of course, was that this type of structure, whether slight in construction or more heavyweight, was not free-standing and consequently had no footprint of its own that

[3] This mill burnt down spectacularly on 25 June 1906 at midday amid full photographic coverage of the event.

would appear on a map. The odd thing is that the photographic record has not uncovered more of these so-called barn-top mills than it has, given that this sort of device would have had many potential uses. The example at West Ashling was of fairly substantial construction and carried four normal sails along with a fantail (Plate 36). As with all such structures, the drive was taken down into the supporting building where, in this case, three pairs of stones were placed for normal usage. Even if, as has sensibly been suggested, no more than two pairs of stones could have been worked at the same time due to the size of this mill, this still represents a serious milling operation. In the visibility stakes, barn-top windmills that are of comparable size to this example at West Ashling score heavily and they thoroughly deserved their recognition by the Revised New Series.

Name of Mill	Depiction at 1:10,560 (six-inch scale)				Revised New Series depiction
	Sheet	Revised	Annotation root word	Ground plan	Sheets 316, 331 revised 1893-4
Hampshire mills					
Hambledon	67NE	1895	*Windmill*	circular	no recognition
Denmead	68NW	1895	*Windmill*	circular	✓ *Windmill*
Chalton	68NE	1895	*Windmill*	circular	✓ only
Barn Green	68SW	1895	*Windmill*	circular	✓ *Windmill*
Wicor	75SW	1895	*Windmill*	circular	no recognition
Langstone	76SE	1895-6	*Windmill*	circular	*Mill*
Brockhurst	83NW	1896	*Windmill*	circular	no recognition
Stubbington	83NW	1896	*Windmill*	circular	✓ *Windmill*
Upton	91SW	1896	*Windmill*	composite	✓ *Windmill*
Bembridge	96NE	1896	*Windmill*	circular	✓ *Windmill*
Wootton (pump)	90SE	1896	*Windmill*	circular	no recognition
Sussex mills					
East Wittering	72SE	1896	*Windmill*	circular	✓ *Windmill*

Table 5.2. Windmills shown at 1:10,560 (six-inch) for areas of Revised New Series sheets 316 and 331.

Circular (plan) indicates a tower mill that is either free-standing and sited completely alone or one that is dominant within a complex of buildings, but where the circularity of its ground plan is very obvious. Composite (plan) indicates a tower mill incorporated into another building where no part of the joint plan is easily identifiable as being that of a windmill. ✓ indicates the use of a windmill symbol on the one-inch Revised New Series (NS-2). All mills are listed alphabetically for each six-inch quarter-sheet in county number order. Any pumping windmill is noted at the end of a county listing.

The idea that the 1890s marked a watershed in the fortunes of the windmill is endorsed by both of the areas under discussion. In Durham, Egglescliffe Grange Mill – which met neither of the criteria for inclusion in Table 5.1 – burnt down in 1897 having only recently had its sails removed and been converted to make use of steam power. Close by, the mill at Mandale (which had been mistakenly considered by Wiggen to be a Durham windmill) had been destroyed by fire in 1890, and so therefore failed by a small margin to be recognised at any scale.[4] The southern mills too were failing fast during this decade. East Wittering Mill in Sussex allegedly lost its sails in 1896 just after ceasing work. All the mills listed in Tables 5.1 and 5.2 were tower mills and many, when visited by the one-inch revisers, will have been in possession of their sails or been in a condition where working by sail might have seemed possible. But primarily, the impression one gets from both tables is that of windmills falling into terminal decline. The evidence of four Revised New Series sheets taken from two areas at opposite ends of the country is interesting since the failures to recognise a windmill at the smaller scale correlate with those mills known to have been very derelict less than ten years later – mills that were probably well on their way to dereliction in the 1890s including one at Hutton Henry in Durham and one at Hambledon in Hampshire. Some non-recognition of windmills by the Revised New Series can be explained away on other grounds, such as the mix-up between the two mills at Elwick or between those at Mount Pleasant and Portrack Lane not far from Stockton. This might suggest that the divergence in windmill recognition between the larger and smaller scales *was* intentional, and that while the remit of large-scale mapping was to show topographical detail in all its minutiae, that of the small one-inch scale was to balance the prominence that a feature had in the landscape with its contribution to the topography of the map – resulting in a lesser emphasis being given to watermills on the one-inch. The capacity that a windmill annotation had under the Revised New Series regime for suggesting a mill to be *Old*, such as was done for many mills in Norfolk, has manifested itself here for more than half of the windmills in the northern sample that had some form of recognition but not for any in the southern sample, even though on a neighbouring sheet the annotation *Old Windmill* is used by this edition for the tower mill at Bursledon. There is little point in trying to tease out an analysis for the use of this annotation using a combined total of just 26 mills except to note that, visually, one type of structure will probably have prevailed within both groups of windmills. While some of these mills were still working at this point – known examples include the mills at Greatham and Hart in the north; those at Barn Green and Upton in the south – a majority of the 26 mills would have been candidates for being described as *Old* by virtue of existing as full-height towers with just the remnants of their caps in place. How are we able to say this? History tells us that when a tower mill falls into dereliction, the sails (if not cannibalised for another mill) decay relatively quickly leaving at most the sail stocks in place. The weatherboarding of the cap is also vulnerable to the elements and can quickly disappear. What then remains is the cap-frame: this consists of two large side timbers known as sheers running from the front of the cap to the back that support two cross timbers – the weather beam at the front and the tail beam at the back. Bearings set into each of these two beams in turn support the rotating windshaft, and it is this iron windshaft (wooden ones were rare in tower mills by this time) to which the sails are attached, and to which is fitted the first and largest gear wheel in the motive power train that delivers energy from the sails to the stones. This combination of four static timbers and a large wooden and cast iron brakewheel together with an iron windshaft makes for a really impressive assemblage of wood and iron that can sit on top of a brick tower, impervious to the elements, sometimes seemingly for an eternity. Many derelict tower mills survived for a sizeable portion of the twentieth century in this condition and their cap-frames might still

[4] Harrison, *Eight centuries of milling in north-east Yorkshire,* 262.

be in place today had not the salvage value of their windshafts been recognised, or had the mills not been converted into houses.

It thus seems safe to assume that windmills in this condition would have been common when the one-inch revisers were going about their business in the 1890s. It is certainly an accurate description of how Langstone Mill in Hampshire appeared when photographed in 1895 and how also, a few miles away, the mills at Bursledon and Owslebury both appeared at the time of the Third Edition when each is described as *Old Windmill*. Langstone Mill had worked using both water power and wind power but at the time of the Revised New Series, when the tower was without sails, it seems reasonable that the simple annotation *Mill* was accorded so as not to emphasise a defunct windmill at the expense of a surviving watermill. Looking at the windmills in the north, the knowledge we have of them supports the notion that those in the same sort of condition in the mid-1890s as Hampshire's Owslebury Mill were a lot less likely to be recognised by the Revised New Series, even though they were still recorded on the six-inch. This could explain the failure of the Revised New Series to show seven mills from the northern sample and four from the southern sample. This thinking is more than reasonably sustained by the evidence, such as it exists, for other windmills from Durham and Northumberland. One windmill survived until 1913 in the hamlet of Cowgate sited a few yards outside the north-western municipal boundary of Newcastle upon Tyne. On the evidence of the first edition 1:2500, this mill is assumed to have been working in 1858 but it had become completely derelict by the early 1890s – manifestly so in 1896 when photographed alongside the equally derelict headgear of Fenham Colliery. At that point, the photographs of Cowgate Mill and Langstone Mill, taken within weeks of each other, show these mills to have been in an identical condition: each consisting of a full-height tower with cap-frame. Both were recognised by the six-inch but the Revised New Series, in failing to record either windmill, and others suspected of being in a similar condition in both north-eastern and southern-central England, was most likely complying with a different set of criteria for depicting windmills than the larger scales were adhering to. This seems to have changed for the Third Edition when Bursledon Mill and Owslebury Mill (and no doubt many, many others) were recognised even though by that time they were in a very derelict condition. Also on the Revised New Series, the criterion for acknowledging a windmill to be *Old*, or occasionally *Disused*, probably stemmed from such mills having become derelict, yet not so much so as to be beginning to lose their visual character and the essence of what it was to be a windmill. Taking this conjecture one step further, a loss of prominence in the landscape could then result in a mill completely forfeiting any recognition by the Revised New Series.[5]

A thread running through earlier chapters, and now this one too, is the hardly surprising

---

[5] Further evidence for the condition of the 26 mills in our two marginal areas for windmill presence comes, of course, from the six-inch annotations accorded each mill and particularly any use made by them of *Old* or *Disused*. None of the southern mills was described as *Old* but of the four described as *Disused* – at Brockhurst, Chalton, Langstone and Wicor – only Chalton Mill challenges this new thinking. The lack of annotation for this windmill on the Revised New Series incidentally is possibly because its symbol was placed within a *Windmill Hill* annotation that was fairly prominent. Four mills from the north – at Easington, Fishburn, North Ormesby and Thorpe Moor – are described as *Disused* and a further two – at Hutton Henry and Yarm – as *Old*. Of these six mills, four are not recognised at all by the Revised New Series; the other two are acknowledged to be *Old* but only one is accorded a symbol. One might reasonably think that this six-inch evidence supports our new hypothesis for the Revised New Series quite well.

idea that what a map edition such as the Revised New Series recorded was naturally linked to what was visually on offer to an observer. What an OS surveyor found himself looking at and how it related to a preconceived notion on his part of what a windmill should look like, or how it corresponded to some putative OS field manual on feature recognition since lost to us, is of real interest. The idea that windmills merited serious status as landmarks on the one-inch was discussed in the context of the Dorington Committee deliberations, but the question of what actually constituted a notifiable windmill structure and what types of mill would be categorised by the Revised New Series using an *Old Windmill* annotation could do with some further clarification.

What the surveyor saw

The problem of knowing what to acknowledge as a windmill begins with a situation where a plethora of mill-related buildings were once common, many of them presenting valid claims to be considered as wind-powered structures. Under such conditions, the representation of windmills inevitably became subjective. No protocols for the notification of wind-powered structures have survived beyond those already discussed, if indeed any that understood the need for more clarification were ever written. One plausible factor underpinning a decision of whether or not to record a windmill was the broader contribution that it was making to its community, rather than any impact it might have been making on the landscape. Such an idea dramatically widens the parameters for appraising any kind of feature representation, and is tantamount to saying that the validity of any map edition could well be immeasurable. Happily, in the case of windmills this idea can be discounted since only a minor proportion of mills in the landscape were upholding central roles in their local community economies at the time of the Revised New Series. The majority were in situations of relegated utility, so it is far more conceivable that the importance of a windmill as a local landmark held greater significance.

But the variety of circumstances the survey teams had to contend with was surprisingly broad. In the apparent absence of detailed instructions as to what constituted a recordable windmill, it is hardly surprising that survey teams working in different parts of the country would use different criteria in making that judgement. Even if the entire country was to be surveyed by only one team, then regional changes in mill typology alone would cause some degree of inconsistency. Any anomalies found across the scales can in theory be attributed to this lack of direction or uncertainty about the criteria applicable at each scale for judging what was to be considered a windmill. But the circumstances of most mills were such that this line of thought can be considered somewhat disingenuous or lacking in common sense. Yet the fact remains that under these survey conditions, the range of situations stretching as a continuum from an obvious working mill to a structure that had some tenuous historical, visual or functional connection with 'windmills' absolves the surveyors from at least some of the blame for OS sheets sending out mixed signals. With perhaps little to go on except common sense, individual surveyors will have fallen back on their individual preconceived ideas of what a windmill looks like. These ideas may have come from childhood familiarity with mills or through some awareness of the popular role of mills as pastoral icons. In the Victorian art world, mills were popular as 'sublime' subjects for photographers and painters. Whether all surveyors (from whatever part of the British Isles) would have had exposure to this type of image, though, or even fictionalised woodcut images from popular magazines, is somewhat in doubt. For the few who were prepared to learn from faithful representations of windmills, the images did exist, even idealised ones belying the harsh reality of working conditions that mills laboured under. A mere glance at the painting of a post mill made by Thomas Miles Richardson (senior) would have sufficed to familiarise any surveyor with the basics of post mill morphology (Plate 37). In this particular image, the rusticity of the mill is exaggerated with its body given an unlikely waisted shape, but the technical features of this windmill have been portrayed to good effect. The dishonesty of the pastoral ambience aside – this was supposedly of a mill near Gateshead – the contrast between the set of common sails mounted on a wooden windshaft shown in the painting, for example, and the normal arrangement in most places by 1890 of a cast-iron windshaft with integral canister mounting for the sail stocks would have been instructive, but only to a very careful observer.

Remaining with post mills, and progressing from what would have been common and unambiguous situations to those that were uncommon or contentious, the simplest scenario is of a post mill that simply gave up working and fell into disuse. The sequence of events for such a mill has already been outlined – the sails would be adapted for another mill or, if they remained static, they would simply decay. But even with its sails gone, a post mill – either an open-trestle one or one with a roundhouse – remains identifiable as a windmill. At some point, though, the body of the mill will become derelict and it is then more likely to attract a *Disused* or *Old Windmill* annotation. It now seems that, possibly dependent on whereabouts in the country a mill was, the recording of a tower mill in the 1890s by the Revised New Series is an indication that this mill was a working one or, if derelict, then not to such an extent as those at Cowgate or Langstone. It could also be that more leeway was given by the Revised New Series to the recording of derelict post mills than to derelict tower mills, since post mills retained their essence as mills rather more so than tower mills did when both were in comparable states of ruination. In a fine balancing act of recognition, this was the case even though the height of the shell of a tower mill still accorded it landmark status. But maybe, as already suggested, thoughts about this cartographic neglect of tower mills, and the consequent loss of their usefulness as landmarks if not appearing on maps, seem to have altered between the Revised New Series and the Third Edition, so bringing more of these structures to cartographic recognition. It is also worth observing that the revisers associated with each of these editions did not necessarily bring the same levels of skill to their work.[6] Tellingly, the very old post mill at Madingley in Cambridgeshire had been recorded by the Revised New Series, and then again by the Third Edition despite its parlous condition in 1905 (Plate 38).[7] Even more surprising is the case of Burton Mill in Cheshire whose circumstances by the same date had advanced to a point where it takes a practised eye to realise that this ivy-clad structure had once been a post mill (Plate 39). Yet, it too was recorded by the Revised New Series and then again by the Third Edition

[6] Hellyer and Oliver, *One-inch engraved maps of the Ordnance Survey from 1847,* 56, 67.

[7] It is interesting that the revisions of 1901 for sheet 39.16 at 1:2500 and then sheet 39SE at six-inch give no indication that this windmill is anything other than a working one. More in tune with its appearance of 1905 is the fact that the last census to identify a miller (and the millhouse as containing a family) is that of 1871, when the miller is Edward Litchfield. In 1881 the millhouse is unoccupied and in 1891 and 1901 a poultry farmer is in residence. Subsequent map editions show this mill to have gone. The mill shown in Plate 38 should not be confused with a surviving post mill re-erected here after being brought from another site.

in 1904. Admittedly both mills were described as *Old* by both editions but this treatment still seems generous, particularly in the case of Burton Mill and the Third Edition. Small wonder then that other post mills, which were no longer working but gently rotting away, were treated quite charitably by the Revised New Series. The large post mill at Clacton in Essex looked forlorn when it was photographed *circa* 1905 (Plate 40). It has lost two of its sails as well as its fantail; the two remaining sails are largely without shutters and some windows are broken. It is recorded by the Third Edition with the use of an unannotated symbol. This is interesting as its depiction on the Revised New Series had been the usual one of a symbol annotated with *Windmill.* But in 1904 the Survey clearly felt unable with its Third Edition to allow the impression to be given that this windmill was still working and so erased the annotation, leaving bare the space it had occupied. This space could have accommodated a slightly longer *Old Windmill* annotation without much effort but the Survey chose not to take advantage of this mode of depiction here – through laziness it must be supposed – and took a risk in offering a depiction open to misinterpretation. Luckily, this ploy of using an unannotated symbol to denote a derelict windmill was not used very often other than in built-up areas where accompanying annotations were not possible. Compare Clacton Mill with the less sophisticated post mill at North Hykeham in Lincolnshire that was still at work when photographed in 1907 (Plate 41). Soldiering on with just two common sails and no glazed windows or fantail to have either broken or gone missing, this mill is recorded by the Third Edition in the usual way for a working mill, yet it is clearly showing signs of distress if the roundhouse roof is anything to go by. Once again, a point in the text has been reached where a selection of examples could be paraded before the reader – this time of semi-derelict post mills – for their turn-of-the-century appearances to be compared with their depictions on the Third Edition, or the Revised New Series if the photographs are of earlier date. As ever, this could be fun but it risks the bigger picture of the depiction of derelict mills getting lost among a host of interesting smaller ones. So one more ailing post mill will have to suffice to confirm the surveyors' generous attitude to recognising such mills. Skirlaugh Mill in the East Riding of Yorkshire had also lost a pair of sails when it too was photographed in 1907, though the sails that it still has appear to be fully-shuttered and the condition of mill body seems sound and the roundhouse newly-tarred (see page 239). By the time this photograph was taken, working with less than the number of normal sails was unusual but not particularly rare and, unsurprisingly, this did not get in the way of the windmill being recognised as a working one. Visually this post mill offers no obvious sign of decline except for the loss of two sails. The two remaining sails were to keep the mill in business only for a further two years, but it is at least fitting that their apparent non-alignment in this view of the mill was not misconstrued as a deterioration by the Third Edition reviser when he visited the area in 1905.[8]

---

[8] This visual oddity was possibly an unintended consequence of the general practice followed in the north of England whereby individual sails were attached to a sail-cross at the end of the windshaft. The other option was for the windshaft to have a poll-end or canister that required opposite sails to be attached to the same sail stock – the practice followed in southern counties. With the sail-cross arrangement there was always a slender possibility that opposite sails could slip out of alignment and it would be understandable to think that this is what has happened here. More importantly in the case of this particular mill, its sails were not set to revolve in a flat plane, but had an element of 'dish' to them. This feature was not commonly seen in windmills but other pictures of this mill reveal it quite clearly. It was probably the outcome of adapting secondhand components

Compared to the visual essence of a post mill, that of a tower mill would have been easier to categorise *circa* 1900. The range of visual circumstances that a tower mill could offer, between one of appearing to be in full-working order and one of simply being a brick or stone tower with all internal machinery seemingly gone, was not large. Visually, only a small part of a tower mill is of wood and hence perishable. The tower is durable but once such a mill falls into neglect and its sails have perished, it can quickly lose the cladding to its cap and so then deteriorate to become the sort of mill seen at Bursledon, Cowgate, Langstone and Owslebury. By this point what remains will have lost its critical quality of being a windmill. It will still have a presence in the landscape and most people would still recognise it as a mill tower and understand why it had been built. But the durability of brick and stone and the perishability of wood meant that patterns of decay in post mills and tower mills were truly not the same, and the point at which each loses its reasonable claim to still be thought of as a windmill consequently differs. As a result, the fewer scenarios of tower mill dereliction meant that the Survey had no thorny issues to deal with beyond the specific circumstances of windmills such as the one at Langstone, which retained the remnants of its cap-frame but had been outclassed in the recognition stakes by its conjoined watermill. This mill, and undoubtedly many others, were perhaps poorly served by the Revised New Series and then not reinstated on the Third Edition in spite of a more accommodating view of this edition towards tower mills that were on the point of losing their uniqueness. Of the other Hampshire mills mentioned, Hambledon Mill was one that by 1900 had become a derelict shell (Plate 42). With no cap-frame or other wooden parts on show at all and with a brick tower that was structurally unsound, it was maybe asking a lot that it should continue to be recognised on the one-inch scale. Unsurprisingly, it was ignored by the Revised New Series and then by the Third Edition too. But the larger scales continued to depict what was fast becoming an ivy-clad pile of masonry: eventually the shattered mill tower became unrecognisable for what it had once been and, after decades of remaining in this state, it was cleared away. The notifiability and depiction of this once prominent windmill, which overlooked the most quintessential of English villages, was not unreasonable as far as the revisers for both one-inch editions were concerned. But it is ironic that this mill, whose remains are still remembered, was only ever recorded at one-inch by the Old Series. This early acknowledgment of the mill at Hambledon would have been from a survey made before 1810 – when sheet 11 of the Old Series was published. Even then it is quite possible that what the surveyor recorded was a post mill that preceded the building of the tower mill.

Tower mills also differed from post mills in that they were able to be refashioned and put to other uses. Presumably what was visually then left for the OS surveyor was not of great interest if the concern was to recognise windmills *per se*. If that interest, on the other hand, was to record notable landscape features for use as landmarks, then a one-time mill tower still had visual clout and could offer *some* reason for being recorded but, if lacking a full complement of sails, any impact made on the landscape would be somewhat reduced. In stark contrast, roundhouses of post mills and bases of smock mills survived for decades

to this windmill with its existing (large) roundhouse. There is an element of conjecture in what is being said here, but the real significance of these comments is that they serve as another reminder that some familiarity with the workings of windmills would certainly have been necessary if the OS surveyors were to record them as objectively as possible.

finding use as places of storage or even as homes after all remnants of their wooden bodies had long disappeared. This happened frequently but these buildings were rarely recognised as windmill remains even by the large scales, though in some extreme cases a few courses of brickwork were provided with an acknowledgement, but such cases are very unusual. As in the early decades of the twentieth century, so it must have been the case in the later decades of the nineteenth century – if not more so with larger residual numbers of mills and lower standards of acceptable housing – that these windmill-derived structures existed, but their presence is anything but obvious from looking at OS large-scale mapping. But, compared to post mill roundhouses, tower mill shells still embraced *some* visual identity of the mills they had been built to become an integral part of, and so their brickwork or stonework purely on its own merit held *some* element of a mill's essence. While this may go a long way to account for tower mills converted into houses being recognised by today's equivalent of one-inch maps, the normal practice of the Survey at the end of the nineteenth century was to ignore truncated and even full-height mill towers that had found other uses, unless there were very good reasons for doing otherwise. This was certainly the case for the one-inch scale, but the larger scales could afford to show discretion. A tower mill by the shoreline at Aldeburgh in Suffolk was converted in 1902 and combined with an equally impressive new house. It even retained its cap and continued to look much like the mill it had recently been, only without its sails (Plate 43). In that sense, it looked no different to hundreds of similar structures that were still receiving recognition by the Third Edition, not to mention the larger-scale sheets, yet it was, quite sensibly, not recorded on the one-inch edition. But two following revisions for both the six-inch and 1:2500 did each annotate this housing complex with *The Old Mill.* This sort of annotative throwback, memorialising the former existence of a windmill, was used regularly by the larger scales in the first two or three decades of the twentieth century. They usually referred to tower mills that were full-height house conversions like Aldeburgh, but such annotations were later eclipsed by innocuous ones such as *Mill House.* This kind of annotation can still be found on modern mapping, providing faint traces of one-time mills. Even where windmills were not a common landscape feature – or maybe even because of that rarity – distinctive round ground plans with a couched annotation are found today that flag up what had perhaps for years been a truncated mill tower incorporated into a house. Brownhill Mill at Birstal in the West Riding of Yorkshire is an extreme case. This tower mill had been dismantled, reduced in height, given a sloping roof and had its ground floor area greatly enlarged by new extensions built around it: all of this had happened before *circa* 1905 when it was photographed in its new state and became the subject of a picture postcard. But it was remembered as having been a windmill by the local community since, like many cards of that time, the caption given is *The Old Windmill.* Not to be outdone, the Survey accurately recorded the composite ground plan of the house on the large-scale sheets of the time, but annotated the plan with *Brownhill Windmill* and then continued to apply the same annotation on later revisions up to the 1930s. If it had perhaps acknowledged that this mill was disused or old, this would still have been an overly generous recognition for a converted mill tower stump. As it is, the depiction used over many years for a one-time windmill site was utterly misleading. Mercifully, neither the Revised New Series nor the Third Edition chose to make the same mistake and, in any case, this degree of aberration is rare at any scale.

The myriad different ways in which windmills could be found depicted on maps at the different scales was not only due to loosely-agreed protocols for notifiability of mills that generations of surveyors and revisers were working to in compiling their field notes, but also to drawing office staff all doing as they individually saw fit within their codes of practice. Given that an ill-considered annotation could have a lifetime that stretched over many editions, as certainly happened at Birstal for example, no single person can be held responsible for a long-term error. It must also be assumed that individual surveyors were working to some form of informal guidelines for minimising the chances that they might misconstrue what they were seeing in front of them. The last few sentences have maybe begun a process of unravelling those guidelines, particularly as they related to windmills, but mistakes inevitably crept in. Was the surveyor who cast his eye over Reigate Heath Mill in Surrey for the first revision of the large scales in the early 1890s – when the mill had been inactive for two decades – content to note it as simply being disused or did he inquire further into the matter? As it happened, the six-inch sheet of 1897 had the mill annotated with *Windmill (Disused)* despite the 1:2500 sheet from which it had supposedly been reduced annotating the mill simply as *Windmill.* To make matters worse, the Revised New Series would have us believe that the mill was working, or capable of working, *circa* 1896 since it offered a *Windmill* annotation. But this may have been a consequence of the mill still having shuttered sails in 1891. After that the Third Edition altered the situation by applying an *Old Windmill* annotation in 1901. None of these annotations was especially appropriate as the mill had stopped working *circa* 1870 and in 1880 its roundhouse had been converted into a church. This may well have posed a problem to a succession of OS revisers who were well aware of the change of circumstances at this mill, but either could not decide how best to handle it, or did not think it sufficiently important to amend the sheets. Or, there may have been a continuing ignorance that lasted for many years until it was eventually resolved by having both a windmill symbol and 'Ch' – a standard way of depicting a church – on much more modern sheets. But ambiguous situations involving mills such as Reigate Heath Mill were never in short supply at a time when considerable ingenuity was being demonstrated in adapting redundant windmills for other uses.

A tower mill overlooking the Avon Gorge at Clifton near Bristol was working as a snuff mill when in 1777 it apparently caught fire, so putting an end to its milling days. It was later converted into an observatory and remained like this for decades. This one-time mill tower is still in place having been home to a camera obscura for the last century or so. The point of interest here is that, while locally appreciated as having once been a windmill, this edifice has, understandably enough, never been recognised as such by the Ordnance Survey.[9] Mill towers found other uses too; they could, for instance, be requisitioned into serving as lofty supports for water tanks, yet would still present themselves to a surveyor as windmill-like.[10] Smock mills could be adapted to sit on existing barns or other convenient buildings as well as the more normal custom-built polygonal brick bases. It is questionable whether or not, in

[9] This awareness, moreover, existed at the time of the early nineteenth-century survey for the Old Series. This is made quite clear by the writings of the novelist Fanny Trollope who described the view from Widow Barnaby's lodgings in Clifton as '.... [looking] out upon the windmill and the down ...'. Trollope drew upon her childhood and early adulthood memories of regency Clifton for her knowledge of this mill. *The Widow Barnaby,* volumes 1-3, London: Bentley, 1839. See also Teresa Ransom, *Fanny Trollope: a remarkable life,* Stroud: Alan Sutton, 1995, 7-8.

[10] Surviving examples include Askham Richard Mill near York. Interestingly, a water-tower structure built *circa* 1927 at South Dissington in Northumberland was contrived, in folly-like manner on a whim of its then owner, to resemble a windmill. Since then this water tower has regularly been recognised as a windmill by the Survey at the one-inch and 1:50,000 scales.

a case such as Darn Crook Mill, we should expect windmill status to necessarily be accorded to an appropriated and very nondescript base if the smock is then removed but leaving much of the original mill machinery still at work there. Common sense and all the evidence suggests probably not. But the reverse was true in the case of a post mill body that relinquished the use of its supporting trestle and was then mounted on a truncated mill tower. This did not occur very often, as it entailed some fairly serious and heavy-duty millwrighting work, but it would have been an attractive proposition where a post-mill trestle was rotten and a tower-mill shell was going begging.[11] Many of the more important windmill counties could each boast a handful of these so-called composite mills, but both visually and functionally there was no reason at all for them to be treated any differently from other categories of windmill – and they weren't.

Beyond these none-too-awkward issues of definition for the surveyor or reviser, there were other cases of unique or very unusual structures that were windmill-associated, maybe clearly so and seemingly working conventionally, but which must have induced uncertainty. One extreme example would have been the case of a post mill that, instead of 'surviving' as just a roundhouse, found itself having its sails and supporting trestle removed and its body, complete with machinery, lowered to the ground. Such a mill would then be able to work by a steam or gas engine. What was the Survey to make of a structure such as this? Flint, in his gazetteer of Suffolk windmills, suggests that maybe five mills here were reconfigured in this way by *circa* 1900: all except one had become disused as conventional mills and their bodies were each then moved complete – no mean task, one might think – to be placed alongside other working windmills. Sensibly, as far as the Survey was concerned, they simply became part of the milling enterprise at their new locations and, without sails and looking to be no more than glorified workshops, they went unrecognised. Equally, the fifth mill to be moved – from Hoxne to a new location at Worlingworth that had no prior association with milling – failed to excite the attention of the Survey at any scale.[12] Other curiosities from late in the nineteenth century included various types of structure that would gradually homogenise into one standard class of structure, which later turned into the annular-sailed metal windpump. But, before this sort of metal structure became ubiquitous, its precursors could appear very ramshackle and makeshift-looking, and they came in many forms – largely made of wood rather than metal. Where mounted on buildings they could be sophisticated barn-top mills similar to West Ashling Mill discussed earlier. Where found at ground level they could be very crude devices – some no higher than a few feet – used for pumping water: these were once found in modest numbers across the eastern flatlands of England (Plate 44). But they could also be more ambitious structures, maybe still used for pumping or possibly adapted to operate a sawmill or any one of a number of light industrial schemes (Plate 45). In short, there was a variety of hybrid wind-driven structures that could be seen in the landscape but, collectively, their numbers and size never threatened the dominance of the classical windmill structure, and in large measure the Survey felt able to ignore them in light of their perceived impermanence.

---

[11] In an associated development, it became fashionable in central Europe towards the end of the nineteenth century for doing away with the trestles of post mills and setting the bodies to rotate instead on small circular metre-high brick 'towers'. It says something for the different milling cultures of there and England that it was considered worthwhile to go to the expense of doing this as late as 1939.

[12] Flint, *Suffolk windmills,* 129-43.

The complications for windmill recognition that arise from the variety of circumstances these mills were offering can become easier to understand if a key piece of evidence is made available. A fortuitously-taken photograph, for example, often serves to clarify exactly what was seen and then recorded, but can also be misleading. When the windmills at Shrewley in Warwickshire were discussed in the context of the details of depiction from the first edition 1:2500 being replicated on the six-inch scale, it was accepted that the inference drawn by the OS surveyors in 1885 of a working tower mill and a just-disused open-trestle post mill, with both mills retaining their sails, was accurate. By 1900 both mills had lost their sails. The post mill was to continue on its course of dereliction and it eventually collapsed in 1937, but the tower mill continued to work using steam power.[13] Interestingly, a photograph of both mills in their pre-1910 state captures the visual essence of the gently decaying post mill, which by this period would have been thought picturesque, but, judging from the caption, ignores the working though rather prosaic tower mill in the background (Plate 46). This prejudice and lack of caution on the part of the postcard manufacturer was perhaps echoed by the Survey, in that the second editions of both large scales, carrying a revision date of 1903, record two disused windmills, each by an open circle. The photographic image is sufficient to repudiate its own caption; whether it allows us to be critical of the description of the condition of the tower mill on these second editions as well is something of a hard question to answer.

The windmills of Essex

Having just considered two small parts of the country to appreciate how marginal windmill populations were recorded by the Survey *circa* 1900, the ploy of comparing small- with large-scale maps can now be brought to a mainstream area of windmill presence. Again this will examine the effectiveness of windmill recognition by the one-inch Revised New Series, but this time using a very much greater number of mills. Ideally, the area chosen for this needs to be one where there is near-congruency in revision dates for the Revised New Series and for the second editions of the large scales. One county area matching this criterion is Essex, where the revised six-inch sheets are dated 1895-6 and the Revised New Series conveniently dates from slightly earlier – 1893 and 1894.[14] The extent of the dissension between smaller and larger scales for the area round Stockton-on-Tees and eastern Hampshire was such that a subtlety of method was not really required. The greater number of windmills involved in studying Essex encourages – and an apparent proclivity on the part of the Survey to record them comprehensively at all scales requires – us to bring a further level of sophistication to our inquiry. One concern might be that the gap in time between revision of the small scale and that of the larger-scale editions, even if this only amounts to a year or so, opens up the possibility of a windmill's status changing in that short interim. For counties where the gap between the dates of revision for the Revised New Series and the 1:2500 (and thus also the six-inch) is greater, there will be increasingly larger numbers of windmills whose dates of

[13] Wilfred A. Seaby and Arthur C. Smith, *Windmills in Warwickshire: a contemporary survey,* Warwickshire Museums Publication, 1977, 21-2.

[14] 1895 and 1896 undeniably saw the revision of the vast majority of six-inch quarter-sheets covering Essex with the northern half of the county having to wait for the southern half to be finished first. Strictly speaking, the span of years taken for this task extended from 1891 to 1896 as the border area with Middlesex was revised earlier than the rest of the county (1893-4) and even earlier Admiralty charts were consulted for some coastal strips, though this was only for information about mudflat areas.

demise – however for the moment this is defined – fall within the intervening period, and they will justifiably be recognised very differently by the two scales. It can also happen that where the Revised New Series and the larger scales are in disagreement, it is the latter rather than the former that is then identified as being in error: this may sound improbable, but it is not impossible. What is consequently needed, to be able to assess the Revised New Series accurately when these sorts of uncertainties are thought to exist, is for verifiable evidence to be abstracted from the historical record to establish dates for when the mills that appear on one map edition, but not the other, disappeared from the landscape. Regrettably, this need to use the historical record is coupled with an awareness that, by and large, the secondary windmill literature is unable to offer much of a shortcut in this process, and that for most windmills across the country this need can only truly be met by a direct petitioning of the primary sources.

Consideration of the windmill as a subject for historical study really dates from the turn of the twentieth century. Earlier tracts and treatises had tended to think of the windmill as a contemporary mechanical object with a need for ever more finesse of design, but then, with the decline of the windmill as modern technology, the later very different treatment of mills as a subject for romantic whimsy within the landscape found a sympathetic and escalating readership. In the 1930s, the travelogue style of writing from that period was adopted for nearly all the books published with windmills as their theme. The treatment at this time by Coles Finch of windmills in Kent, and a companion volume for the windmills of Sussex, were exceptions to the general rule in that they gave historical accounts for each individual mill. Better contextual and factual volumes eventually followed: the 1950s and 1960s saw the publication of generalist texts and county-specific windmill-biography texts.[15] During the last few decades the mill literature has expanded enormously, with its coverage of topics extending to include, for example, medieval mill studies. But the sort of book where a writer selects a county and discusses it on a mill-by-mill basis has proved an enduring one and the exemplar, without doubt, of this approach – the study by Farries of the windmills of Essex – is unlikely ever to be bettered.[16] Realistically, these regional windmill studies are of variable

---

[15] It would be too space-consuming to note here all the admirable texts that appeared at this time and slightly invidious as well to have to compare them, and in any case the reader does have a bibliography in the present volume to fall back on. I will settle for mentioning just three: Wailes, *The English windmill* and Freese, *Windmills and millwrighting* between them opened up a world hitherto unknown to most people. Very different in style, the first is still widely considered to outweigh any other in importance for the study of windmills. This derives in part from the large measure of esteem in which Rex Wailes is held for his mill studies (dating from *circa* 1924) and for his promotion of the conservation and welfare of mills. A mechanical engineer by education, Wailes was also acutely aware of the contribution made by millwrights to the pantheon of Victorian technical achievement. The second text is equally informative and written in a far more homely style that reflects the passion of its author. Coming somewhat later; John Reynolds, *Windmills and watermills,* London: Evelyn, 1970, provided an erudite blend of historical, architectural and engineering approaches to the study of mills. It too should be seen as one of the more influential mill texts. This is a convenient place also to point out the large and often overlooked contribution made around the middle of the twentieth century to the field of windmill historiography by H.E.S. Simmons, who corresponded prodigiously with anyone he felt could be of assistance in unravelling individual windmill histories. In the case, for example, of the mills of Durham (often disregarded by Simmons's contemporaries), one of his local contacts was a Mr Oxberry whose information had been gleaned in earlier correspondence with a Mr Angus, the grandson of someone who had worked one of the mills on Windmill Hills and who had taken oral soundings from those whose memories went back sufficiently in time to remember the mills. The exceptional output of Simmons, largely unpresented to the mill community in his own lifetime, can be found in manuscript, but now bound, in the Science Museum Library at Kensington.

[16] Farries, *Essex windmills millers & millwrights,* volumes 1-5.

format and scholarly rigour. Only in relatively few cases are fully-referenced primary sources resorted to, not simply to put into context the practice of windmilling and millwrighting in a county, but also to sustain a biography of every mill mentioned in the historical record, no matter how brief the existence of that windmill might have been. For other counties there are competent windmill biographies which describe the distribution of mills in their chosen county and the technological traditions of the area, but which then give individual narratives only for the mills that remain and some that have vanished, while settling for a gazetteer to mention very briefly other windmills that have long disappeared. Otherwise, the secondary literature might consist of short pieces of writing having as their theme the windmills of an area; these might offer brief but reasonable accounts of the later mills known to have been present in areas where few were ever built, or a broad outline of the windmilling scene in a area where there were many, using sample mills by way of anecdotal illustration. For many types of work, the reference sources used can be rudimentary and some studies may be no more than succinct reliable gazetteers for existing windmills written in pamphlet or article form. All these types of study have their place of course: all are the outcome of somebody's enthusiasm and all have something to add to the larger picture of the historiography of the windmill. It is just that not all are of direct use in the sort of exercise we are now planning.

By any standard, Farries' account of the mills of Essex excels in its use of documentary evidence to trace out individual windmill histories. The sources used are wide-ranging and include the map record. So, by happy coincidence, the best of all counties one could choose to study for which the historical record has been meticulously exploited, is a suitable one, in closeness of revision dates, for comparing the Revised New Series and the larger scales. Our particular interest lies in assessing when each windmill ceased to work or when its site was cleared, or more generally when it assumed the sort of appearance we now associate with a surveyor revising his opinion of how to record a mill. In reviewing the types of evidence that Farries invoked for his assessment of when each windmill found itself in this situation, we can draw some conclusions about the value of each type for its ability to date the demise of a windmill and to qualify just exactly what that demise might entail. Three considerations arise: first, Farries naturally included the evidence of the map record in his study, though he did this critically. Even so, if we are to ensure that our inspection of the Revised New Series is to be objective, then we need to disregard this element of Farries' gathered evidence, or at least proceed here with caution. Second, there is the extent to which the range of primary sources has been used uniformly for all Essex mills and; third, there is the subjective nature of the importance attached to each of them, which needs to be viewed in conjunction with the writing style used by Farries. The sources of evidence commonly employed by Farries are listed below, where for each there is an indication of how many mills from a convenient selection of 236 one-time Essex windmills they were invoked for, though some judgement has had to be exercised here owing to ambiguity on occasions as to exactly which kind of source is being cited (Table 5.3).[17] It should be noted that only the types of source used to

[17] These 236 windmills are those that survived long enough to be candidates either for actually being recorded by first edition of the six-inch (1862-76) or, if missed by it, for Farries to acknowledge this to have been an error. The use of the six-inch rather than the 1:2500 is deliberate, and the art of arriving at this number of mills has not been a precise one since it includes at least one mill in a similar state to Darn Crook Mill in Newcastle upon Tyne at the time of survey. Selecting just a single example: Farries claims that Woodham Walter Mill is shown on the first six-inch edition, yet he refers to a building in the water- and steam-mill complex here as that 'which must have carried the windmill', which certainly does not imply cartographic certainty. Indeed, no windmill recognition can sensibly

clarify a post-1875 situation are included here; sources of a more medieval flavour used by Farries to unravel histories of earlier mills are not. Each of these sources is self-explanatory, except perhaps the last one, for which some qualification is needed because of the way that Farries chose to write, and because of his trust in the notes of earlier windmill fieldworkers. While carrying out his own fieldwork in the 1950s and 1960s, and accruing his knowledge base, Farries became intimately acquainted with the work of, and friendly with, those who had been doing the same sort of work in the 1920s and 1930s. Accordingly, and justifiably, he had no problem with windmills that survived later than 1920, accepting as substantiated evidence the field notes from that small circle of people. Besides, the photographic evidence for any windmill that managed to survive beyond 1920 was not hard to come by. Typical of the protracted history of the demise of a windmill that was still standing in 1920 is that of one of the tower mills at Belchamp Otten. This mill was reported as having ceased to work by wind in 1918. Its sails were removed in 1921 but it continued grinding by steam till 1943. The cap blew off in 1947 and the tower was burnt out *circa* 1952. This mill was eventually pulled down *circa* 1960. The reference to this windmill ceasing to work by wind in 1918 was derived from a Dr Turner who was an early gatherer of mill information, but who restricted his pastime to Essex. Turner had been in a position to talk to many millers whose mills had disappeared in the preceding thirty years and, as an early exponent of the art of oral history, his work was accepted more or less unconditionally by Farries. Herein lies a problem: for windmills such as the one at Belchamp Otten, the 1918 citing is credible as Turner was on the scene at that time but for windmills that had disappeared earlier the citings are possibly less credible. Worse still, it is not always apparent from the mill histories in Farries when a Turner-derived piece of information or one of more dubious provenance is being offered. Farries is often conscious of this dilemma and he makes the effort to differentiate between, for example Lindsell Mill which was 'said to have been demolished about 1906', and Great Leighs Mill where 'the mill was demolished *circa* 1912', though he offers no references for either.[18] For other windmills, Farries considers the oral tradition to be reliable. For example, the post mill at Great Bentley is declared unequivocally to have been demolished in 1891, again without a reference being given, but with the implication that the knowledge of that demise was got from an impeccable source such as Turner speaking directly to the miller.[19] Certainly many such claims are true and others have a basis of truth in them, but problems do occur such as that posed by Little Stambridge Mill, where the miller informed Turner in a letter dated November 1919 that his mill had been pulled down at least thirty years earlier, in which case the Revised New Series, which shows it, is at fault.[20] This sort of oral history may have something to recommend it but the number of times that an elderly person when questioned about the last days of his windmill, replies 'the mill was demolished about 1900', and which is thought worthy enough by Farries for him to cite is suspicious to say the least.

Another example of a difficulty in using Farries is a seeming inconsistency between mill histories in his use of tax returns and trade directories. For some mills, the date of demise is

---

be attributed to either of the larger scales for this site. A better interpretation would have been that this particular windmill was missed at both the 1:2500 and six-inch scales.

[18] Farries, *Essex windmills millers & millwrights,* 4, 70-1.

[19] Farries, *Essex windmills millers & millwrights,* 3, 36.

[20] Farries, *Essex windmills millers & millwrights,* 4, 116.

deduced from a miller ceasing to operate at the mill: this is probably reasonable if supported by other evidence, or even if it simply does not disagree with other reputable findings. For other mills, however, including some where no alternative evidence is offered, this tax and directory information is not discussed. There may be a very good reason to account for this such as the partial survival of these records, but what this chiefly highlights is the difficulty of achieving comparability between mills when using the historical record. These comments are not meant to detract from Farries who conducted his studies with an obvious scholarly precision, but to voice a concern lest it be thought that comparability can exist between the blend of evidence for one windmill and that of another, when in reality the circumstances

Type of source cited as providing evidence for the demise of a windmill	Number of mills
Newspapers: capable of providing detailed reports on local events including the aftermath of storms and litigation involving millers.	66
Directories, notably Kelly's and White's (see note below): Kelly's directories provided newly-acquired information in respect of millers at work and became available every four years from 1870 – interim volumes being derivative.	195
Land tax returns and records.	110
Poor rate lists and books.	45
Auction documents and sale notices – those appearing in newspapers as well as supplementary ones.	42
Estate records.	6
Electoral registers and census returns: primarily used to track the movement of millers, implying in some cases the cessation of trade at a mill just relinquished.	39
Trade invoices of millwrights.	17
Contemporary dated photographs – sometimes with a record on the back of a mill's destruction.	13
Contemporary travelogues and antiquarianism.	71
Maps: notably Ordnance Survey.	225
Oral / Recorded local history.	135

Table 5.3. Sources of evidence cited by Farries for the windmills of Essex.

For inventories of the directories available by geographic area, see Gareth Shaw and Allison Tipper, *British directories: a bibliography and guide to directories published in England and Wales (1850-1950) and Scotland (1773-1950)*, second edition, London: Mansell, 1997. For earlier directories, see Jane E. Norton, *Guide to the national and provincial directories of England and Wales, excluding London, published before 1856*, London: Royal Historical Society, 1984. For other types of source tabled here, refer to county record offices. In addition, the national coverage of past newspapers is held as the BL Newspaper Collection.

surrounding each mill have to be gauged individually. Any investigative framework used for weighing up the evidence gathered by Farries for individual windmills has to be mindful of these points when considering mills that vanished before 1920. A system of five categories for windmills can be devised, each defined by the type of evidence that Farries offers for assessing the date when a mill fell into disuse, or perhaps disappeared completely from the landscape. In order of degree of certainty, these categories can be defined by the availability of:

A a written contemporary, or near-contemporary, account of the destruction of the mill, or of it being disabled beyond economic repair and used for salvage;

B an auction notice for the mill specifying removal or dismantling;

C a sequence of trade directories, or similar in terms of integrity, corroborated by at least one other, preferably non-map-based, source that can identify the date of demise to within three years;

D a variety of sources of evidence that, collectively, can only identify the date of demise between a range of dates;

E a single uncorroborated source, such as a trade directory, that allows the date after which a mill ceased to work, or fall into disrepair, or disappear from the landscape altogether, to be surmised, or an uncorroborated claim for the date of demise.

The categories are self-explanatory. Category A includes a few mills in a position similar to that of Great Bentley Mill where Farries can be taken on trust. Otherwise the great majority of mills found in this category are there due to the survival of contemporary reports of their destruction. They are thus similar to the windmills in Category B whose dates of destruction can also be presumed reasonably safely. The mills in Category C are more problematic due to considerations already mentioned, and in construing a windmill's history from the diverse types of source, a harsh line has been taken to ensure that any reliance placed upon maps is minimised or, where it has to occur, it is made very apparent. It is important to identify the mills for which, in Farries' opinion, the map evidence outweighs the value or availability of other sources. Moreover, since the Revised New Series is to be our prime focus of attention for misrepresentation, the evidence from it has been totally disregarded. Category D is more problematic yet again, and Category E caters for all mills for which no reliable information is offered by Farries concerning when each one met its end.

The handy, though arbitrary, sample of 236 Essex windmills consists of all those *in situ* at the time of survey for the first edition six-inch: they can all be considered as fair game for possibly having survived another twenty years until the era of the Revised New Series. This list of windmills excludes six known windpump sites, but does include mills known to have been overlooked by the six-inch and, equally, the windmills that were recorded in spite of potentially having disappeared years before. It also includes Little Dunmow Mill, which was moved from near Sawbridgeworth in Hertfordshire *circa* 1875, as well as mills that because of boundary changes are either now in other counties or have been brought into Essex.[21] In

[21] County windmill biographies are normally careful to inform readers that the mills they consider to be of their county will have been decided by reference to the 'historic' pre-1974 arrangement of county boundaries, so much so that this arrangement has become a universal referent for determining which county any windmill is in, or should be thought of as having been in. This is misleading as the boundaries of counties and parishes have always been subject to alteration and, until comparatively recently, counties commonly had detachments

exploiting all the source material cited for each mill by Farries, though, we have not escaped the thorny issue of defining what exactly is meant by a mill's demise. In line with previous thinking, we have to fall back on the assumption that it is the time of a windmill ceasing to be a landmark, as best as that can be ascertained. For a post mill we have seen this to be the date of removal of the wooden body in those cases where the mill was not demolished in its entirety; for a tower mill it represents the date when, if not completely demolished, it was at least severely truncated, leaving the equivalent of a post mill roundhouse. It is perhaps just as well that one difficulty we will not have to face is fixing a date of demise for the windmill at Belchamp Otten. It was only many years after its recognition by the Revised New Series that this mill headed into its prolonged period of decline, during which a valid case could be made for the year of its demise being anywhere between 1918 and 1960! Looking ahead to the comparison of large-scale sheets with Revised New Series sheets for congruence of their windmill evidence *circa* 1895, potentially we could want to know how any of the 236 sample mills was then faring. But many of these mills, such as the one at Belchamp Otten, were still standing in 1920. This does at least reduce to 170 the number of mills whose recognition (or not) by the Revised New Series and final circumstances may be such that we are forced to seek the help of Farries. Sorting these 170 mills into the five categories (A-E) is helpful: the mills whose dates of disappearance are known to within a reasonable margin – all those perhaps in Categories A, B and C – have been assessed primarily using sources other than maps. Since the supporting evidence cited by Farries for 30 of the Category C mills is of a non-map-based nature, it can be claimed that of the 97 mills occupying categories A, B and C, our knowledge via Farries for as many as 76 of them has been gained without resorting to the map record – demonstrating that this type of evidence was by no means a dominant one for Farries (Table 5.4).

If we now compare the Revised New Series sheets of Essex with those of the second edition six-inch for the whole windmill sample, a number of mismatches are revealed. To reiterate; the comparison exercise exposes instances of the Revised New Series omitting to take account of a windmill that it should have recorded, or omitted to delete the depiction of one that no longer deserved recognition. These are signposted by the appearance of a windmill on either the one-inch or the large-scale editions, rather than on both or neither. Also idiosyncrasies on the six-inch, or even the 1:2500, perhaps relating to the roundhouse of a dismantled post mill or other obscure windmill remains will be highlighted. There turn out to be 30 such mismatches (excluding two where a windpump is involved) between the smaller and larger scales for Essex – instances where either the Revised New Series or the second edition six-inch alone records the presence of a windmill. All these occurrences of a

---

or exclaves embedded within neighbouring counties: this was a legacy of age-old patterns in land ownership – notably ecclesiastical ones. This situation was rationalised in the nineteenth century with the use of tidying-up legislature: primarily, the Counties (Detached Parts) Act of 1844 but also the Local Government Act of 1888, which made significant boundary changes that led, *inter alia*, to large numbers of windmills from south-eastern counties being taken into a new County of London. This happened between initial publication of appropriate New Series sheets and their Revised New Series equivalents. Some windmills were thus forced to change their county allegiance and the current discussion for Essex highlights those whose status altered through changes to the borders of Essex with Cambridgeshire and Hertfordshire. Elsewhere, a sprinkling of windmills appear in one county on the New Series and then in another on the Revised New Series. Tilbrook Mill was one such mill, 'moving' in this case from Bedfordshire to Huntingdonshire; another close to the village of Netherseal 'moved into' Derbyshire from Leicestershire (Plates 47 and 48).

mismatch need careful scrutiny before the Revised New Series can be definitively assessed for this decidedly non-marginal county for windmills. If the simpler condition is applied of only noting the second edition six-inch depictions unmatched by earlier Revised New Series depictions, then the mismatch count reduces to twelve. Some of the 30 mismatched mills disclose the errors of representation being sought. Others are genuine in their depictions at both scales and are cases where the windmill disappeared in the interim between the dates of revision of each scale, small as it was, and for some there is insufficient evidence to draw any conclusion. This last position could be improved, though, by taking a more lenient view of the validity of certain types of evidence for the date of demise of a windmill. But to reach a ready conclusion, the category into which the evidence for the demise of each mill best fits – unless it is a windmill that survived beyond 1920 – together with the date when each mill is thought to have disappeared, is given below. Also offered is an assessment of the Revised New Series for each of the 30 mismatched mills, which, it could be suggested, creates a first impression of containing a very mixed bag of individual verdicts (Table 5.5).

Category of Mill	A	B	C	D	E
Numbers of Essex windmills (236 in total) (66 remaining in 1920)	41	5	51	36	37

Table 5.4. Numbers of mills by category of sources of evidence for their demise according to Farries.

Before judging what these 30 mismatches mean for the Revised New Series, some extra background discussion for the OS treatment of Essex and its windmills may be helpful. It took fifteen sheets of the New Series to cover this county and almost one hundred six-inch sheets. The New Series sheets exhibit all the problems of this edition in depicting windmills and are dismissed by Farries with the remark that 'little heed was paid to windmills': he then declines to mention them again, which seems a trifle harsh given that their recognition of windmills is not consistently uninformative.[22] The variability of New Series sheets for Essex stems, of course, from the changes in policy occurring in the late 1880s for acknowledging windmills. Unlike the southern sheets of the Thames estuary that only offer a handful of annotated ground plans, the mills shown on, for example, sheets 222 *Great Dunmow* and 223 *Braintree*, although still indicated by *Windmill* or *Mill* annotations, are at least comparable in number to the mills shown on the Revised New Series. While it is true that the 'improved' NS-2 sheets reveal some windmill gains, they also suffer some losses over the earlier NS-1 sheets. NS-1-O:223 *Braintree* is a useful sheet for making two further observations: first, for the equivalent area of an Old Series quarter-sheet, its time of survey was as much as twelve years (1873-85), which is a reflection of the survey pattern for the constituent 1:2500 sheets and over that length of time many changes would be happening to the windmill population. This is being somewhat disingenuous in that this lengthy period of survey can be put down

[22] Farries, *Essex windmills millers & millwrights*, 1, 45.

to the fact that sheet 223 *Braintree* – like so many – straddles a county boundary, in this case the one between Essex and Suffolk. The years of survey for Essex were reasonably limited to 1873-6 but publication of this sheet was delayed until 1887 by the late survey of Suffolk, meaning that, as the second observation, this coincided with the point at which the decision

Name of Mill	Category	Demise of windmill	Comment on the Revised New Series
Windmills recorded by the Revised New Series (NS-2):			
Aveley	C	1915-17	Correct *
Blackmore	A	1875	Non-Deletion
Boreham	D	1880-96	Inconclusive (RH)
Colchester: Distillery Mill	D	1889-96	Inconclusive
Fordham	D	1885-95	Inconclusive
Great Bromley	A	1895	Correct
Great Burstead (Bell Hill): West Mill	E	*circa* 1892	Inconclusive
Little Sampford: East Mill	E	1886	Inconclusive
Little Stambridge	D	1886-95	Inconclusive
Little Warley	A	1866	Non-Deletion
Pebmarsh	C	1894-6	Correct
Purleigh Barns	D	1873-95	Inconclusive
Ramsden	A	1873	Non-Deletion
Saffron Walden: Copthall Mill	C	1894-6	Correct
Shenfield: East Mill	E	1878	Inconclusive
Shenfield: West Mill	E	1870	Inconclusive
Stebbing: West Mill	A	1895	Correct
Steeple Bumpstead: North Mill	C	1893-4	Correct
Windmills not recorded by the Revised New Series (NS-2):			
Broxted	–	*circa* 1953	Omission
Finchingfield: Village Mill	–	still standing	Omission
Great Tey	A	1865	Not Incorrect *
Helions Bumpstead	E	1896	Omission
Hempstead	D	1898-1904	Omission
Matching Tye	C	1880	Not Incorrect * (RH)
Stansted: Mill Lane	–	still standing	Omission
Stapleford Abbots	–	1923	Omission
Steeple Bumpstead: South Mill	E	1896	Omission
Takeley: East Mill	C	*circa* 1897	Omission
Wicken Bonhunt	–	1920-9	Omission
Wimbish: Lower Green Mill	E	1896	Omission

Table 5.5. Essex windmills with conflicting depictions at one-inch and six-inch scales (circa 1895).

* indicates an error on the second edition six-inch
(RH) indicates roundhouse survival

was made to reinstate the windmill symbol. This led to the neighbouring, but slightly later, sheet 224 *Colchester* of 1889 being different. On this sheet, the mills that are recognised are limited to those in the south-east quarter of the sheet despite the presence of many other windmills in the other quarters: they are all, except one, shown with unannotated symbols. The exception is the post mill recorded as *Wivenhoe Windmill* but there is no obvious reason for this mill to have received such treatment. Further north, the New Series sheets covering the border of Essex with Cambridgeshire and Suffolk record windmills in the conventional manner of post 1888 using unannotated symbols, and they do so extensively.

The fifteen New Series sheets were all revised in 1893 except for sheets 239 *Hertford* and 259 *Foulness* whose revisions were made the following year: all were then published in 1895 or 1896. These Revised New Series sheets record 163 supposed windmills, the vast majority by symbol and *Windmill*. Only six windmills are recorded with unannotated symbols and of these, four are in congested areas where any kind of annotation would not have been viable. Three of these four confined symbols record mills, and the fourth – at West Ham records a windpump whose exact details are unknown, but which vanished shortly afterwards. The other two unannotated symbols recognise Walton-on-the-Naze Mill, which, its inaccessible coastal position aside, had no obvious feature to warrant its cartographic uniqueness – there was certainly sufficient space for an annotation – and lastly, an unidentified structure as far as the literature is concerned that was sited at Potters Hall, Great Yeldham. This was almost certainly not a conventional windmill and the symbol probably denotes a pumping facility. Owslebury Mill in Hampshire has featured throughout this study. Mention was made in the first chapter also of the brick-towered windpump that stood alongside it. At Colne Engaine in Essex, a virtually identical structure, also carrying an annular sail, was built ten years after its Hampshire counterpart to work for the local sewage works. The similarity between the two extends no further than this as the newer pump was not accorded a symbol but merely annotated as *Tower*. Elsewhere, structures of a smock-mill-type construction or tower mills of modest dimensions were built in Essex to pump water, but the only example recorded as a windmill on the Revised New Series is at East Tilbury. In Essex, this kind of structure was never as prolific as it was, for example, in Norfolk. The annotation *Old Windmill*, which can often appear on Revised New Series sheets elsewhere, appears only once in Essex, where it records Purleigh Barns Mill on sheet 241 *Chelmsford*. This is curious since many mills would by most definitions of dereliction have qualified to be so labelled; the mills at Fordham and Stebbing (West), for instance, had both been disused for years. It is possible that describing mills in this way was considered new and untried at the time except that sheet 259 *Foulness*, revised a year later than all but one of the other sheets, still does not take advantage of this facility to affirm that Foulness Mill had stopped working three or four years beforehand. Purleigh Barns Mill itself was a strange hybrid windmill, having been rebuilt from parts of an earlier mill that had been demolished. This in itself was not unusual except that in this case its appearance may have been disconcerting to an untrained eye. By 1893 it might have been in a disused state for twenty years if the first edition large-scale maps are anything to go by, but a late photograph shows it to have been in no more parlous a state than others at the time of one-inch revision. Other details of representation include two occasions where an annotation of *Windmills* is shared by two symbols and, at Great Dunmow, the windmill is recorded in unique fashion by a symbol annotated with *Tower Windmill*. Almost predictably, no conceivable reason offers itself for why this tower mill should have been favoured in this way over the many other tower mills, not only of the vicinity but across the country. A last oddity on the Revised New Series is the lone occasion when the word *Windmill* is engraved but, in the fashion of earlier New Series sheets, there is no symbol alongside it to record the mill's precise location.

The survey dates for the first edition, and then those of revision for the second edition of the larger-scale maps differ from sheet to sheet as one would expect. The survey that led to the first edition of the 1:2500 and six-inch for Essex took place between 1862 and 1876, consistent with that for Hertfordshire – after all, they share the same county meridian – but earlier than those of Suffolk and Cambridgeshire. Even if it could be established that across the county boundaries the differences between survey dates were small and that the survey progressed incrementally with no discontinuities, working with these scales cannot, without adjustment, provide a meaningful account of the mills of this whole area as a group. Solely within Essex, the fifteen-year span that the survey took distorts the message on offer about the status of the windmills of this county at any single point in time. As with anywhere else, the six-inch sheets as reductions of the 1:2500 sheets can be difficult to read on account of print size, and the annotations are not as helpful as on the 1:2500. But the actual windmills are recorded tolerably well on the six-inch even though the annotation is often simply *Mill* and instead of a symbol, of course, the ground plan of the mill is shown at near-true-scale. Of all the windmills that warranted a mention at the time of this first survey, only two are wholly missing from the six-inch – Harlow (Potter Street) Mill and Woodham Walter Mill – leaving 233 mills to be recognised by some combination of annotation and ground plan.[23] However, given that the survey dates spanned the years 1862-77, it can be understood why the smock mill at Little Warley, despite disappearing in 1866, did appear on this edition, yet other mills which outlasted it quite legitimately did not. Luckily, the edition is fortunate not to have compounded this distortion by recording Little Dunmow Mill (re-erected *circa* 1875) as this would have implied that this post mill had been present in the landscape at the same time as Little Warley Mill, which was never the case.

The second editions of the larger scales were the result of revision that took place really in the two years 1895-6, and consequently they escape the difficulty of misrepresenting the windmill count as all sheets can be assumed to be contemporaries of one another. The drop in the number of mills on the revised second editions to 154 highlights the decline that had occurred in little more than twenty years. But even though the second editions were capably revised in a short space of time, they too are not without their uncertainties of depiction. These, at a point just before the new century, included one of how the Survey was to deal with dismantled post mills whose roundhouses had been left intact. This dilemma does not seem to have existed with the shells of tower mills on account of their continuing landmark status, but innocuous single-storey roundhouses did for a while pose a problem. On the one hand, Bell Hill (West) Mill at Great Burstead, which was taken down *circa* 1892 but leaving its roundhouse behind (to be shown by a symbol on the Third Edition ten years later), had no annotation at either of the larger scales. On the other hand, Great Tey Mill, described in 1865 as having burnt down completely, was annotated as *Old Windmill* presumably because

---

[23] The natures of these two omissions differ slightly. The mill at Harlow is shown by an annotated grey square at 1:2500 – indicating a wooden structure, *ie* one with no roundhouse – but is then later excluded from the six-inch. Woodham Walter Mill on the other hand, and as seen earlier (Note 17), is rendered totally invisible at both larger scales. As a late arrival into Essex, Little Dunmow Mill is the other missing mill from the sample of 236.

some brickwork remained. The anomaly is made more noticeable by the cases of post mills standing in a more or less complete condition, or even working, at the time of revision, but which then attracted no map annotation. Aveley Mill, which was demolished in 1916, was one example and Copford Mill, whose body was taken down *circa* 1900, was another. Just as there were omissions of mills, so too there was a non-deletion in the form of Matching Tye Mill, demolished *circa* 1880 yet still recorded with the annotation *Mill (Corn)* over a decade later. The national position on this prospective move to record roundhouses on the second edition six-inch seems to have been quite different to that of the first edition, maybe as this was a time when sentiment was beginning to colour peoples' perceptions of windmills. In nearby Suffolk, when the body of one of the three post mills that stood in a row at Swefling was removed *circa* 1900 and its roundhouse turned into a dwelling, it retained its symbol at the one-inch. This may have set a precedent for roundhouses to retain their one-time mill status in the eyes of the Survey, but this was in stark contrast to mills that had been turned into accommodation only a few years beforehand, and that had tended to be deleted from maps and only rarely later reinstated. But, in any case, the recording of roundhouses at the turn of the century was not done evenly and it proved to be a short-lived phenomenon, as was implied a little earlier in the chapter. Two points of interest can be made in conclusion: first, the precept that the evidence of the six-inch is not always an absolute reflection of the 1:2500 is endorsed, but this had been due only to one single case, on the first edition at least – the mill at Harlow (Potter Street) – and second, the decline in numbers of mills between the first and second large-scale editions – from 236 to 154 – is really quite staggering.

Returning to the 30 instances of a mismatch in windmill depiction between scales, and remembering that the evidence of the Revised New Series should be disregarded when we assess the evidence gathered by Farries for when each of these windmills met its end, nearly half of the 30 mismatched mills still highlight this edition as being at fault. This remains so, even when all the criteria for judging the evidence are in favour of this edition being correct, and the proportion increases if we take a slightly harsher view of the validity of some of the sources of evidence used. Perhaps the fairest interpretation for nine of these 30 windmills is that an inconclusive verdict has to be delivered as to whether their depiction by the Revised New Series was correct or not, given that the evidence for their dates of demise falls into either Category D or Category E. One of these mills was the post mill at Little Stambridge, which was considered by Dr Turner to have disappeared before 1889. This was based on a much later discussion between Turner and the former miller; in addition, a trade directory refers only to water power and steam power being in use at this site in 1886. But are these crumbs of evidence really sufficient to decry the Revised New Series depiction of just a few years later? From a general realisation of the number of times that each type of source can isolate the date when a windmill disappeared from the landscape, a number of points can be added to those already made. Given that the aim is to establish the date of a mill's demise, by which could be meant the probable loss of landmark status, the usefulness of sources such as directories and land tax records is questionable. While they can be used to pinpoint cessation of trade and may help to confirm a local oral-history citing for demolition, they do not help in those cases where a mill languishes for years in a derelict state. Perhaps it was with this in mind that Farries saw no need to consult full runs of directories for some mills. It certainly appears that, of all the sources he used, directories in particular were often only consulted in a partial way. Compared to this, the availability from newspapers of reports of such untimely events as the destruction of a mill or its severe disablement, or auction and sales notices in their advertising columns, have provided the most date-specific evidence of all. But even here, sales notices can vary in the degree of optimism with which they suggest that future trade may be viable. Colchester (Distillery) Mill was sold in 1889 very much as a going concern. Little Sampford (East) Mill, on the other hand, was probably on its last legs when sold three years earlier in 1886. But whatever the source, the importance attached to it, together with the level of discretion shown when assessing its usefulness, is crucial, since ascribing the highly individualistic cocktail of sources of evidence for any windmill into one of just five categories is a highly sensitive decision: each decision requires a fine degree of judgement. That Farries' study has embedded within it his own particular value judgements does not detract from its value. His careful treatment of the sources in uncovering the fate of mills has more than sufficiently drawn attention to a fallibility of the Revised New Series. Moreover this has happened for a county where one might have been forgiven for thinking there would be safety in large mill numbers and that, at least here, the Revised New Series might have been accurate in its representation of the windmill.

Even if one wanted to be charitable about how the Revised New Series fared in Essex, this edition failed to record ten windmills. Irrespective of whether these ten mills had had a tenuous existence, and disappeared immediately after being recorded on the second edition six-inch, or whether they took the process of being recognised at that scale in their stride and survived well into the next century, the Revised New Series was at fault here. It erred by simply not recording windmills at Broxted, Finchingfield, Helions Bumpstead, Hempstead, Stansted, Stapleford Abbots, Steeple Bumpstead, Takeley, Wicken Bonhunt and Wimbish. Admittedly, two of these windmills – at Stapleford Abbots and Wimbish – had undoubtedly stopped working by the mid 1890s, but each would still have merited a symbol alongside an *Old Windmill* annotation. At the opposite end of the spectrum, two of these ten windmills – at Finchingfield and Stansted – are still standing. The windmills that had disappeared by the time of the Revised New Series, but which were nonetheless recorded by it owing simply to not being deleted from an earlier map, provide a second, smaller, list of faulty depictions. These were the mills at Blackmore – burnt down in 1875; Little Warley – blown down in 1866; and Ramsden, which was blown down in 1873. To this list of three mills, the two at Shenfield can perhaps be added, even though a real lack of evidence for their demise has hitherto rendered them as inconclusive for this analysis. But, unlike any of the other mills whose evidence was inconclusive, these two were considered by Farries to have disappeared as many as fifteen years before the one-inch revision took place. If both these groups of wrongly-depicted windmills are plotted, it is very obvious that each forms an unmistakeable cluster. The first is a broad arc that follows the north-west county boundary and the second approximates to a circle whose centre lies to the east of Brentwood. The implications of this are very significant. On the face of it, either human error was limited to two relatively small areas, in which case the specific surveyors or revisers allotted to handle those areas may be held responsible, or some explanation based on the processing of the survey information for these particular tracts of land may provide a reason. Either way, this unexpected finding, if it were extended across the entire country, would point to the existence of cartographic 'hotspots' of undue fallibility, while everywhere else would be near-perfect in its Revised New Series statement. One intriguing question that Farries asks, but makes no real attempt to resolve, is related to a decay curve he derives for the disappearance of Essex windmills over a long period of time. The curve is a smooth one, as would be expected, except for a sharp and inexplicable discontinuity that occurs in the mid-1890s – the time of one-inch revision leading to the curious clusters of omissions and non-deletions. The two clusters are unlikely to be related and each is open to further inquiry, but the findings for Essex should not lead us into assuming that any failings of the Revised New Series found elsewhere are also going to be concentrated: we need to be cautious before presuming to extrapolate these findings into a broader national picture.

From the types of misrepresentation that the Revised New Series presents for Essex, a major influence on the accuracy of this edition, or lack of it, might be thought to stem from a dependency on what the survey for the first edition 1:2500 had recorded – this eventually leading to the variable mill representation on the New Series. The one-inch revisers carried out their work by making necessary deletions and additions on the most convenient larger-scale map – six-inch sheets that were based on the same early survey.[24] As this was the case, it may be thought that the accuracy of the Revised New Series would depend somewhat on how many years had elapsed since the first large-scale survey, the non-derivativeness of the Revised New Series notwithstanding. In this respect, some other counties are as bad if not worse than Essex, being surveyed for the first editions up to forty years before the one-inch revision, although the convenient corollary of this is that the survey for the second editions of such counties often, as is the case here, only slightly postdates the Revised New Series. Another corollary is that because of the high attrition rate for windmills in Essex between the two large-scale editions – from 236 down to 154 – the revisers would surely have been aware of the foolhardiness of relying too heavily on the first edition six-inch sheets they were carrying and have been forced to do what was expected of them – a full field revision of the one-inch. Importantly, the use of our comparison exercise to investigate the mills of Essex and their declaration by the Revised New Series reveals all misrepresentations except possibly one. Unbelievably, a post mill at Copford eluded recognition by both the one-inch Revised New Series and the second edition six-inch if Farries' evidence for the survival of this windmill until *circa* 1900 is to be believed. Farries elects to be ambiguous on this point, acknowledging as he does an interview by Turner with the windmill's last miller, who had assuredly pinpointed the date when his mill had stopped working. Farries, in acquiescing to the probity of Turner, went on to state that 'the body, bereft of sails, is thought locally to have survived until *circa* 1900'. Perhaps, on this occasion, it is still the maps that should be believed.

Further mention should be made of OS town plans. They add an extra dimension to the story of nineteenth-century windmill representation, specially for areas of urban growth that were first surveyed mid-century. They can also help with our understanding of mills in towns where the 1:2500 and six-inch scales offer two different stories. Often though, the geographic extent of the early town plans is such that, by today's reckoning, only the most central mills of a town are shown. The plan, for example, of Colchester in Essex, surveyed in 1875, shows only one windmill (Butt Mill) but happily the first editions of the six-inch and the 1:2500 scales, covering not just the old town but also its modern suburban area, are in agreement with one another, with each showing five windmills. Given that the position

---

24 Duncan A. Johnston (ed), *Account of the methods and processes adopted for the production of the maps of the Ordnance Survey of the United Kingdom,* second edition, London: HMSO, 1902, 99; Hellyer and Oliver, *One-inch engraved maps of the Ordnance Survey from 1847,* 57.

of urban mills was usually one of relative cartographic jeopardy, Essex is probably atypical in that only one of the 30 mismatched mills – Colchester (Distillery) Mill – was an urban windmill. But, though by today's calculation its site would count as being in Colchester – it was one of the five mills just referred to – that observation would have been a lot less safe in 1890, and Distillery Mill most certainly would not have been thought of as an urban mill *circa* 1875. As all of Colchester's windmill sites of eighteenth-century origin had lost their mills by the early nineteenth century, what was thought of as the town of Colchester at the time of the survey for the OS town plan had thus only boasted Butt Mill. This windmill was auctioned in 1881 for demolition, leaving no other what would now be termed inner-city urban windmills in Colchester, or indeed Essex, for the Revised New Series to slip up on. But the situation in the 1890s for built-up areas in other counties could be very different as we have seen, and an extended use of the comparison exercise would need to be particularly mindful of the case of urban windmills that were already hemmed in by 1890.

So, the comparison of small- with larger-scale maps, linked to careful use of the sources that can best establish dates of windmill demise, can be the basis of fruitful inquiry into the mill content of one-inch sheets. In the case of Essex this was helped by the interval of time separating the revisions of the two sets of OS maps being a short one. For many counties, publication of the Revised New Series and of the second editions of the six-inch were also sufficiently close in time for the comparison exercise to be equally useful. Also helpful for the realisation of good assessments of the Revised New Series is the observation that if the simpler condition is applied of noting only the depictions on the revised six-inch sheets that are unmatched by an earlier one-inch revision, then the number of mismatched Essex mills reduces to twelve, but that as many as ten of the thirteen windmills definitely misrecorded by the Revised New Series are included in this figure of twelve. Even for counties where the one-inch revision rather more equally divides up the span of time between a county's initial large-scale survey and its first revision, the comparison exercise is still very useful. It simply means that rather more work has to be done with the primary sources to establish the dates when individual windmills no longer merited recognition, since a greater number of mills will be involved. An assessment of the Revised New Series windmill coverage for Norfolk, for example, would be much more of an undertaking than the analysis conducted for that county in the last chapter, since there we were able to take advantage of Apling's work, but his publication only covered surviving windmills. The revision of Norfolk's one-inch sheets occurred in 1897 but the larger-scale revision was not complete until 1906 – nearly a decade later. Not only here, but elsewhere in East Anglia, and also in counties such as Lincolnshire, assessments of the comprehensiveness of windmill depiction on Revised New Series sheets are feasible, but they might present more of a challenge than Essex did simply because there we had Farries to help us.

Remaining in East Anglia a while longer, the first large-scale survey of Suffolk had taken until 1885 to complete and revision here was not considered appropriate until between 1900 and 1904. The pace of change in this county was slower than in many others, and this had resulted in the relatively late survival of a large number of windmills from the uncommonly large nineteenth-century mill population. The high proportion of mills, not just surviving but working here *circa* 1895 from those of, say, fifty years beforehand, means that this is a county able to offer more evidence for the depiction of windmills by the Survey than most, simply due to the high residual number of mills involved. A brief study of the Revised New

Series treatment of the sizeable population of Suffolk windmills can usefully draw attention to two aspects of the comparison exercise in its reckoning of the worth of the NS-2 edition. First, the results achieved for Suffolk are comparable to those for Essex – so demonstrating that the convenient use of Essex did not, by some hideous stroke of bad luck, lead to an atypical outcome in respect of the Revised New Series and windmills; second, that the ease with which lessons can be drawn from comparing sets of county sheets does, indeed, differ between counties. Using the same large sample of 204 first edition six-inch quarter-sheets of Suffolk first cited in Chapter Two (when looking at the variety and incidence of the types of windmill-associated annotation), the number of windmills acknowledged by them as a result of a survey lasting from 1880 to 1885 is 263.[25] This number reduces to *circa* 192 mills by the time of the second edition in the mid-1900s, thus revealing a proportional decrease in mill numbers between the two editions less severe than that seen in Essex, and one that reflects the pattern of survival for windmills in 'remoter' counties such as Suffolk. Revision of the one-inch in the southern part of the sample area took place in 1893 and in the northern part in 1897, thus neatly dividing the gap between initial survey and revision of the larger-scales. The recognition by the Revised New Series of exactly 200 windmills within the sample area suggests that this edition may not have been as remiss here as in Essex. Applying the simple condition of noting those mills recorded by the second edition six-inch, but not the revised one-inch up to as much as a decade earlier, ten cases are revealed. These include two where there was a problem of space on the one-inch – Saxmundham Mill, which was dismantled in 1907, and Woodbridge (Theatre Street) Mill. Also included in these ten instances are two tower mills – at Oulton Broad (Lady Mill) and at Wrentham – which were not demolished until 1932 and 1964 respectively; a windmill at Blythburgh that was demolished in 1937 and an open-trestle post mill at Stonham that was wrecked in 1908.[26] Alongside other anomalies such as the two mills shown by the first editions of the larger scales at Thorrington Street and Polstead being one and the same mill – it evidently travelled faster than the survey team recording each site – the Revised New Series is again shown to be fallible. But not, one may think, to an unacceptable extent, *except* that a lot more work would need to be done here to guard against the windmills that would have still been standing in the mid-1890s, but which then disappeared in the ten-year gap before the revision of the larger scales. For other parts of the country, considerable research may be needed to uncover the evidence for using our comparison exercise or, depending on numbers of windmills, maybe not. What is plausible

---

[25] In its completed version a six-inch quarter-sheet was contoured but, as soon as possible after survey, the Survey issued, as an interim edition, a 'First Edition without Contours' version of the sheet, with the definitive contoured edition following some years later. Both editions of the sample sheets have been inspected to verify that the representation of windmills is identical between the two. These sample sheets offer 286 sites where the root word *Windmill* forms part of the annotation but this number reduces to 263 if those depicted as supporting pumping mills are ignored. Another 35 sites can be presumed with varying degrees of confidence both from their annotations and their topography to be supporting a windmill. Lastly, the use of a set of four dots to denote the brick piers of an open-trestle post mill and, in a separate instance, the appearance of a surrounding circular track on which ran the tram to the ladder of a post mill, each coupled with an annotation of *Mill House*, tell their stories of windmill presence. But if, for the sake of the argument, these 37 sites, including many of known windmills and even one of a surviving working post mill, are ignored and if, furthermore, complications such as windmills that had been stripped of their sails and moved to new locations to work using other forms of power – like the smock mill body moved to Peasenhall *circa* 1882, but which subsequently went unrecorded – are also ignored, we are still left with the very large number of 263 mills.

[26] Dates of mill destruction taken from Flint, *Suffolk windmills*.

is that carrying out the exercise on counties that sustained reasonable numbers of windmills will endorse the findings seen so far, and that counties with lower numbers of windmills will support the conclusions drawn earlier in the chapter.[27] What can be affirmed with some conviction is that any assumption on the part of the Survey that windmills, as point features in the landscape, would offer consistency of form and appearance, or of utility and purpose, was sadly mistaken. This must have undermined any policies for the representation of mills that the Ordnance Survey may have formulated and, demonstrably, it led to a sequence of one-inch editions losing their authority as definitive reflections of ground-truth accuracy, at least in respect of windmills.

An assessment for the one-inch scale

The findings of the last two chapters and now of the present one need to be drawn together to offer a quantitative analysis for the one-inch scale during the nineteenth century, as well as something of a qualitative one too, though this will depend on discussing new ways of thinking. The initial-state sheets of the Old Series, later sheets of the New Series, all those of the Revised New Series and the Third Edition (where symbols were used to denote mills) all exhibit, it seems, much the same range of percentage errors in their recognition of mills. The errors are mainly the total non-recognition of windmills or the continued depiction of mills safely known no longer to exist. The scale of these errors is statistically significant in that, typically for the area studies of windmills carried out, each displays an error of around five per cent. This observation is a statistic. It says nothing about the success or otherwise of other types of map representation. It neither insinuates any verdict of success or failure, nor does it imply that the Survey viewed this as an acceptable level of misrepresentation or even that it acknowledged the level to exist, as each area study exhibits an individual mixture of defects in how it depicts mills. Since these defects range from the Survey misconstruing, perhaps quite understandably, minor idiosyncratic features of mill construction, to making blatant, inexcusable errors of omission, it is not so simple to uncover underlying causes. It would certainly appear the case that there is no single overarching reason that could account for even a reasonable minority of the individual cases of errors made at the small scale. Any ambition of recording windmills accurately is also hardly one that the Survey can claim to have paid much attention to during the many decades of updating its Old Series sheets, as it showed total disregard for changes in the landscape other than those related to new railway construction. The simple fact would appear to be that the Survey was unable or unwilling to

[27] One source of evidence that has become much easier to access in the years since Farries was involved in his research is the collected census information, available now (2012) up to and including the 1911 national census. It has become quite simple, for example, to unravel the story of the members of the Bellwood family who worked a post mill (discussed and illustrated earlier) together with its near neighbour (another post mill) at Thorne in Yorkshire between 1851 and 1901. Moreover, tracing the familial links between the windmillers of an area and tracking the movements of a miller in his career between different mills can be done reasonably easily. The census information is useful in other ways. One uncaptioned early picture postcard (a good source of mill photographs from *circa* 1900) offers a clue as to the whereabouts of this mill – it is postmarked Wem in Shropshire (Plate 49). Careful use of 1:2500 sheets to find a windmill that corresponds topographically with what the image shows and, more significantly, the use of census data from 1901 to identify the original sender of the card, suggest that this is an image of Loppington Mill. In that year, this windmill was run by two elderly brothers (William and Thomas Kynaston, listed as millers) together with the son of one of them who is listed as the carter at the mill – surely the three people seen in the photograph.

record the fluctuating circumstances of the windmill population in the later decades of the nineteenth century with any great accuracy. Whether that inability or unwillingness was the consequence of procedural default, or whether there were other forces at work, is essentially unknowable. If this seems to be a harsh judgement, it is grounded in the cultural idea of the Survey being infallible. The Revised New Series did much to reinstate the value of the one-inch scale at the very end of the nineteenth century and even though this edition registered a five per cent error in its representation of windmills, this does still equate to being ninety-five per cent successful!

The business of the Ordnance Survey during its first one hundred years can be seen as a schedule of work that only really grew to become a large undertaking in the 1850s once the decision to adopt the 1:2500 scale had been made. This then also became the basis for issuing sheets at the 1:500 scale along with derivative six-inch sheets, and one-inch sheets too, which were needed as long overdue replacements for Old Series sheets that were taking a protracted length of time to complete. The expenditure of effort that went into mapmaking at all these scales after 1860 was massive and the output from it – measured in thousands of different maps – ranks alongside other Victorian programmes of achievement where some regulating of nature or society was under way. This was not carried out in a cultural vacuum: sustaining the Ordnance Survey at all points along the way was a changing political and economic climate and, more broadly, an outlook on life based on the philosophical grounding of the age. Of the many influences moulding the social and intellectual milieu out of which nineteenth-century OS mapping evolved, and that may have affected its nature, the notion of geometric form offers an interesting example. The cultural transformation whereby faith and reason were finally disassociated from one another can be dated to the beginning of the nineteenth century, at which time scientific thinking had come to be seen as the only legitimate means for the interpretation of reality. The myth of a divine nature, which had sustained metaphysics as a philosophical tradition, was redundant by this stage but in its time its guidance of human sensibilities had promoted the ideas of geometry and number as having a transcendental importance for an understanding of symbolism in the arts and for architectural style. Suddenly, in the wake of increasing industrialism and the emergence of science as the validating authority for advances in technological endeavour, these abstract ideas of geometry and number fell from their position of intellectual vitality, and the finesse accruing to them was submerged within the new philosophies that came to govern nineteenth-century lifestyles. This weakening of geometry, from a position of having had an almost mystical importance attached to it, to one of simply being a formal discipline with neither intrinsic value nor symbolic meaning, had implications for maps. The potential for a less rich significance being 'read' into nineteenth-century maps, compared with before, follows this decline in the meaning of geometric form. In contention with the very powerful drive towards explaining society and rules of nature along rationalist lines throughout the nineteenth century, successive movements in art sought to escape this prescriptive embrace by retaining and amplifying their use of symbology. In the language of theoreticians of maps and mapping, the term 'symbol' is assumed to be synonymous with, or at least analogous to, the terms 'sign', 'sign-vehicle' and sometimes 'icon', but within broader cultural parameters the term 'symbol' has always had many dimensions of meaning.[28]

---

[28] MacEachren, *How maps work: representation, visualisation and design,* 234. For a further explanation of some of the

Art movements came and went throughout the nineteenth century but the importance of symbolic expression in the world of art would eventually prompt the development of the dominant movement associated with the late 1880s, the 1890s and the turn of the century – which, indeed, was labelled Symbolism. This name had its origins in a letter to the Parisian newspaper *Le Figaro* in which it was suggested as an appropriate tag to describe the culture of a contemporary school of French poets. The name stuck and came to be seen as a fitting label for the broader cultural movement that artists of all kinds were becoming engaged in across western Europe. From its beginnings in France (and by French standards) much of this artistry was considered mildly degenerate; by the time it crossed the channel its practice was thought positively decadent. Indeed, Symbolism in England came to be associated with the formally named movement of Decadence from *circa* 1890 onwards, an alliance much reinforced by the perceived behaviour of adherents to both such as Aubrey Beardsley and Oscar Wilde. This *fin-de-siècle* feeling of decadence continued alongside the artistic style and language of Symbolism into the new century and provided a cultural backdrop to, perhaps, the most innovative decade of all in the nineteenth-century history of the OS one-inch map series.

The Ordnance Survey in the nineteenth century was, in the minds of all concerned and for better or for worse, guided in the creation of its one-inch maps by the military protocols to which its executive staff had a natural affinity. It is not in any doubt that a tension existed in the Survey between its civil remit and military exigency. This may have been exacerbated by the large-scale work consuming most of the resources available to the Survey from 1840 onwards, resulting in funding available for the one-inch being disproportionate to the public awareness and appreciation of it as well as to the military requirement for maps at this scale. The one-inch certainly found military approval – by Colonel Wilson among others – but the presumption that mapping at this scale was made at the behest of the War Office principally to be of use to military commanders in the field is not an altogether safe one. True, when it became expedient to do so, the one-inch was hailed as 'military' mapping; at other times this was less the case and the attitude towards the one-inch could be one of indifference. After the reverses of the South African War (which severely unsettled the military establishment), another small scale – the half-inch – began to find military favour and also chimed well with a society that was increasingly mobile. But the one-inch did not disappear; it was to maintain a longevity that survived from the Napoloenic era to *Entente Cordiale* and well beyond. More than a century after the event, what may now appear to some people as condescension in the testimonies of army personnel to the Dorington Committee and in the correspondence from earlier Director-Generals with, for example, the Treasury, was at the time designed to promote the one-inch scale in a forthright, if not necessarily diplomatic, manner. Attitudes needed to change and, though the governance of the Survey was to remain in military hands for much of the next century, the premise that military concerns could override any other soon lapsed. But *circa* 1900 the design of the one-inch map with its use of signs and symbols was still being achieved within a military setting. Within such a specialised context as this, it would have been natural for the practice of map design to have developed following its own teleology, whereby it progressed towards an end state that favoured, say, the use of symbols

wider cultural shifts raised here, see Alberto Pérez-Gómez, *Architecture and the crisis of modern science,* Cambridge, Massachusetts: MIT, 1983.

over any other system of depiction. The development from Old Series to New Series, from there to the Revised New Series and on to the Third Edition is matched by an escalating tendency towards the use of conventional signs. The representation of windmills – from depictions of individually-engraved symbols on Old Series sheets to standardised punched symbols on New Series sheets – is as illustrative of that trend as is the representation of any other type of landscape feature.

The idea that maps gradually made greater use of a language of symbols (becoming free of unnecessary annotation) is unmistakably sustained by the unfolding plot of the OS one-inch New Series. The reintroduction of the windmill symbol in 1888 can be seen as sensible, not only from a need to save space on a map, but also in according a 'modern' legitimacy to a map by offering a better consistency of depiction. Seen in this light, the simple annotation of perhaps *Windmill* added to Revised New Series sheets was maybe unnecessary, especially since a characteristic sheet or legend was given linking the symbol to a written identification. But, despite instructions to revisers to note the use to which each windmill was being put, no other information was given. Since also, in more than a significant number of cases, the status of a windmill was wrongly recorded through adding an inappropriate annotation or causing confusion through having none at all, this suggests that three different designs of symbol (with none requiring annotation) might have been a better option to introduce onto the Revised New Series as a way of recording a windmill, an old windmill or a windpump.[29] That this option was not taken up in the early 1890s can be considered odd when elsewhere on the map a trend towards greater use of map symbols was in vogue – as shown by the new use of different symbols for churches. It can also be noted that the military mind was ahead of the game in this respect, as annotating a map to show tactical military information was regularly done using a sophisticated, codified array of geometric patterns. Perhaps it is even the case that our knowledge of windmills on the Revised New Series through the use of annotations has been offered as a token gesture towards resisting the 'foreign commodity of decadent Symbolism', which was, after all, regarded at this time with 'a certain frisson, as something caught from France', and quite possibly therefore viewed by many at the Survey as a thoroughly unwholesome thing![30]

A concern to depict windmills at the one-inch scale, irrespective of success or otherwise in achieving that aspiration, may simply have been traditional, where for a long time no-one saw any reason to discontinue the practice or it may have been a policy subject to ongoing review, either for this one landscape feature or for it and a raft of other features. The failure to update the Old Series sheets and the less than effective recording of mills on the earliest New Series sheets probably needs to be interpreted as a manifest lack of a considered policy for the representation of windmills, as well as for other features too. That the Survey could condone its maps showing windmills whose presence had not been field-checked in half a century, alongside other mills whose upgrading depended purely on proximity to a railway line, shows scant regard for the general idea of consistency on a map, and none whatsoever

[29] This was an option the Survey eventually introduced for its one-inch scale in 1952, with the use of two distinct windmill symbols – each unannotated. One of the symbols was an iconic representation of a post mill complete with sails – this at a time when the great majority of windmill survivors were tower mills – while the other symbol mimicked a post mill bereft of its sails.

[30] Andrew Wilton and Robert Upstone (eds), *The age of Rossetti, Burne-Jones & Watts: symbolism in Britain 1860-1910*, London: Tate Gallery Publishing, 1997, 15-16.

for the notion of accurately recording the windmill population as a specific entity.[31]

In time, however, the sheets of the New Series began to demonstrate that thought was being given to the representation of windmills. The decision in 1888 to reinstate the symbol for them was made at a time when the need to be consistent in this regard was becoming urgent. The preparation of sheets for the eastern counties was underway, but these counties had large mill populations and there is evidence to suggest that senior officers at the Survey were sympathetic at this stage to the importance of windmills as conspicuous landmarks in the landscape. Of all the considerations that have been mooted as being possible grounds for deciding how an individual windmill's inclusion on a sheet would have been ensured, the idea of it possessing a status due to its prominence and facility for providing a navigational fix fits the observed map data better than any other. Not only would this have been a vital military consideration – and the one-inch was still very much a military map *circa* 1890 – but the Admiralty too was interested in the portrayal of coastal windmills. This proposition is supported by the case of Stiffkey Mill in Norfolk with its unique annotation, but a good deal more convincingly by that of Bidston Mill sited at the northern end of the ridge dominating the Wirral. This windmill is still a prominent hilltop landmark overlooking the River Mersey and its estuary. It is recorded by a Admiralty chart of *circa* 1860 that sets out the line of sight to the windmill from a key location at sea for shipping entering the estuary. For the whole coastline from Formby in Lancashire around to the north of Wales, only one other non-maritime feature has the same role.[32] It may not be sensible to generalise about Admiralty policy from a single example – especially as a contemporary chart of the River Tees and its estuary elects not to show any of the equally visual coastal windmills that existed between Hartlepool in County Durham round to Redcar in the North Riding of Yorkshire. It does, however, lend *some* weight to the general idea that windmills, particularly tower mills, were of value as beacons. For all the concern to maintain the recognition of windmills at the one-inch expressed by the Baker Committee, and given the range of circumstances that the OS command structure should have realised that windmills can occupy, good instruction as to precisely how windmills should be recorded was still not being provided. Instead, included in a broad instruction to surveyors and revisers dated 16 March 1896 was a single line of text that simply stated 'all windmills are to be shown, and the purpose for which they are used should be noted'. In sharp contrast, the following subsection dealing with churches uses more than six times as many words to spell out the intricacies of depicting churches depending on whether they have spires, towers, or both or neither.[33] The recognition of

[31] It can definitely be declared that not only was the Survey unclear on policy here but, from an organisational point of view also, it was manifestly not desperate to be consistent in its mapping. The job of revising Old Series sheets on the ground was on some occasions put in the hands of military personnel of the lowest rank. Given the strict association of responsibility with rank in any military establishment, having a lone sapper – comparable to a private in the infantry – on railway insertion duty, as happened *circa* 1847-53 (see staff returns in TNA PRO WO 55/963 & 964) was clearly not attaching any wider significance to this task beyond literally reassessing the narrow corridors of land through which the railways passed, though undoubtedly the soldier concerned was, like most of his kind, conscientous and able in his work. It should perhaps be explained since this is not the first time that the word 'sapper' has appeared in this study that, while in this instance it refers to an army rank, as a title it can also be applied to anyone, irrespective of rank, who is serving or who has served in the Corps of Royal Engineers.

[32] Coastal chart 'Liverpool Bay', first published by the Admiralty in 1859, resurveyed 1875-9, corrected to 1881.

[33] Later instructions – the one dealing with windmills in particular – were briefly mentioned in the last chapter (Note 68). The timing of this instruction on how mills should be depicted together with its lack of clarification – leading, no doubt, to the lack of compliance that the instruction engendered – is pivotal to our understanding of

windmills was not being ignored, it is true, but unless it can be supposd that queries about their recording were being raised at field level and going unanswered, then throughout the Survey there was a culture of allowing their recognition to be insufficient and, therefore, less than reliable.

Other issues raised in passing have included a concern for the precision with which mill symbols were placed on maps. In some cases a symbol was unnecessarily positioned on the wrong side of a road such as at Chattenden in Kent but the error in distance terms is rarely other than small and, from an extensive perusal of sheets from both small and large scales, obvious mistakes such as these were unusual. Much more frequently, instead of siting a mill symbol – which, at approximately a tenth of an inch in height and corresponding to almost 200 yards on the ground at one-inch – so that it is centred on the exact location of the mill, it is moved so that any slight overlap with surrounding detail is avoided. In this compromise between legibility, technical accuracy and even aesthetics of the map, it is the sensible desire for clarity that usually rules the day, though there can be exceptions to this as was illustrated earlier (see Fig. 3.3). There is also the suspicion that retention of the root word *Windmill* for an annotation on successive editions possibly led to the later annotative addition of *Old* not always being applied where it might have been appropriate.

All of these different concerns expressed over the use of symbols at the one-inch scale and all of the findings for each of the studies carried out over five chapters can together be a basis for evaluating the Ordnance Survey of the nineteenth century. Quantitative analyses have understandably predominated: anyone with an interest in windmills and who has read thus far will want to know how good each one-inch edition was at recording ground-truth accuracy for windmills, and so how valuable a guide to the past it is. The larger scales have been shown as more than reasonably dependable if 'read' with care. A modification of the small scales' earlier language of signs and symbols came to be seen as the way forward for the one-inch at the beginning of the twentieth century, and this opened up the prospect of landscapes now being 'readable' and 'mappable' in new inventive ways. Accompanying this enhanced use of symbols for depicting windmills on maps is the understanding that there can also be an approach to assessing the Survey's windmill representation that tries to be qualitative rather than quantitative. This can embrace some surprising ideas, including the suggestion that the actual numbers of windmills recorded plays a secondary role. The use of symbols as a wider cultural phenomenon has been described as:

> '... a profound human need and indispensable for the perpetuation of culture. Man's humanity depends on nothing less than his ability to come to terms with the infinite in terms of the finite, precisely through symbols, whether totems or magnificent churches. Symbols are part of the visible world, but are also outside the world. Goethe wrote that they made visible the invisible and expressed the inexpressible.

---

how windmills were treated at this crucial stage in their declining years; mention of it thus bears repetition. For the large organisation that the Survey had become by the middle of the nineteenth century, there would clearly have been a welter of memoranda concerned with all manner of day-to-day minutiae that a (military) headquarters had to send to its dispersed divisional offices. These were formally labelled as 'Southampton Circulars' and the more important of them covered matters of direct mapping policy. Our knowledge of the 1896 instructions is via this route (Book 3, ff 127-32). Ironically, our awareness of these Circulars is due to a surviving set maintained before partition by the Ordnance Survey's Irish division at Phoenix Park, Dublin. See 'The one-inch revision instructions of 1896', *Sheetlines* 66 (2003).

Like human knowledge and perception itself, symbols are ambiguous; they possess an eternal, ahistorical dimension and a dimension determined by a specific cultural context.' [34]

In the context of reading a map, how map symbols are viewed perpetuates that ambiguity and otherworldliness. In a modern culture, the mental operation for 'reading' something is habitually one of analysis but, as Schwaller de Lubicz argues, this was not always the case. In a translator's introduction to a text of his, Schwaller de Lubicz is credited with recognising that an analogical mode of 'reading' texts has been lost where instead of symbols just being 'conventional designations, abbreviations or literary metaphoric devices' as they are today, they can also be 'the means for transmitting precise suprarational knowledge'.[35] In a detailed narrative and appraisal provided for map signs and symbols by MacEachren (who would be reluctant probably to endorse fully the argument of Schwaller de Lubicz), he acknowledges that 'when a sign-vehicle has visual or structural similarities to its referent, it can be easy to metaphorically transfer the apparent meaning to another context'.[36] MacEachren also feels able to quote Muehrcke, who declared that 'the human mind seems quite adept at letting the map serve as a surrogate for reality for convenience purposes, sometimes even to the extent that the symbol becomes, for most intents and purposes, the reality itself'.[37] These last few lines in the narrative may appear to some to have drifted away from a discussion centred on windmills and the Ordnance Survey but, taken together with an earlier attempt at providing a scene-setting account for the (parts and wholes) use of symbols in a mapmaking capacity, they have been included to substantiate the idea that the 'reading' of symbols on a map is a

---

[34] Pérez-Gómez, *Architecture and the crisis of modern science,* 323.

[35] Robert and Deborah Lawlor, translators' introduction to R.A. Schwaller de Lubicz, *Symbol and the symbolic,* Rochester, Vermont: Autumn Press, 1978, 9. (Text first published as R.A. Schwaller de Lubicz, *Du symbole et de la symbolique,* Paris: Dervy-Livres, 1949.)

[36] MacEachren, *How maps work: representation, visualisation and design,* 324. Instances of this happening have included cases where a pictorial image of a windmill has been used to signify the presence of milling activity by watermill in an area. This was regularly done by Speed in the seventeenth century as his treatment of Newport in the Isle of Wight clearly demonstrates. His depiction of the island (1610-11) shows only five windmills, three of which are near-identical in their size, aspect and degree of pictoriality. Of the other two, much the smaller is used to indicate the presence of a mill on the west bank of the River Medina at Newport. In the usual practice of Speed's county maps, this key town is the subject of a supplementary larger-scale map provided as an inset within the border of the main map. However, at this scale no windmill is signified but two watermill symbols are shown instead. This undoubtedly is the topographically more correct interpretation for what mills there were physically on the ground. Obviously not thinking there to be a contradiction between his two maps, Speed is evidently using the compact visual appeal of the windmill symbol to signify on the smaller-scale map the prevalence of milling (as it happens here) by water rather than by wind. But this practice was not limited to as long ago as the seventeenth century: the one-inch Revised New Series (NS-2) was quite capable of the same cartographic sleight of hand. At Ellerker in the East Riding of Yorkshire, a watermill and a windmill less than a hundred yards apart were both still working at the end of the nineteenth century. As had long since been usual for the one-inch, the Revised New Series gives no indication either by annotation or by distinctive ground plan that there is a watermill here. What sheet 80 *Kingston upon Hull* does do, however, is to record each of the mills by using two windmill symbols and then compounding the deception by adding an annotation of *Windmills.* Given that this is potentially a real problem if trying to assess numbers of windmills on Revised New Series sheets, it comes as something of a relief to know that the watermill and windmill sited together at Cottingham, just eight miles east of Ellerker, received the conventional treatment – use of a single annotated windmill symbol and complete disregard for the watermill. In fact, cases such as Ellerker are very rare on the one-inch New Series.

[37] P. Muehrcke, 'Beyond abstract map symbols', *Journal of Geography* 73 (1974), 35-52. See MacEachren, *How maps work: representation, visualization and design,* 339.

hugely difficult act to grasp intellectually, and that for many people when looking at symbols on a map as part of their ordinary experience, there is a naive and innate sense in which the image of the symbol is taken to represent or to be like its referent. Putting this another way, the depiction and the real object being depicted blend into one another. Map symbols, it can be suggested, act as triggers, so everyone 'reads' a map according to a set of values that they have individually developed. This mirrors the idea of landscape appraisal being very much a personal activity, as other writers have suggested. Simmons, as just one example, places high value on ones lifeworld *(Lebenswelt)* as the arbiter for how anyone views a landscape before them, and within this schema discusses at some length the role that the arts play in shaping how landscape is appreciated.[38]

Maps, like all pictorial and artistic representations of landscapes, can be seen as aesthetic artefacts and it is known that mapmakers down the years have linked the attractiveness of their products to commercial gain. The idea that a map can primarily be a thing of beauty is an important consideration and the promotion of this through careful use of symbols has been discussed as a serious subject. Such representation, it is suggested by Hacking, is not generally intended to say how things are but, as importantly, to offer them as 'portrayals or delights'.[39] The nineteenth-century OS large-scale sheets for towns, the 1:2500 and the six-inch scales are doubtless broadly reliable in their representation for the landscape features they recognise, but they are also attractive artefacts, as too are the one-inch sheets of both the Old and New Series. Who would deny that the usefulness of an OS sheet and the ease of using it is not enhanced by factors such as its attractiveness? To some, it is only one step further to argue that this attractiveness can be achieved, if thought necessary, at the expense of full quantitative accuracy. Yet, regardless of any such consideration, by its own standards the Ordnance Survey would surely have been critical of the failure rates that are manifest in its representation of windmills at its smaller scales, the Revised New Series edition included. Conceding that a map is a static *tableau* where each part 'stands for' a dynamic, differentiated and complex element in the landscape so that the full intricacy of a landscape can never be totally and objectively replicated on paper is one thing. For the Ordnance Survey to admit that it had a golden opportunity to record all windmills at a crucial time in their history, but that it somehow fell short of the mark is quite another.

---

[38] I.G. Simmons, *Interpreting nature: cultural constructions of the environment,* London: Routledge, 1993, 76 ff.

[39] Ian Hacking, *Representing and Intervening,* Cambridge University Press, 1983, 138.

6

Epilogue

The story of how the windmill found itself represented by the Ordnance Survey does not, of course, end with the completion of our narrative for the one-inch Third Edition. From this point in time (around 1912) onwards, there remains much of interest to relate. Over the past century, successive Director-Generals of the Survey have each left their mark on how the depiction of landscape features might be achieved, and there have been two massive wartime upheavals each profoundly altering the plans and capabilities of the Survey. Above all, the technical process of mapmaking was evolving in ways that would have been unimaginable to earlier generations. In short, the story of the Survey during the twentieth century is every bit as riveting as that of the nineteenth century but there is for this study, as previously spelled out, one very big difference. The rate of decline of windmills that had gathered pace in the 1890s did not ease off until after the end of the First World War, and then only did so because the damage to mill numbers had been so great that the relatively few that were left got a new lease of life in the postwar economy. But even though life in rural areas continued in much the same way as it had done before the war, for all windmills the writing was on the wall. Some continued to work by wind but many more were using ancillary power instead. Time was simply not on their side and increasingly they fell into neglect, either after structural failure or just through becoming too uneconomic. The process of decline in the windmill population continued until well after the Second World War. In the 1950s, one of a very small number of authors writing about windmills felt moved to dedicate his book to 'the remaining thirty millers who keep the sails turning'.[1] This story of a collapse in mill numbers has been raised repeatedly, but it needs saying again to justify providing a much shorter narrative here for how windmills were depicted after 1912. This concern for the decline in mills has, as its corollary, the idea of a change in their status from one of workaday 'invisibility' to one (eventually) of cherished icon, so clearly the Survey was put in a position of needing to alter its stance.[2]

[1] Freese, *Windmills and millwrighting*. Even then, Freese's estimate of the number of windmills still at work by wind in the mid-fifties was somewhat inflated. Similarly, the major work by Rex Wailes from the same period includes many photographs of seemingly pristine mills all carrying the caption 'still at work'. But, by the time of publication, this was sadly no longer the case for many of these mills.

[2] The idea that windmills became more iconic as the twentieth century unfolded or emblematic of a former (more 'golden') age has been mentioned a number of times. The status of the windmill in the first decades of the twentieth century was enmeshed in a raft of wider cultural movements that sought to redefine the values of the countryside in the aftermath of one appalling war and with the prospect of another increasingly likely. This complex position is discussed in David Matless, *Landscape and Englishness*, London: Reaktion Books, 1998.

A real distinction should be made between the time when mills were commonplace and their essence was marked out as one of everyday utility, and much more recently when profound shifts in our culture and material prosperity have led them to be reconstituted as museum pieces. This is not to decry the valiant efforts being made today to preserve individual windmills – far from it – but it is a mistake to suppose that the interpretations people put to windmills during the nineteenth century are in some way comparable with those that people put to them today in the twenty-first century. For all these reasons this narrative for the representation of windmills after the outbreak of war in 1914 is so much shorter than that for the years before. In truth, it is offered as an epilogue to the study of an era when depicting windmills on OS maps was a more serious business. Nonetheless, it does explain for those who wish to understand it, how the story of the recognition of windmills unfolded once these mills were heading into their twilight years, and how they became caught between an 'authentic' nineteenth-century existence and their modern-day one of redundancy in the landscape and of being actors in the heritage industry.

Progress of the one-inch map

Following the third edition of the one-inch it was inevitable that a fourth edition would be contemplated following a third national revision of the New Series and, indeed, this is what happened. But this newest edition – still using the 360 small-sheet configuration – was discontinued after the issue of only a few sheets.[3] The reasons for this happening need not concern us unduly since the revision data that had been gathered, but then just used for the only seven sheets to have *Fourth Edition* inscribed in the top left hand corner – and all of the east of Kent – eventually contributed after further revision to the next generation of maps, the so-called Popular Edition. The short-lived Fourth Edition that sought to supplant the Third Edition was not the only hindrance to a smooth transition between the Third Edition and the 'Popular Maps' that we associate with the 1920s. The First World War, as can be imagined, caused huge turmoil and a redirecting of priorities at the Survey that effectively caused the domestic mapmaking programme to stagnate. An end to hostilities eventually came, of course, and with it a desire to initiate a new scheme of domestic map production. But before considering the Popular Edition as an important contribution to twentieth-century mapping, a glance backwards at the few sheets of Kent that had been published before 1914 can still be of interest since, stylistically, NS-4 sheets resonate far more with what had gone before than what would come later. In a way, they provide an ending to the story of the one-inch as it had developed during the nineteenth century. So, just as the Revised New Series was an updated version of the New Series and later the Third Edition was of the Revised New Series, so the association of this Fourth Edition with the Third Edition is quite apparent despite the fact that there are differences in both the detail, and the way that it is shown. One such difference seems to be that all annotations attached to windmill symbols have been removed. Taking sheet NS-4-O:305 *Folkestone* as just one example: seven windmills had been shown by the Third Edition, all of whose symbols were annotated, but the number of windmills later recorded by the Fourth Edition falls away to five and none of their symbols are now annotated.

---

[3] Hellyer and Oliver, *One-inch engraved maps of the Ordnance Survey from 1847,* 67-75.

Whatever benefits the small number of Fourth Edition sheets bestowed on the area of south-eastern England they covered, the remainder of the country was forced to wait until after the war before sheets of the Popular Edition became available to replace those of the Third Edition. Thus seen by the map-buying public as a creation of the 1920s, the normal coverage of a 'Popular' sheet was an area measuring 27 miles by 18 miles – the same format as had been used for most sheets of the Third Edition Large Sheet Series.[4] But the look of these new maps was very different to that of earlier ones. One significant change was the introduction of a cover that was attractive in its own right – a marketing concept that has persisted up to the present day. The standard cover for Popular Edition sheets is coloured red with a buff surround and features an image of the sort of leisurely activity that buying the map would facilitate. As with earlier editions, each sheet of the Popular Edition would then be issued in a number of states, with each giving full details of provenance. So it is that sheet 132 *Portsmouth and Southampton* tells us that the original survey for the area of coverage – the 1:2500 survey – was carried out in 1855-76 and that publication of all relevant New Series sheets had taken place in 1876-7. Also, that later revision of the one-inch sheets (to provide NS-2) had occurred in 1893-4 and that further revision (for NS-3) had taken place in 1901. Lastly, and in keeping with earlier mention of wartime delay, we are informed that the third revision had taken place in 1913 but that it was not until 1919 that this sheet was first printed, with minor corrections (for the state inspected) being carried out up to 1932. As informative and interesting as these notes of provenance are, Popular Edition sheet 132 shows only seven of the windmills that this area had once sustained. The wealth of information that the 'Popular' sheets collectively contain would make for a fascinating story but it is not one that needs to be told here as this has already been done, both stylishly and definitively.[5] But, because these were the maps that were used by early fieldworkers in the study of windmills such as Wailes, then perhaps a few words on the portrayal of windmills by this generation of maps might be useful.

Across all sheets of the Popular Edition, one convention that seems to hold up well is how windmills were depicted. The windmill symbol and the windpump symbol are kept on from the Third Edition, but the windmill symbol now either appears on its own or in conjunction with just one form of annotation – *Old Windmill.* The conclusion one might sensibly draw is that unannotated symbols are now being used to denote windmills that can work or that at least appear to be in reasonable order; others that appear neglected or to be derelict are denoted with an *Old Windmill* annotation. This assumption can be tested against what is known with some certainty to have been the case on the ground, but now this only needs to be done for a modest selection of areas since, with reducing numbers of windmills, the scope of the data is correspondingly reduced. Popular Edition sheet 116

[4] Unlike the Old Series, and the New Series in both its small- and large-sheet formats, all of which have had schema showing their sheet lines presented here, the Popular Edition and each of the one-inch editions that followed it provided the equivalent information on the back cover of each folded map, irrespective of which part of the country one was interested in. For this reason, it has not been felt necessary to provide the sheet lines here for these later one-inch editions. But, for readers who are particularly addicted to relating parts of the country to sheet numbers for all OS small-scale map editions, the comprehensive guide is Roger Hellyer, *Ordnance Survey small-scale maps: indexes 1801-1998*, Kerry: David Archer, 1999. Much of the same information can also be found in Higley, *Old Series to Explorer: a field guide to the Ordnance map.*

[5] Hodson, *Popular maps: the Ordnance Survey popular edition one-inch map of England and Wales 1919-1926.*

Chatham and Maidstone covers the same substantial area of northern Kent as Third Edition Large Sheet Series sheet 117 (NS-3-LC:117 by the Oliver notation) had done a decade or so earlier: it shows fifteen windmills, five of which are annotated with *Old Windmill*. This Popular Edition sheet had first been printed in 1921 from a revision carried out between 1914 and 1919 – once more, a length of time that conveys something of the upset caused by the war. The state used here however carries a print code of 1932 and a note that the sheet had been corrected to the previous year. But, since it would appear that counts of windmills on Popular Edition sheets were left unaltered between states, then the wartime years of revision would seem to be the sensible period for wanting to know something about the condition of these fifteen windmills. From Coles Finch, three of the five mills recognised with an annotation were certainly in no condition to work at the onset of war, nor had they been for some time: Lower Stoke Mill not since 1905, Perry Wood Mill not since 1910 and Shorne Mill not since 1870 – though this mill had later been used as an observatory. But the other two windmills from this group of five are not so supportive of our hypothesis. Throwley Mill was still working in 1914 shortly before it was pulled down in 1915 and Higham Mill was still at work in 1919 before it too was pulled down in 1921. With the ten windmills depicted on Popular Edition sheet 116 simply by a symbol, and conjecturally still capable of work, we are on surer ground. Five of these mills were still working by sail in 1920. Another had worked by sail until 1914-15 and then by steam – Preston Mill; yet another had worked by sail until 1917 – Rodmersham Mill – and Oare Mill had worked until at least 1919. Of the remaining two windmills, Milton Regis Mill appears to have been active until around 1914, thus leaving Whitstable Mill – thought not to have worked since 1905 – as probably the one mistake. Even here the Survey can be absolved from blame, since a new suburban road layout surrounding the mill needed to be shown, and this will have taken precedence over any inclination to annotate a symbol that merely denoted a redundant windmill.

These fifteen mills were all that was left from twice as many mills standing in this area at the time of the Revised New Series. Even sheet NS-3-LC:117, which was published in 1909, shows ten windmills that later went unrecorded on the Popular Edition sheet little more than a decade later. Some of these ten 'mill removals' were justified: two – Ospringe Mill and Upchurch Mill – had burnt down in 1910; Challock Lees Mill had disappeared as early as 1906 when it had been destroyed in a storm, and Eastling Mill was destroyed *circa* 1912 due to its condition having become so severe as to cause concern over its imminent collapse. The evidence for the fate of another two mills neither supports their departure from the Popular Edition nor condemns it. Hoo Common Mill close to Chattenden had become derelict by 1905 but Coles Finch is silent as to when it subsequently disappeared from the landscape. He is similarly vague about the fate of a mill standing on Broom Hill in Strood so that the Popular Edition may or may not have been right to remove it from the map record. Another two mills should never have been on the Third Edition in the first place, or even the Revised New Series for that matter: one had been pulled down around 1890 and the other had burnt down in 1887. The Popular Edition was also right not to record a windmill at Lynsted as this mill had become disused by 1895, although it did survive as a much-altered truncated stump before becoming overgrown by trees. The tenth loss of a mill symbol between the Third Edition and Popular Edition also reflects well on the latter. A pumping mill at Copton waterworks near Faversham had first of all been ignored by early large-scale sheets. Then it was wrongly accorded a windmill symbol and *Windmill* annotation by the Third Edition. Later, more correctly, it was recognised by the use of a windpump symbol on the Popular Edition and, further to the credit of the postwar map, two working windmills that had been ignored by the Third Edition were later recognised – one survivor from the group of windmills that worked at Boughton Street and Delce Mill in Rochester, which was still working when Coles Finch published his windmill book in 1933. This recognition of Delce Mill was, incidentally, not just an improvement on the Third Edition but on the two earlier New Series editions and the Old Series too. The New Series can partly be forgiven as its relevant sheet was published long before 1888, but for the Popular Edition effectively to be the first one-inch map to record this windmill highlights the part that serendipity played in the recording of mills as well as the serious inadequacies of earlier mapping.[6] Yet sheet 116 of the Popular Edition was not perfect. As well as the flaws already outlined, it too missed mills that earlier maps had failed to note, most likely on account of their less than robust large-scale depictions. One example was Kimmins' Mill at Frindsbury that was not demolished until 1931, but which had continuously been ignored by a succession of one-inch editions. This was a substantial mill and at the time of the Popular Edition revision it was still in a state that warranted recognition.

These findings for Popular Edition sheet 116 and for its antecedent one-inch sheets – Old Series sheet 6, sheet 272 from each of the New Series, Revised New Series and small sheet series of the Third Edition, together with sheet 117 of the Large Sheet Series Third Edition – can certainly be considered as reliable an account for the windmills of an area as the findings for anywhere else can offer; this is due to a number of factors that work to the benefit of the Medway area. First, William Coles Finch was born in Rochester in 1864 and brought up there. He would have been well acquainted with the windmills both here and in the surrounding area at a time when they were being recognised, or not, by

---

[6] Quite conceivably in this instance, the congested map space surrounding this mill suited the depictional style of the Popular Edition, whereas earlier the practice of incorporating an annotation into the depictions of active mills could have been the reason for this mill's exclusion from the Revised New Series and the Third Edition – if the Survey had wanted particularly to avoid the compromise of an unannotated symbol here. Or possibly, this oversight arose from this mill being in a sorry state for a few months after having burnt down – in 1872 according to Coles Finch – and before an opportunity to rebuild it had presented itself. This may just have coincided with the OS surveyors' visit to the area but this seems unlikely. In any case, the preparation of the Revised New Series – intended to be the outcome of field revision – was clearly remiss over this windmill together with a neighbouring mill at Luton. This second mill – listed by Coles Finch as a Chatham mill – had apparently burnt down in 1887 but continued to contribute to a confused story of OS windmill depiction for its immediate vicinity. The Revised New Series records what might have been sizeable remains at a site shown on the large-scale first editions as being occupied by Luton Mill, but for which the second editions show just a *Mill House*, only this site is four hundred yards east of a site that more closely fits Coles Finch's description for Luton Mill and where, crucially, the second editions, revised 1895-6, record a seemingly working windmill. It is very unlikely that Coles Finch missed a second mill here as he seems to have placed great faith in the recall of his informant for this area. But it could be that the oral evidence was flawed to the extent of confusing a destruction through fire with partial destruction and subsequent removal over a short distance. If this was the case, then this was a much-travelled windmill. Much earlier – *circa* 1850 – it had been brought to the Medway from as far away as Salt Hill (Slough) in Buckinghamshire. Alternatively, it could simply be that the Survey got its recording of one windmill in this small area near Chatham horribly wrong. In any case, the capacity of late-nineteenth century OS mapping to cause confusion for map users was to fade with the decline in numbers of mills. Popular Edition sheet 116 *Chatham and Maidstone*, almost needless to say, does not recognise a windmill at either of the two putative mill sites for Luton.

Old Series later states. He is generally to be trusted on the windmills of Kent (he misses only one – at West Malling) and especially so, one might think, on those of the Medway area. In a professional life that led to him becoming the manager of the local waterworks, he ended up taking responsibility for having to destroy one of the windmills he discusses. Second, Chatham (Brompton) was (and still is) the home depot of the Corps of Royal Engineers. If extra care was ever going to be taken over the mapping of anywhere in the country, then with the exception maybe of Southampton as home of the Survey, the area surrounding the mouth of the Medway would have been that place. Third, there is always the possibility that thought may have been given to subjecting the depiction of windmills to censorship on account of their potential role as aids to navigation – which would have helped to ensure that depictions on 'master copies' were correct in the first place. While this would have been in keeping with the non-depiction of the substantial fortress works in this area, by the early 1920s this may have been a rather outdated way of thinking and, in any case, it is somewhat thwarted as a hypothesis by the continuing recognition of, for example, churches.[7]

This *Chatham and Maidstone* sheet of the Popular Edition is of an area that had by any measure once been rich in windmills, yet by *circa* 1920 the number left to record is only in the order of fifteen. Sheets of the Popular Edition showing other areas also once rich in windmills depict similarly small numbers, but what can vary between these sheets is the proportion of unannotated symbols to annotated ones. Sheet 84 *Bedford*, for example, has only three annotated symbols but twelve that are unannotated – probably offering a fairly accurate account of the situation on the ground. In contrary fashion, sheet 132 *Portsmouth and Southampton* has six annotated symbols for mills that, in most cases, were mere shells by this time with one of them completely shrouded in ivy. The sheet shows only one mill – Gale's Mill at Denmead – by just a symbol. This mill, conjecturally still at work when it was surveyed in 1913 (probably true), was pulled down in 1922 three years after the initial printing of sheet 132. If we venture to parts of the country that only ever had a sprinkling of windmills, then by the time of the Popular Edition their numbers are severely reduced. Sheet 112 *Marlborough and Devizes* provides a single annotated symbol at Chiseldon, and a single unannotated symbol for the mill at Wilton, which given that it still had its sails in the 1920's seems very reasonable. Elsewhere though, areas that supported small numbers of windmills are fairly unevenly recorded: sheet 110 *Cardiff and Mouth of the Severn* manages four annotated symbols, as well as two unannotated symbols for Somerset windmills that were supposedly still active on Worle Moor and in Brockley Wood. Worle Vale Mill still carried sails into the 1920s and maybe warranted its depiction, but the love affair between Brockley Wood Mill and the Survey was doomed to continue well into the second half of the twentieth century despite this one-time mill by then only consisting of an inaccessible pile of stones and, even by the most conservative of estimates, probably having been out of work for well over a century!

If the message so far on offer for the Popular Edition has been one of sadness at the reduced numbers of windmills apparently still in place for it to depict, then something of

[7] The depiction of military works here continued to reflect successive phases of a thawing in international relations (essentially with France) and then increasing tensions. Hence, the first revisions of the large scales (1895-6), as with earlier sheets of all scales, do not show the military installations here though later these were restored to the one-inch sheets after 1907, only then to be removed again after 1925.

a corrective should be offered. Some areas of the country – and thus some sheets of this edition – did still have fairly reasonable numbers of windmills to record, but they are the exception rather than the rule. Counties such as Lincolnshire where the tower mill took hold in large numbers could still offer many windmill structures in the landscape and sheets of the Popular Edition reflect this fact. But this does rather overlook the reality that these tower mills were falling into disuse, ultimately to end up as empty brick shells. However, these shells increasingly retain their recognition as mills with this remaining the case when some were altered into houses. Sheet 33 *Hull* covering parts of Lincolnshire and the East Riding of Yorkshire illustrates this well. As many as 27 annotated symbols can be found on this sheet, nearly all recording tower mills – the post mill at Skirlaugh being the exception. In many cases, these mills were defunct and, even if still at work using ancillary power, they were certainly no longer able to muster a set of obviously workable sails. In contrast, the number of unannotated symbols is much smaller at eight. This includes one for the tower mill at Skidby that was still working and which, in due course, would became the last mill in the East Riding to work by wind. But for the other seven mills supposedly still at work in the traditional way, there is a lack of easy evidence to confirm that they were indeed doing just that at the time of revision in 1920-1.

Lastly, there were still pockets of the countryside where economic conditions fostered the continuing use of windmills though, as suggested earlier, major damage, particularly to the sails or to other costly parts of machinery could result in sudden death for a mill and cannibalisation of its remaining components to keep other mills going. One of these pockets was East Suffolk where the tradition of milling by sail remained strong into the 1920s and even later, but inevitably for a diminishing number of mills. Popular Edition sheet 87 *Ipswich* is one of the few sheets of this edition not only able to record a sizeable number of windmills but also to depict a significant majority of them as supposedly still working. Based on a revision lasting between 1914 and 1919, this sheet was published in 1921 and a later state (used here) was updated with minor corrections to 1933. A total of 38 unannotated symbols appear on this sheet in contrast to just four annotated symbols, but the picture that these numbers portray is not quite as simple as it seems. In an area that supported mainly post mills, but with healthy numbers also of both tower mills and smock mills, the way in which the post mills particularly are depicted indicates something of the nature of the last years for windmills in this area. The four mills that are recorded using symbols annotated with *Old Windmill* include a post mill at Earl Soham that had stopped work in 1917, but which still carried sail frames into the 1920s and continued to look fairly pristine, another post mill at Eyke that had been dismantled in 1910 and two other mills demolished *circa* 1919 (Bredfield Tower Mill) and in 1922 (Ashfield Place Post Mill). These last two mills are thus credibly depicted by this sheet. Less so the previous two. Turning to the much larger number of unannotated symbols appearing on sheet 87; these include a post mill at Debenham that was demolished *circa* 1917, the smock mill at Leiston demolished in 1917 and the post mill at Wetheringsett demolished in 1919. When these windmills had been surveyed just months previously, each could still have been at work and the map cannot therefore be condemned on their account. Many other mills in this area, indicated by the Popular Edition as working, met their end not that much later. Markin's Mill at Snape was demolished in 1922; Station Road Mill, Aldeburgh was pulled down in 1924 and Mount Pleasant Mill at Framlingham came down *circa* 1927 – all post mills. Others from the group of 38 windmills depicted by unannotated symbols will have disappeared by 1930 and, regardless of the quirk that some mills shown as disused were to outlast others depicted as still working, this was certainly a time when the survival of any mill, particularly a post mill, could hang by a thread. One can only feel very sorry for a mill enthusiast who bought a copy of this sheet in 1933 intending it to be the basis for exploring the windmills it recorded. But even so, there would have been compensations. Inevitably (as by now it must seem), windmills were missing from this sheet just as they were from others; here, they included a tower mill working up to *circa* 1920 (Tricker's Mill at Woodbridge). Previously, in 1922, a visitor to Suffolk would have been rewarded with the sight of a post mill body being moved wholesale over some distance and refashioned as a windpump. This operation would have been reminiscent of the heavy-duty work done by millwrights a generation earlier and, to a casual observer, only the non-circularity of the newly-built 'roundhouse' would have suggested that this was not a construction of some new cereal-grinding windmill. This was still a time and place where windmills were working 'authentically' and moving this mill body to its new home at Thorpeness made it no different, cartographically speaking, to the dozens of tower-mill structures that served hereabouts as windpumps; it was only somewhat later that the uniqueness of Thorpeness Mill was to become celebrated (Plate 50).

Before going any further with this twentieth-century story of OS one-inch editions, our attention can briefly turn to other mapping that had been an innovation of the early to mid-1900s – the half-inch. Authorised initially in 1902, early sheets at this scale were derived from the Revised New Series and issued in the familiar 18 inch by 12 inch format that contemporary one-inch sheets were following. The idea was that there should be 103 of these sheets but not all were issued. Just as the Third Edition found itself re-issued in larger sheets, so too did the half-inch in 1906 with the introduction of a new 'large' 40-sheet series, now mainly based on the Third Edition. There was an obvious intention to depict windmills at this new scale and this happened, unsurprisingly at such a small scale, by just using symbols. Remaining in East Suffolk close to Saxmundham and the coast, it is insightful to compare, for example, the common coverage of sheet 88 of the Third Edition Large Sheet Series (NS-3-LC:88) of 1908 and sheet 25 of the '2 miles to 1 inch' scale of 1911. This reveals a close correlation of mill symbols but not a perfect one. On the evidence of three omissions from the half-inch scale – mills at Blythburgh, Leiston and Snape that only vanished in 1937, 1922 and 1917 respectively – something is amiss, but by the evidence of mills omitted at Stratford St Andrew and Coldfair Green, which had been dismantled or demolished in 1905 and 1908 respectively, one may even suspect the Survey of having done some last-minute field verification. Sadly this is improbable as it now chose to substitute a windpump symbol for the windmill symbol that had depicted the second of these mills – a substitution that had no justification whatsoever – and this was not the only windpump symbol that crept onto the smaller-scale sheet. The post mill at Saxtead Green is also treated in a similar way – ironically so given this mill's survival to become today's classic example of a preserved Suffolk post mill. With the arrival of the one-inch Popular Edition, the chance was taken to use it to provide an updated half-inch sheet for the area. As a result, a copy of sheet 25 of the half-inch, still dated 1911 though acknowledging 'periodical corrected reprints' but in fact able to be dated by its print code to 1936, closely mirrors the windmill representation of Popular Edition sheets, including sheet 87 *Ipswich* cited earlier. This is achieved even to the point of adding annotations to those symbols that had been given one on the one-inch sheet, though in each case it now appears in a slightly abbreviated version as *Old W.Mill.* A few discrepancies also appear: absurdly, the post mill at Peasenhall is depicted using a windpump symbol in place of the earlier windmill symbol and to rectify the earlier omission of a post mill at Snape, not just one but two symbols appear on the map at the site where this mill had been demolished in 1922. Less strange, one must think, is the failure to erase a symbol from an earlier half-inch sheet: the post mill at Sutton had been blown down *circa* 1917 and it had not been recorded by the Popular Edition, so its later appearance on the updated half-inch sheet was in all likelihood simply an oversight.

Looking briefly elsewhere, the windmill depictions on NS-3-LC:33 *Hull* are replicated fairly well on sheet 10 (1907) of the half-inch. The symbol count is the same for each but three mills shown at one-inch are not then cascaded down to the smaller scale and, oddly, there are three additions to the half-inch that do not appear on the Third Edition. Half of these anomalies are sited in or very close to the town of Barton upon Humber. A tower mill right in the centre of the town, where space may have precluded the use of a symbol, and a tower mill by the railway station were each missed by the half-inch. Both mills are still standing. Another tower mill on the south-eastern edge of the town, demolished *circa* 1980, is shown on the half-inch sheet yet had been missed from the one-inch sheet. A failure to record all urban mills at half-inch can hardly be seen as blameworthy if, when earlier comparing the larger scales to the Revised New Series, it was acknowledged that sometimes the Survey had no option other than not to record some mills at one-inch if they had been encroached upon by urban development. This had occurred any number of times at one-inch, so accepting this practice for the smaller half-inch was inevitable. The complexities of depiction engulfing the windmills at Brigg in northern Lincolnshire have already been debated. Sheet 14 of the half-inch scale (1908) provides a very credible account of the mills of this part of Lincolnshire yet it fails to show any of the three mills at Brigg that the Third Edition had recorded despite, on this occasion, there being ample space in which to have done so. In the city of Lincoln, two of the three mills shown by the Third Edition are precluded due to lack of space from appearing on the half-inch, but maybe the miracle here is that symbols for these mills were shoehorned into the one-inch sheet in the first place. Lastly, returning to Kent, principally to the Medway towns, four mills found on the Third Edition within this area reduce to three on the half-inch with the non-recognition of the last of the windmills that had stood on Broom Hill. But also on sheet 40 (1906) of the half-inch, a windmill at Whitstable that had not appeared on the Third Edition is resurrected from the Revised New Series – it had survived until 1905 when it was pulled down. It stood close to the railway station about a mile north of the other (surviving) mill in the town, and the dependence of the earliest half-inch sheets on the Revised New Series rather than on the Third Edition was probably the reason that a mistake was made here. Sheet 40 also reduces the number of windmill symbols used at Drapers Mills near Margate from three to two but, as a *quid pro quo*, it does also correct the symbol used to recognise Copton Pumping Mill from one of a windmill, as seen on the Third Edition, to one of a windpump – as the Popular Edition was to echo much later. All in all, it can be acknowledged that the Survey made a reasonable job of depicting windmills on their early half-inch sheets. At a scale that found increasing favour with the military, this serves to emphasise the perceived value of windmills as landmarks. There were, of course, omissions in the reduction from the Third Edition, but given that only a quarter of the map space was now available for symbols that were not proportionately reduced in size, this represents a fine deal for windmills. Furthermore, for those people who bought maps at this scale (or even those today who collect them), they represented value for money in terms of depicted windmills per shilling spent! Most of Kent appears on sheet 40 of the half-inch and, since it depicts well in excess of one hundred windmills, having access to a copy of this sheet is enough to offer a comprehensive, if not definitive, account of the windmills in this part of the country at the turn of the twentieth century.

If the 1920s was the decade of the Popular (substantive fourth) Edition of the New Series one-inch scale, by the early 1930s consideration was being given to a replacement fifth edition. This point in time marks a sea change in how sheets at this scale were to be prepared: now, instead of copper plates or lithographic stones, the basis of maps was to be photographic negatives of pen and ink drawings.[8] The rendering of the landscape on this next edition thus looks decidedly bolder than it had done on Popular Edition sheets, but sadly only a part of the country was destined to be catered for before this edition was superseded. The sheets issued for it had covers declaring them to be of this Fifth Edition and they echoed the style of cover used for the Popular Edition. This had incorporated a picture of a wholesome-looking fellow gazing out on a pastoral scene. He was still clearly benefitting from participation in the outdoor life after the cover had been redrawn for the Fifth Edition. Still with an unfolded OS map, he has moved with the times in the new image; his dress style now embraces a sleeveless jumper and rolled-up sleeves rather than his earlier less casual attire that had still conveyed a slight whiff of the Edwardian era. His one-time cap has been discarded too but his pipe is still firmly clamped between his teeth. More to the point, the windmill that he can see in the far distance on the Popular Edition cover is still in place for the Fifth Edition. Of the less than a dozen Fifth Edition sheets covering areas of reasonable windmill presence, sheet 96 *Hertford and Saffron Walden* takes in part of the border between Hertfordshire and Essex. It continues the earlier practice of distinguishing between windmills that are recognised only by symbol – twenty in this case – and those with a symbol annotated by *Old Windmill* – two mills, at Little Hadham and Manuden. One noticeable change that paves the way for later depiction of windmills at this scale is the nature of the symbols themselves. Most of those used on this sheet are of a new stronger style, with particular emphasis given to making the sails prominent in a way that had not happened before. See particularly, for example, the depiction of Guilden Morden Mill. Yet symbols of the older type appear on this sheet too, notably a cluster of four at Aythorpe Roding, Great Dunmow, High Easter and White Roding. This suggests that, as had been the case in earlier decades, changes in policy for how features were to be depicted were implemented without having to wait for the launch of a next edition. A decision must have been made *circa* 1933 during preparation of sheet 96 to strengthen the windmill symbol and the part of the sheet already finished (the south-eastern quarter) was not brought into line with the other quarters of the sheet. This sheet was published in 1935 and was then subjected to periodic and undated corrections including the removal of the annotation for Little Hadham Mill on the 1938 printing. It, together with other

[8] Richard Oliver, *A guide to the Ordnance Survey one-inch Fifth Edition*, London: Charles Close Society, 2000, 4.

sheets of the Fifth Edition as well as the many ageing sheets of the Popular Edition, was then replaced by a sheet from a new one-inch edition that, in a renewed abandonment of the numbering notation, was called the New Popular Edition.

The genesis of this next edition, and the premature demise of the previous edition, in 1938 was linked to the decision to superimpose a metric grid onto the face of each sheet. One outcome of this was that the standard New Popular Edition sheet size represented an area on the ground that measured 40km in an east-west direction by 45km in a north-south direction. But, importantly, the scale remained at one-inch-to-one-mile. This New Popular Edition went on to be completed within a decade, the upheavals due to a second major war notwithstanding. A major disruption caused by the war was a need to provide two 'emergency' one-inch editions – the War Revision and the Second War Revision. No doubt, the last thing on the minds of those whose job it was to provide these editions was the guarantee of an up-to-date record for windmills. Interestingly though, later sheets without legends from the second wartime edition have either *Popular Edition Style* or *5th Edition Style* printed at the top of each. These headings indicate what the source mapping had been for each sheet and the nature of the conventional signs used, but this practice of using Popular Edition material and Fifth Edition material also later underpinned the New Popular Edition. So, sheets of this next edition form two groups – those based on Fifth Edition or Fifth Edition style material and those using Popular Edition material.[9] Being a new one-inch edition (the *de facto* sixth edition), we can expect numbers of mills shown on it to have altered despite the overall lack of new survey material other than that which never got used for the short-lived Fifth Edition. As with the Second War Revision sheets that overtly declared the provenance of their revision material, the precise style of windmill representation on sheets of the New Popular Edition reflects whether a sheet's provenance lay with the older Popular Edition or with the later Fifth Edition. Both make use of the newer stronger style of symbol having prominent sails, though not exclusively so since there are instances of the older-style symbol surviving as non-deletions twice over, first from the Popular Edition and then again from the Fifth Edition – the group of symbols near Aythorpe Roding included. On sheets based on Fifth Edition material, it would seem that no *Old Windmill* annotations are used but, on sheets based on Popular Edition material, they are. Putting to one side these complexities of provenance for the detail of the New Popular Edition – whether it was based on Popular Edition material or Fifth Edition material, all of which went through wartime revision – the decline in numbers of windmills recognised across these editions is relentless.

As ever, individual sheets of the New Popular Edition can be inspected in some detail for their accuracy of windmill depiction. Such is the larger coverage of each New Popular Edition sheet that all of east Kent appears on sheet 173 *East Kent* published in 1945. This sheet was based on a full revision of 1936 intended for the Fifth Edition, and later states of this sheet were subsequently issued showing major corrections to the road network, in 1947 for example, as well as for other minor detail. For the area covered by this sheet, 49 windmills had been declared by the half-inch sheet of 1906 but only 21 of these survived

---

[9] This explanation for how the New Popular Edition evolved out of the curtailment of the Fifth Edition and the need for special wartime measures is necessarily brief; seriously so given the complexities surrounding the introduction of the New Popular Edition. For a fuller account of this period, see Richard Oliver, *A guide to the Ordnance Survey one-inch New Popular Edition*, London: Charles Close Society, 2000.

to be recognised by the New Popular Edition, though this is a little harsh as windmills at both Eastry and Canterbury should also have been depicted. Even so, in something less than forty years the number of windmills in the area had been reduced by more than half. Other areas and other sheets tell the same story of diminishing mill numbers, of the sole use of the bolder symbol, and of one or two mill omissions. Adjacent to sheet 173 of the New Popular Edition, sheet 184 *Hastings* was prepared in 1940, again based on a revision of 1936. A later state for this sheet declaring the road system to have been corrected to 1946 and incorporating other non-specific alterations is, however, found to be remiss. In spite of the fieldwork done to be able to make these corrections, this state fails to record Northiam Mill, which was not taken down until 1949, and it continues to show two mills at Woodchurch, one of which – Upper or Black Mill – was almost certainly dismantled before the war. Elsewhere, even in counties that only half a century before had supported many windmills, their losses as revealed by sheets of the New Popular Edition were even more dramatic than they had been for Kent. Sheet 162 *Southend-on-Sea* published in 1945 shows just six windmills. This includes a mill near Chelmsford that had been demolished in 1932 and two mills that had possibly disappeared sometime during the war. Doubtless other significant remains were still *in situ* such as the full-height tower at Rayleigh (a mill since restored – in 1974), but this was still a time when it was thought unusual to record a structure at this scale unless it retained more than just vestiges of being a windmill. Even so, six windmills represents a poor survival rate for the forty that the Revised New Series had shown for the same area only fifty years before.

Rather than seeking to consolidate its New Popular Edition for any longer than it did, and not swayed by years of immediate postwar austerity or indeed the rationing of paper at this time, the Survey instead went ahead in 1947 with a replacement one-inch edition – the Seventh Series.[10] The welcome change introduced onto the Seventh Series, as far as windmill enthusiasts are concerned, was a dual-classification mode of depiction which differentiated between windmills that had lost their sails and those that still carried them. This would be used for the remaining life of the one-inch scale, changing only with the launch of metrication in the 1970s. In the 1950s the number of remaining windmills was giving grave cause for concern to those interested in their welfare. The number of voices being raised on their behalf was still small and it would be another decade before local initiatives and public funds combined to save something of what was left. Consequently, the inexorable decline in mill numbers can now be witnessed by comparing sheets of the Seventh Series with those of the preceding New Popular Edition. Comparisons are made easy as sheet lines and sheet numbers stayed the same so, for example, the areas covered by sheets 173 *East Kent* and 184 *Hastings* common to sheet 172 *Chatham & Maidstone* had supported nine mills on the earlier edition. This reduces to merely two windmills for this common area by the time that sheet 172 of the Seventh Series is published in 1957 – the outcome of a full revision in 1954-5. Mills at Headcorn, Sinkhurst Green and Smarden had either collapsed or been demolished between 1952 and *circa* 1955, with Sissinghurst Mill vanishing a little earlier. Woodchurch is now correctly credited with having a single

---

[10] Some readers may well be wondering why this series was not referred to as a 'Seventh Edition' in keeping with earlier sheets that were labelled as belonging to the Third Edition or to the Fifth Edition. The answer to this lay with the need to comply with recently-agreed NATO mapping protocols. But the Seventh Series did at least stay mindful of its position in the numbering sequence of one-inch editions.

'disused' mill – the 'White Mill'. Its partner – the 'Black Mill' – had almost certainly gone before the war (see above) as had another mill at Bethersden, in 1937. These two pre-war removals meant that these mills had barely scraped in on the New Popular Edition and they are absent from the Seventh Series. Gone cartographically as well is a mill at Lenham Heath that had actually been dismantled as early as 1925. However, as a encouraging sign for the future, the smock mill at Cranbrook – then in the process of being restored to full working order – is shown as having sails, an unchanging position since Cranbrook Mill is today regarded as one of the foremost remaining examples of a traditional windmill in the country.

Returning ever so briefly to the story of the half-inch scale; a last mention of it can be made at this point. It will be remembered that the initial edition of this scale was in some measure based on the Revised New Series and that subsequent printings had been based on updated survey data as it became available. 'Half-Inch Map' sheet 51 *Canterbury* carries a nominal date of publication of 1956 and includes the coverage of its near-contemporary one-inch Seventh Series sheet 172 *Chatham & Maidstone* of 1957. Any expectation that the windmill depiction of the half-inch might mirror that of the one-inch, however, leads to disappointment and serves as a cautionary note to anyone deciding to extract information from a half-inch map. Of the seven mills just discussed as having correctly been removed from the one-inch, five are still shown on the half-inch including Bethersden Mill, blown over in 1937. Moreover, of these five windmills, four – including the one at Bethersden – are depicted using the 'windmill-with-sails' symbol (this half-inch edition having adopted the dual-symbol style of depiction introduced on the one-inch Seventh Series). Elsewhere on half-inch sheet 51 *Canterbury*, the post mill that stood at Ash until it was blown down *circa* 1954 is shown and there are other windmills recorded that had disappeared from the landscape by the mid-1950s. There is good reason for all of this: the compilation diagram of this sheet – dated 1956 – actually shows the sheet to have been based on pre-war Fifth Edition and New Popular Edition material. So, the incorrectness of the sheet then shifts to windmills still standing in 1956, for example, at Canterbury, Eastry and Hawkinge not being shown. In the earlier days of this scale, it had given a good account of the nation's windmills considering the limitations of how much could be squeezed into a sheet and, half a century later, not all sheets of this scale were as flawed as sheet 51. Nevertheless, the use of the half-inch to gather windmill-related information of this period is not to be recommended – and the quarter-inch scale even less so – particularly if one-inch sheets are easily available.

Moving from Kent to anywhere else in the country that once had tower mills – even if in modest numbers – there is a good chance of finding that the Seventh Series recognised the remnants of such mills even though their sails were maybe long gone. The recording of mill towers, even those that were truncated, was something the one-inch now appears increasingly to have wanted to do. Having acquired its new facility for showing mills that, if not completely disused, had indeed lost their sails, it could now record these old towers without badly misrepresenting the situation or taking up space – given the new symbols were unannotated. In light of this, sheet 165 *Weston-Super-Mare*, published in 1958 from a full revision of 1949-56, was able to show nine such tower mills though the reality on the ground for most was not much better than the pile of stones that Brockley Mill, as one of the nine, had become. The sheet also recorded a mill at Chapel Allerton that was about to be restored, by using the alternative symbol (that has sails) but missing was the full-height tower of Felton Mill that stands today in a dominant position overlooking the airfield at Lulsgate, south of Bristol. The point being made here is that by the 1950s Somerset had long since ceased to be known as an area that supported a windmill population, and other parts of the country were following suit. Yorkshire by this time was down to its last mill still really capable of working by sail – the elegant, now restored, example at Skidby. So, this too in the popular imagination was fast losing its standing as an area for having had windmills. Sheet 97 *York* was published in 1960 after a full revision in 1958: it has many 'windmill-without-sails' symbols. They were used here to denote derelict mill towers and even stumps of mills including one that, before its shortening, had supported a very odd composite-style mill at East Cowick. Holgate Mill in York, long since encroached upon by new housing, is accorded one of these symbols. The long-term loss of its sails and its alien residential setting had increasingly made it little more than a faint reminder of the legacy of windmilling all but forgotten in this area. The welcome irony in this instance is that this mill is not only still standing but undergoing restoration – on a mini-roundabout in the middle of a large housing estate. Could it even be that this mill's present condition has, in some very slight measure, been helped by an unbroken history of recognition at a popular map scale?

Yet the cartographic survival of derelict mill towers was not always as comprehensive. Seventh Series Sheet 150 *Ipswich*, revised in 1953-4 and later published in 1956, is fairly parsimonious in its offering of 'windmill-without-sails' symbols. The long since converted full-height tower of a former windmill standing amid other tall buildings at Aldeburgh is understandably not recognised, but the disregarding of another full-height mill tower at Woodbridge (Tricker's Mill) is not so easily justified. However, it is the notable vestiges of post mills in this area that are the key casualty of this and other Seventh Series sheets. Surviving roundhouses are simply not given the same status as mill towers or even tower mill stumps. Thus, surviving post mill roundhouses at Gosbeck, Grundisburgh and Otley are not recognised, though in fairness the one at Swilland is. None of the six symbols on sheet 150 is of a mill 'with sails', yet Buttrum's Mill at Woodbridge was in a similar state to Chapel Allerton Mill at the time of sheet revision in that it was undergoing restoration to quasi-working order. Even so, the mix of windmill types is represented well, even by just six 'without sails' symbols. Other than the roundhouse at Swilland, one derelict post mill, two derelict smock mills and two tower mills are recorded. On neighbouring sheet 137 *Lowestoft*, the situation is different. An understandable disinclination to recognise post mill roundhouses means that many go unrecorded including, for example, the prominent double-storey example at Earl Soham. However, some other less grandiose examples are shown, such as at Ubbeston where the mill body had been taken down in 1924 leaving an unremarkable roundhouse set back from a small road. But, on this sheet, some post mills are recognised as being 'with sails' despite not just being derelict, but in some cases totally gone by this time. Examples include windmills at Wrentham (Carter's Mill) demolished in 1955, Pettaugh dismantled in 1957 and Worlingworth (New Mill) that was demolished in 1952 while others such as the post mill at Peasenhall were dismantled (*circa* 1957) leaving a roundhouse. All of this was on a sheet published in 1954 but reprinted with corrections to 1962. If all this is thought to be harsh on the Survey given that revision for this sheet was completed by 1952, it can be pointed out that a tower mill at Debenham, last worked *circa* 1934, is also shown to be 'with sails' as are both a post mill at Framsden, which also stopped working in 1934, and Saxtead Green Post Mill which worked on until 1947. The survival of sail remnants on these three mills, at least for a few years, is not disputed but it does seem more than slightly disingenuous on the part of the Survey to choose to use a symbol with four symmetrical sails – so implying a mill in pristine condition – for a mill that is instead barely standing and carrying the most meagre of sail remains. The irony in all of this is that if the corrections to 1962 had thought to include an accurate assessment of whether the mills on the sheet still had their sails or not, then the depiction of Saxtead Green Mill would be accurate as this windmill was totally reconstructed by the Ministry of Public Buildings and Works – starting in 1957 and having finished by 1962 (as this writer can happily attest). Looking beyond these post mills, the depiction of tower mills, smock mills and redundant towers on this sheet mostly all conform to the tolerable standard of accuracy that the Seventh Series can, as a rule, offer though this assessment is influenced by the relative durability of tower mills. Other Seventh Series sheets across the country will unquestionably offer narratives for windmill representation that differ slightly, but to explore them all is not the task here. One last fragment of mill-related depiction found on a Seventh Series sheet will have to suffice. Sheet 146 *Buckingham* has six 'without sails' mill symbols and just one 'with sails' symbol for the post mill that survives at Brill. One of the six 'without sails' symbols, strangely, has the annotation Windmill written alongside it. The use of this gothic script was reserved by the Survey expressly for when it chose to record antiquities and it had done this since the 1830s, largely at the larger scales, to note features such as tumuli, long barrows and their archaeological like. Using exactly the same script to annotate the symbol that depicted the old post mill at Pitstone displays an approach to the task of recording windmills that many may find surprising in that it redefines, for one thing, the parameter as to how old something had to be before it could be regarded as an antiquity. At the very least, the Survey is conceding that this windmill is not cutting-edge technology! But why the sole choice of the mill at Pitstone for this style of recognition? Even on sheet 146, there were at least two other post mills at the time of revision in 1950 that would equally have merited 'antiquity' status, either as a very decrepit mill in the last throes of dereliction (Arncott Mill) or simply as an aged and venerable mill survivor (Brill Mill). The answer may lie in the fact that, at the time of this sheet's publication in 1954, Pitstone Mill was being fêted by one author on windmills as the oldest dated example in the country.[11] It would have been a huge coincidence for the Survey to have selected this windmill for its special type of depiction and at the same time to be ignorant of this mill's significance. That the Survey was not just aware of this assertion for Pitstone Mill but felt moved to signify it on sheet 146 is noteworthy.

New Popular Edition and early Seventh Series sheets had come in covers that, in the aftermath of war, looked rather dowdy. Coloured buff and red in equal measure, they had lost the pictorial image used on the Popular and Fifth Editions. The look of the Seventh Series covers was gradually improved: first, by being made an attractive cream and red colour and then by being laminated in white and red. In 1969 all-red covers were brought in, but the one-inch scale was nearing its end. Providing a metric equivalent of this scale

---

[11] Freese, *Windmills and millwrighting*, 145.

became the new order of the day and thus was born the 1:50,000 that remains with us in the form of *Landranger* maps, and whose covers include a colour image of somewhere to be found on the chosen sheet. The detailing on maps went through a period of transition in the move to metrication, but the point of interest for the representation of windmills was that the dual-classification style of depiction was dropped in favour of a single type of symbol that could be used to denote any form of windmill. There were anomalies; the two symbols for depicting 'windmills-with-sails' and 'windmills-without-sails' survived metrication at first, simply due to the fact that early metric sheets were straightforward enlargements of the one-inch sheets. But, even as late as the early 1980s, with the process of redrawing completed, some of the old symbols were still appearing on maps. As one example, sheet 100 *Malton & Pickering* provides a mixture of one old-style symbol – a mill 'without sails' – at Old Malton together with two of the newer symbols. Moving ahead in time and with ever-reducing numbers of windmills to record, we arrive at the present day. Unannotated mill symbols are easy to find on current *Landranger* 1:50,000 sheets covering areas where windmills were once plentiful. Nearly always, these symbols depict redundant mill towers or ones that have been converted into homes; post mills together with smock mills have all but vanished from the landscape. The exceptions are those that have been restored and, if this has happened, or if a surviving mill is sufficiently well known in the neighbourhood or to the national windmill-conservation fraternity, the symbol depicting it may be annotated with a local place name. Sometimes too, the cover photograph used for a *Landranger* sheet is one of a local mill. Skidby Mill, mentioned just a little earlier, is only one example of a windmill that has been the subject of both types of supplementary recognition.

Progress of the larger scales

The story of the one-inch scale from the early days of the Old Series until its replacement by a metric near-equivalent has now been told, and throughout the telling this small scale has received mixed reviews. But, as far as the Ordnance Survey's overall representation of windmills is concerned, its saving grace has been highlighted by the interwoven narrative for the larger scales that show these to have been near faultless in their depiction of mills. What of the larger scales since 1914? The practice of periodically undertaking a complete revision of the 1:2500 on a county-by-county basis was discarded in the 1920s. From this point on, revision at this scale becomes rather piecemeal and hence our capacity for being able to gather windmill data from it to take to a comparison with the one-inch is reduced, largely due to the diluted utility of the six-inch. In place of these once useful large scales, and inextricably bound up with the demands of pre-war and then wartime mapping, new larger scales were introduced, but one in particular – the 1:25,000 – was to emerge as the most useful to those who wanted to search out windmills in the immediate aftermath of war. But, for early sheets of the new large scales, it is likely, sadly, that the provenance of their windmill data was the evidence of the one-inch, meaning that the postwar story of the larger scales is not of so much use or interest to us as the corresponding story from before 1914. Given also the diminishing windmill population, the evidence for mills on later sheets, continuing up to the digitally-assembled ones of today, offer very little extra to what the one-inch sheets can tell us. The wartime years, meanwhile, brought about a barrier of another kind for those using OS maps to search out windmills. In short, this became a proscribed activity. Wartime restrictions, compounded by the non-availability of updated larger-scale maps, meant that anyone innocently trawling round the landscape searching for mills could find themselves in severe trouble with the authorities, as indeed notoriously happened to one diehard windmill enthusiast.

The scale that did become of great use, and which today has in many ways supplanted the 1:50,000, is the 1:25,000 or very nearly a scale of two-and-a-half inches to one mile. The genesis of this new scale dates from before the war as War Office map series GSGS 3906.[12] Sheets of this edition, which were not on sale to the public, only catered for areas deemed to be militarily significant. The environs of Hull were considered to be one such area: sheet 53/44 NE is useful for studying windmill depictions in that it shows Beverley with its neighbouring stretch of open land known as the Westwood. This area has figured before in our narrative as home to five mills in the mid-nineteenth century; three towers apparently survived to the time of this sheet being issued in 1936. All presumably were of full height as each is marked as a 'Trigonometrical Station on a Building' with annotations of *Black Windmill*, *Union Windmill* and *Westwood Windmill* (Plate 51). Complete recognition as named mills was given for all three mills in spite of the fact that at least one of them – Black Windmill – had survived for a long time without any sort of residual cap structure. This was possibly done to differentiate between three trigonometrical stations that were in such close proximity to one another, though quite why the Survey would need stations so close together is debatable. It also does not explain why nearby Thearne Mill, recorded too as a trigonometrical station, should have the more appropriate annotation of *Windmill (Dis.)* when the degree of disuse here was no more obvious, and indeed a lot more recent, than was the case on Beverley Westwood. Elsewhere on this sheet, Walkington Mill and Skidby Mill are recognised by filled-in circles and each has an annotation that includes its place name – unlike Thearne Mill. Despite Walkington Mill being shown as disused, these mills still had substantial if not complete cap structures in place in the mid-1930s and it can be assumed that this might have prevented the setting up of survey instrumentation at either mill. This raises the interesting retrospective question as to whether the selection of a windmill in the nineteenth century to be a trigonometrical station suggests anything of its status and ability to work at the time. No surveyor would presumably want to erect costly equipment on some tall structure that was going to collapse around him if it was already derelict, vibrate if still at work or generally interfere with his lines of sight. It may have been the case that windmills were only used as 'up-stations' or 'intersected stations' so that observations were made *to* rather than *from* them – this happened, after all, with church spires. It also does not have to be forgotten that the permanency of a mill tower could offer a convenient site for a benchmark, useful for the process of contouring, but the Survey would acknowledge these in a different way. To finish the story of windmill-related depiction on sheet 53/44 NE: the mill tower at Bishop Burton was at some time shortened and incorporated into a new house, yet it too is shown on the map as having been a trigonometrical station, only this time the annotation added to it was *Mill House*, suggesting that this mill tower had already been converted by 1936.

---

[12] Two references are appropriate here: first, Richard Oliver, 'Episodes in the history of the Ordnance Survey 1:25,000 map family', *Sheetlines* 36 (1993), 1-27; also, Roger Hellyer, *A guide to the Ordnance Survey 1:25,000 First Series*, London: Charles Close Society, 2003.

Sheets of the postwar 1:25,000 edition, available to the map-buying public from 1945 onwards, also had drab-looking covers to match those of the one-inch maps, except that these were coloured buff and blue. Later, as with the Seventh Series, the buff colour was replaced by cream. Both the design and colour of the 1:25,000 covers were then regularly updated until, eventually, they evolved into visual counterparts of the modern 1:50,000 covers. As to the content of the actual sheets, a narrative for all of them looking at more than half-a-century of individual sheet histories would be even less riveting, or necessary, than a narrative concerned with every nineteenth-century 1:2500 sheet might have been. Instead, the development of the 1:25,000 since 1945 will be appraised relatively speedily. The first postwar publically-available 1:25,000 edition was titled the Provisional Edition. More than twenty years later this was replaced by a Second Series when the Provisional Edition retrospectively became the First Series: this later series was also then rebranded to become the Pathfinder Series that from 1994 then developed to give us *Explorer* maps. This severe editing of the 1:25,000 story is permissible since there is no dependency on this scale for an assessment of the 1:50,000 in the same way that the 1:2500 was necessary for our nineteenth-century assessment of the one-inch. Furthermore, the usefulness of early 1:25,000 sheets as 'snapshot' records of the landscape has been called into question and, as will be seen, the depiction of mills on these sheets in any case lacked coherency.[13] In their favour of course the early 1:25,000 sheets were able to show detail in a way that the one-inch and later the 1:50,000 sheets could not do. Since the mid-1940s, therefore, sheets of this scale have had their devotees for whom this better ability to represent the landscape has meant an extra outlay but not necessitated the purchase of countless sheets for an area, as even the six-inch had required one to do. For a great many people who are interested in all aspects of the landscape, the 1:25,000 has become the scale of choice.

Almost any early 1:25,000 sheet of a district that had once sustained several windmills will have its own story to tell of how it recognised the remnants of these structures. Only from 1996 were symbols used for depicting windmills at this scale. Before then, however, there was nearly always ample space on a map, especially in rural areas, to apply fulsome, often idiosyncratic, windmill annotations alongside innocuous ground plans, just as had happened on the six-inch in the previous century when numbers of windmills had been much greater. On modern *Explorer* sheets the symbol used for a windmill is the same as that found on *Landranger* sheets and its design is such that it caters for all eventualities of windmill status. But, when used at the larger scale, there is a correspondingly greater map space that offers some chance of an annotation also being used. But returning to a time long before 1996, it is almost as if the depleted numbers of windmills and their perceived irrelevance *circa* 1955 allowed those mills that remained to be depicted in very individual ways. This observation could underpin an interesting study of this period when, for just a few years before the heritage movement took hold, the lack of interest in windmills and their relegation in the collective imagination appear to have provided an excuse for not mapping these archaic artefacts with any precision. Such a study would have its appeal, but it is not for us at this point. It is enough to note that the renewed use of annotated ground plans for recognising windmills on 1:25,000 sheets did not generate a consistency in mill depiction this time around any more than it had done half a century before for the

[13] Oliver, *Ordnance Survey maps: a concise guide for historians,* 46-7.

six-inch scale – in fact, much less so. To illustrate this point, a few sheets can be reviewed for their fragile representation of windmills during this period. A copy of sheet TL64 of the *Scale 1:25,000 or about 2½ Inches to 1 Mile Provisional Edition* dated 1960 shows the part of the country where three counties – Cambridgeshire, Essex and Suffolk – meet close to the small town of Haverhill. Like all standard 1:25,000 sheets of the time it represents an area on the ground that measures 10km by 10km and this particular sheet carries notes of provenance showing it to have been:

> *Compiled from 6" sheets last fully revised 1901-24. Other partial systematic revision 1946-52 has been incorporated. Made and published by the Director General of the Ordnance Survey, Chessington, Surrey, 1957. Reprinted with minor corrections 1960.*

Such remarks are hardly guaranteed to ensure confidence in the individual depictions of unsettled landscape features such as windmills, and they serve to reinforce the suggestion made earlier that, for a considerable time around the middle of the twentieth century, the reliability of scales larger than the one-inch was certainly not what it had been a hundred years before. Then, or at least in 1890 when the New Series sheet for this area was issued, eleven windmills were declared as present; this then reduced to nine on the Revised New Series. Sheet TL64 of 1960 has just two depictions of *Windmill (Disused)* including one for a mill at Streetly End that neither the Revised New Series nor the Third Edition deigned to show. This mill had lost its sails in 1895 but changing remits between then and 1960, and 1960 and now, for how a windmill has to appear in order for it to be recorded means that not only was the 1960 sheet justified in recording it, but so also is the latest 1:25,000 sheet (2006). Not very far to the north-west of Haverhill lies the point where, this time, Bedfordshire, Cambridgeshire and Huntingdonshire all meet close to the large village of Gamlingay. When sheet TL25/35 was reissued in 1984, the era of the windmill being a forlorn and superfluous feature in the landscape had passed. Three mills near Gamlingay are recorded on this sheet in different ways with the variation due, as much as anything else, to the use of gothic script seen much earlier in the depiction of Pitstone Mill on the one-inch. Of these three mills, Bourn Windmill – the post mill truly considered to be one of the oldest, if not the oldest in the country – appears as **Bourn Windmill**. Another post mill at nearby Great Gransden is also recognised by **Windmill**, but this time *(Restored)* also appears in normal script. Finally, the windmill just south of Gamlingay is recorded using normal script with *Windmill (Disused)*. Clearly, a distinction is being drawn here between a derelict smock mill deemed to be of no significance and on the verge of collapse and two post mills shortlisted for being acknowledged as worthy antiquities. But, for years before this, no such deference to antiquity status had been forthcoming – quite the opposite in fact – and the anomalies and discrepancies seen in mill depictions over years of postwar 1:25,000 sheets can be attributed to a loss of esteem suffered by windmills. Of the many 1:25,000 sheets that could be cited to endorse this claim of inconstancy over a span of at least twenty years after their introduction in 1945, one example must speak for all. Then, to finish the story of this scale, the latest 1:25,000 sheets will be assessed. Issued in 1954, sheet TA07 records two ruinous tower mills from the East Riding of Yorkshire less than three miles apart. One, at Hunmanby, is annotated as *Hunmanby Windmill (Disused)* while the other to the north of Burton Fleming is annotated *Old Windmill (In Ruins)*. There is a further annotation of *North Burton Mill (Disused)* alongside the second windmill that very probably refers to a small complex of outbuildings that may have housed a steam mill at some point, though none of the earlier larger-scale maps offers evidence for this. There is, in theory, an outside chance that all three annotations can be defended and, moreover, that making a distinction between *(Disused)* and *(In Ruins)* in terms of windmill status was sensibly applied in this case. However, given the variable provenance of the data for this sheet – and indeed all contemporary 1:25,000 sheets – and given also the lack of precise depiction on rather too many other occasions, it would take a special breed of optimist to declare that sheet TA07 gives an accurate account of the windmills then surviving in this corner of Yorkshire; the point being that for too many years this scale lacked reliability in its representation of windmills. Nor was this inability to depict windmills reliably limited to the 1:25,000 (which many would classify as a medium scale). The practice of offering information at much larger scales has continued since 1945 with the retained use of the 1:2500 scale together with other new metric ones. But often these larger-scale sheets do not unambiguously indicate the presence of the few windmills that have survived up to comparatively modern times. Normally this is simply the outcome of a scenario we have seen before, namely that ground plans of mills are readily identifiable if it is known where to look on a map, but without a suitable annotation the windmill becomes invisible. It is certainly unsafe to assume that, even using postwar 1:2500 sheets, all windmills will be identified. Scarborough (Harrison's) Mill worked by sail until 1898 when the four sails that it had left – out of an original complement of six – were removed: it was worked by a gas engine until 1927, but then fell derelict. It was restored *circa* 1990 and given a set of dummy sails.[14] When this full-height mill tower was recorded by 1:2500 sheet TA0388, issued in 1968, the outcome was little better than the total lack of recognition accorded this windmill by the one-inch throughout the entire lifetime of this scale (Plate 52).

As already noted, current *Explorer* 1:25,000 sheets make use of windmill symbols and these are often complemented by annotations that combine a place name with *Windmill*. It would be surprising indeed if there were no caveats to add to this observation, or if there were not a whole raft of issues that needed discussion before these sheets could be put to critical use, but – largely skirting this need to be expansive, even though these are the maps that nowadays most people prefer to use when searching for mills – our study of them will be restricted to a few sheets. This is not only in keeping with the remainder of this chapter, but a course of action that will at least still give us a better than cursory impression of what they can offer in the way of evidence. The village of Gamlingay in Cambridgeshire now appears on *Explorer* sheet 208 *Bedford & St Neots* – edition A1/ of 2008 based on a last revision of 1999, but including selected changes to 2005 and minor changes to 2008. On this sheet, the two post mills at Bourn and Great Gransden retain

[14] The penchant for adorning 'restored' windmills with mock sails – very often just lattices of unseasoned and inadequately-sized timbers – to provide a visual impression not so dissimilar to that of a working windmill is a popular one, and it is a custom that has been practised for many decades. Usefully, dummy sails can serve as a metaphor for issues connected with the concept of 'heritage'. While many would argue that sails such as these keep alive for future generations an idea of what windmills once looked like, others view them as an anathema and an insult to the technological refinement that windmills once possessed. This is a huge area for intellectual concern and there is much serious literature for those who wish to explore it. If time is pressing, however, the reader is directed to two key texts; David Lowenthal, *The past is a foreign country,* Cambridge University Press, 1985; Raphael Samuel, 'Theatres of memory', volume 1 (1994), *Past and present in contemporary culture*, London: Verso.

their gothic-script annotations, probably since both belong to an extremely small number of surviving open-trestle post mills. The map also provides place-name annotations in normal script to accentuate windmill depictions, for example, at Guilden Morden, Potton and Stevington. Also, simple unannotated windmill symbols record mills at Sharnbrook, Steeple Morden and Thurleigh. Of further interest is the antiquity-status annotation of **Mill Mound** appearing at Colmworth. The idea that one-time windmill mounds should be recorded by the Survey is not a new one. It will be remembered that as long ago as the 1840s with the first edition six-inch of Yorkshire, the sheet showing Beverley recognised redundant mill mounds. This renewed instance of recording a mound is by no means an isolated one either. Cases also occur on neighbouring sheets: at Balsham and Wimpole Park on sheet 209 *Cambridge*, at Little Thurlow on sheet 210 *Newmarket & Haverhill*, at Ely on sheet 226 *Ely & Newmarket* and, last of all, at Hoxne and Stradbroke on sheet 230 *Diss & Harleston*. Further afield, in today's areas of marginal mill presence, other instances of this trait are found such as at Highbridge in Somerset on sheet 153 *Weston-Super-Mare & Bleadon Hill*. This particular mound is also recognised at 1:50,000, again by an annotation of **Mill Mound**. As windmill-related features, these one-time mounds now regularly appear at 1:25,000 due, it has to be supposed, to their antiquity status but they appear much less often on 1:50,000 sheets. Yet excavated mill mounds that happen to be mentioned in the archaeological literature relating to windmills, small as this is, are not necessarily recorded by the relevant 1:25,000 sheets. Neither is the important example found at Kirkby Bellars near Melton Mowbray in Leicestershire. This particular mill mound was the one cited by Beresford and St Joseph in their study for the analysis of archaeological remains based on aerial photography.[15] This re-emerging feature of the depiction of windmill mounds on OS maps certainly deserves further consideration, both for their patterns of distribution and for the provenance of the data that the Survey has collected for them.

The evidence of mills found on *Explorer* sheet 208 *Bedford & St Neots* can be added to by studying any sheet of an area where it might be thought that the remains of windmills can be found. Returning to the area of Cambridgeshire and Suffolk, specifically sheet 209 *Cambridge* – revised 1999 with changes to 2007 – to see what more can be gleaned of the ways by which this map series records windmill relics; unannotated symbols are used to record windmills at Ashdon, Foxton, Fulbourn, Hildersham, Ickleton, Madingley and Six Mile Bottom. In contrast, symbols are not used for either of the two house conversions at Barrington and Little Wilbraham: instead, annotations of *The Windmill* are applied. But lest it be thought that maybe all windmills converted into houses are so treated, it can be added that the mills at Hildersham and Ickleton are also now homes. Elsewhere, Harston Mill is recorded only by *The Old Windmill* but, hedging its bets, Linton Mill is recorded by both a symbol and an annotation of *The Windmill*. In addition to using this mix of styles

[15] For their consideration of mill mounds see M.W. Beresford and J.K.S. St Joseph, *Medieval England: an aerial survey*, second edition, Cambridge University Press, 1979, 64-6. A distinction needs to be made between two categories of mound: first, those on which complete windmills were placed, where archaeological examination usually only uncovers discarded clay pipes, worn-out nails that once held the sails together and so on. Second, mounds that once enclosed the wooden trestles of post mills. These, viewed from the air, notably take on the appearance of 'hot-cross buns' because the two main horizontal components of a trestle – the crosstrees – are placed at right angles to one another. Encased in earth, they eventually rot and, with the body of the mill long gone, this leaves the earth above them to collapse into the gaps provided, so leading to indentations in the top of the mound.

for depicting the mills lingering here, and in common with other sheets, something of a distant past when these mills were numerous can be gathered from the plethora of such annotations as *Mill Farm*, *Mill House Farm* and *Millfield Farm*. Sheet 230 *Diss & Harleston* – revised 1999 with changes to 2006 – offers a typical mix of styles of depiction, including the recognition of a post mill at Stanton by **Windmill**. There are also thirteen annotations of *Mill Farm*, nine of *Mill House* and three of *Mill Green*, together with a range of similar descriptions all encouraging the user to infer additional sites of one-time mills. Elsewhere in Cambridgeshire and Suffolk, unannotated symbols and annotations without symbols are found in parishes that are next to one another – at Burwell and Soham on sheet 226 *Ely & Newmarket*, for example. Sheet 211 *Bury St Edmunds & Stowmarket* possesses merely a single symbol (for the post mill at Drinkstone) despite other mill remains spread across this sheet. On sheet 212 *Woodbridge & Saxmundham* Saxtead Green Mill is fully deserving of recognition: so what it gets is the full treatment of a symbol annotated by **Saxtead Mill** and, just in case the user is still unaware of the significance of this windmill, the light-blue English Heritage symbol is applied as well. Meanwhile, on this same sheet the full-height mill tower at Woodbridge (Tricker's Mill) continues to be ignored.

Returning across the country one last time to reconsider sheet 153 *Weston-Super-Mare & Bleadon Hill* – revised 1998 with major changes to 2004 – it is quite evident that, in an area that became marginal for windmills generations ago, updating the depictions for the few mill remains that survive here is an ongoing business. In 1998, the symbol for Chapel Allerton Mill was annotated *Ashton Windmill* and the very derelict coarse stone tower of Kenn Mill kept its mill symbol. With the changes made in 2004, the annotation changed to *Ashton Windmill (dis)* even though, if anything, this fairly basic tower mill is in as good a condition as it has ever been even if, technically, it is incapable of working. The symbol for Kenn Mill disappeared to be replaced by recognition of a new light-industrial park. The Survey could easily have been fooled into retaining its windmill symbol here if it was relying on aerial photography or a casual visual re-inspection of the site, since the stone windmill tower apparently survives here. In all probability, this structure was overlooked as it is now surrounded by warehouses. But this is just as well since Kenn Mill was totally destroyed to be replaced by a stone-faced breeze-block structure built over a reinforced-concrete base. A tablet set into the wall declares it still to be the windmill albeit in 'rebuilt' form. The sad assumption made by someone is, no doubt, that this simulacrum of a mill tower lends a sense of historical continuity and bogus legitimacy to the new development of the area. This sort of liberty taken with the remains of windmills aside, the prevalence of windmills is now such that *Explorer* and *Landranger* sheets can only record the minimal numbers of them left, even in areas that were once heavily populated. Windmills are still being omitted from current maps; usually these are the truncated remains of tower mills or the one-time post mill roundhouses that, by now, have possibly had a longer history as a garden store or the like than ever they had as part of a windmill. Full-height mill towers escape notice as well, even though as individual mills they might have been recognised by earlier editions. It is as difficult to offer a critical analysis for today's maps in respect of their windmill representation as we have seen it to have been for different times in the past. This is not only due to the statistical difficulty of there being fewer mills to assess, but also due to the distorting overlay that comes from the involvement of surviving mills in the heritage movement. In many ways, a more meaningful assessment of windmill representation on 'modern' Ordnance Survey sheets can come from revisiting the sheets of the Seventh Series. The numbers of windmills on these one-inch sheets, dating from the third quarter of the twentieth century, is significantly greater than what is on offer on current sheets, and the recognition of mills then was at a time before their status became badly obscured.

What we are left with today. Well-intentioned and, in most cases, careful conservation of the mills that have survived includes the renovation of Thurne Dyke Drainage Mill. This was carried out initially in 1949 and then again in 1962 just before being photographed here from exactly the same viewpoint as seen in Plate 12. The mill has pristine paintwork but is clearly not meant to work since there are no shutters in the sail-bays. It stands mute in a landscape where once it was operated hidden from all but a few. Today, many hundreds of people pass by it each year oblivious to any distinction between cereal mills and drainage mills. Visually very similar to how it looked when at work, its cultural identity has fundamentally altered, leaving it as one of today's most recognisable icons for those interested in traditional windmills.

A personal conclusion

So, this story ends on the same note with which it started – one of mild criticism towards the one-inch for an inability at the end of its life to give an accurate account of windmills. Of course it was asking a lot that sheets of the Seventh Series be accurate in not depicting mills that might only have disappeared from the landscape just months beforehand. But back then, even if time was not a constraint during the long school holidays, my keenness for cycling the length of a county to visit non-existent mills was not limitless, and there were times when my admiration of the Survey became less than total. Furthermore, my introduction to the idea of oral history centred around such gems as 'Oh yes, I remember my father telling me he saw the mill being pulled down. I imagine it will have been before the war'. Discovering later that 'before the war' could mean anything from about 1920 to around 1950 did little to rescue the situation, but all of this did at least prompt me to find the British Library Map Room – then within the British Museum – with its treasure trove of OS sheets of all scales and from all periods. This reduced the need to be dependent on 'modern' maps and my feelings towards the Survey mellowed. But even by the 1960s, an enthusiasm for windmills and searching them out had one real limitation. The number of mills that had survived represented only a tiny fraction of all those that had once existed. Moreover, only a few of these survivors were able to offer any insight into the conditions of working windmills at a time when they had been a common feature in the landscape. It more and more came to appear that delving into the working lives of mills long gone was going to be a more rewarding approach to take to the study of mills. Since the 1960s, this slant has been complicated by having to sidestep the approach to heritage that dominates the modern nostalgia industry – the practice of smothering the past in a layer of pastiche. Moreover, in presuming that we are able to access any past way of life and render it fully knowable today, we are surely deluding ourselves. Even by assuming a critical stance for the use of primary sources, we are only able to go so far down that road. But to those for whom the journey is important, the evidence offered for mills by the Survey over a time-span of two centuries makes its maps a compelling primary source and one that is much enhanced by the visual immediacy of these maps. Accordingly, it is important when using OS material in a context of being interested in mills – or anything else for that matter – that any limitations that might be inherent are clarified. This is what this book has tried to do. Whatever criticisms of the Ordnance Survey can be construed from reading these pages, it does not have to be forgotten that the windmill as a landscape feature set the Survey the sternest of tests for accurate representation. While it is true that the one-inch has teetered on the brink of deserving harsh words at times, we can let the precision that the larger scales once demonstrated in recording windmills perhaps compensate for this. If this seems reasonable, the representation of the windmill over much of the lifetime of the Ordnance Survey can be seen as cause for quiet celebration.

Part Two

Examples of Representation

Introduction

The second part of this book takes a more practical approach to understanding how mills were mapped by the Ordnance Survey. This time, seeing how the Survey recorded mills is meant quite literally as the approach taken is founded *entirely* on the evidence of the maps. So, while Part Two hopes to support and complement the aspirations to scholarliness of Part One, an added advantage of its approach is that the insights it offers will perhaps be agreeable to more people, free as they are of any theoretical content. Indeed, this part can be read separately from the rest of the book and it offers a model for carrying out further research into the mapping of mills. The narrative of Part One wove information from the literature dealing with the history of windmills with historical details of different OS map series, and it took evidence too from other types of source as seemed appropriate. This linking of, hopefully, something of a grasp of the history of the Ordnance Survey to an interest, nay passion, for windmills has not been tried before in book form to the best of this writer's knowledge. Considering this something of a lacuna – particularly for people whose interest lies with windmills rather than OS maps – the first part of the book was written for those with a foot in either camp, but who see the sense in knowing something of the other. To assist this process of linking maps with windmills, this second part takes a different stance: it selects a number of Revised New Series sheets and for each provides a facsimile map demonstrating the windmills that the sheet recorded as well as some that it did not but that had been shown on the first edition of the 1:2500. Also, windmills not fitting either of these categories but which had been recorded by the Old Series are also indicated. All windmills appearing on either the Revised New Series or the first edition 1:2500 for the area of the one-inch sheet, or both, are listed and any later recognition of them on for example the Third Edition, the Popular Edition, the Seventh Series and the current 1:25,000 is also noted. In all of this, it will be evident that the Revised New Series has been accorded a primacy, so implying – in keeping with Part One – that the end of the nineteenth century should remain our principal focus. But a demonstration of how the numbers of recorded windmills varied from early in the nineteenth century to beyond the end of the twentieth century is also an important part of what this second part of the book is trying to achieve for the chosen sheets and, by extension, for the whole country. Each of the facsimile maps makes use of the symbols seen below for signifying windmills belonging to each of the three categories:

- All windmill symbols, whether annotated or not, appearing on the one-inch Revised New Series 
- All windmills recorded by the first edition 1:2500 but then not by the one-inch Revised New Series 
- All windmills recorded by the one-inch Old Series, but not then later recorded by either the first edition 1:2500 or the one-inch Revised New Series 

Predictably, this seemingly simple recipe for the compilation of windmill information on a Revised New Series facsimile sheet where only three types of mill recognition are used, needs qualification. It is assumed that the windmill record of a Revised New Series sheet, as well as that of a first edition 1:2500 sheet, remains unaltered between sheet states. But the record of a one-inch Old Series sheet can be subject to alteration due to the updating of a sheet to accommodate new railway features. Accordingly, where information is taken from an Old Series sheet, a date for the state used is declared. Sometimes the coverage of a Revised New Series sheet falls within the coverage of a single Old Series sheet; at other times their unrelated sheet lines can mean that four Old Series sheets are needed. Other points needing to be aired, if there is to be a consistency across the facsimile maps when applying the three types of symbol, centre around the nature of the evidence required for windmills that are at the margins of recognition, and also the naming of windmills. Each location logged using any one of the three symbols will have definitely supported a wind-driven structure at some point. Most of these will have been normal corn-grinding mills, but some will have been pumping mills. For the majority of grinding windmills, the word *Windmill* is incorporated into the depiction of the first edition 1:2500, and in most cases into that of the second edition also. For a minority of windmills where the annotations instead centre around the use of a name, say *Union Mill*, corroboratory evidence for each mill – that it was indeed a windmill – has been sought before its inclusion in a map's mill list. This evidence may have been in the form of an unambiguous ground plan for a free-standing tower mill sited on high ground or one of the variety of ground plans for a post mill also sited appropriately, or it could otherwise have been the prime evidence of the one-inch Revised New Series sheet where a standard depiction of a windmill symbol with an annotation of *(Old) Windmill* appears. All mill symbols on a Revised New Series sheet, regardless of whether they are accompanied by an annotation, are listed. An unannotated symbol will at times be recording a windpump, as the larger scales are capable of making clear, or it may be unannotated through lack of map space. Where the explanation is not apparent, one will usually be provided in the notes for that sheet. Very often the second edition 1:2500 will be the first larger-scale sheet to record a pumping mill, but this should come as no surprise when one recalls the 'modern' nature of many of these contrivances. On the large-scale second edition sheets, it frequently happens that a mill depiction lacks any annotation and one is simply presented with a shaded circular ground plan, implying that this windmill has almost certainly fallen into disuse and most likely into dereliction as well. In this situation, a windmill is still considered to have been recognised and so the listing details of that mill will reflect this fact. Very occasionally, the sense of a windmill's continuing presence implied by a sequence of ambiguous depictions from the Old Series, the first edition 1:2500, the first edition six-inch, the second edition 1:2500 and, indeed, the Revised New Series can still mean that discretion has to be exercised as to whether or not the mill concerned deserves to be included in a sheet's listing. Where this occurs, it is brought to the attention of the reader.

The only requirement for gathering and collating the evidence for windmills from the chosen sheets has been the patience to scour each one for any depiction associated with a mill. No other knowledge of individual windmills other than their recognition on the Old Series, the Revised New Series and the first edition 1:2500, together with any subsequent recognition by later maps, has been included here. The only exception to this statement is when some occasional point of interest, highlighted by one of the photographs of mills, is explained. Consequently, this is an exercise anyone can carry out for any of the many sheets that do not feature here. In this spirit of leaving the maps to speak for themselves, the listings of windmills for each sheet are by names that appear on the map. This means that a majority of the mills are listed by the name of their parish – a process made easier by the sheets of the Revised New Series showing parish boundaries. This can sometimes lead to mills not being named after an obviously nearer settlement if the mother parish is sited some way off. An example of this occurring is Southill Mill on sheet 204 *Biggleswade*. Sometimes though a mill *is* named after a sizeable hamlet within a large sprawling parish if confusion could otherwise result, or if a place name other than that of the parish was used as part of a larger-scale annotation. Examples of mills where this has been the case include Shepeau Stow Mill in the parish of Whaplode Drove on sheet 158 *Peterborough*, Ward Green Mill in the parish of Old Newton on sheet 190 *Eye*, Martin Mill in the parish of East Langdon on sheet 290 *Dover* and Sissinghurst Mill in the parish of Cranbrook on sheet 304 *Tenterden*. A complication in the naming of a mill close to a parish boundary can arise if the Revised New Series mill symbol straddles that boundary or is even positioned the wrong side of it, and one is only alerted to this by scrutinising the larger scales. This happens to mills within the parishes of Ripple on sheet 290 *Dover*, Wellingore on sheet 114 *Lincoln* and Wetherden on sheet 190 *Eye*. Conversely, a mill sited in an isolated spot just a few feet inside a parish border can be the cause of the border having deviated from its obvious course to go around the mill. Examples of this happening are at Stanfield and Little Snoring on sheet 146 *Fakenham* and at Ashby St Ledgers on sheet 185 *Northampton*. All of these features can be seen on sheet 318 *Brighton* in the case of one windmill – one of a select group of mills that providence has decided should have a high visibility within the mill literature. This post mill still stands in the parish of Durrington in West Sussex at the border with the neighbouring parish of West Tarring. The first edition 1:2500 has an annotation of *Windmill (Flour)* but on the following edition this becomes *Salvington Mill*, which is misleading on two accounts since Salvington is merely an outlying hamlet within this neighbouring parish of West Tarring. Furthermore, this mill has since gone down in history under its adoptive name of High Salvington Mill, attracting to itself a mythology that owes much to it having been a popular place to walk to in the 1920s after a tearoom had been opened in its roundhouse, and the rumour circulated that during the building of the mill its post had been fashioned from a living tree!

Another factor in the choice of names given to the windmills listed for these sheets is that some, if close to the edge of a sheet, take a convenient name from that sheet rather than a more correct name, such as that of their parish, if this only appears on an adjacent sheet. In other words, all mills are identifiable by a name that appears on the sheet under review. The two windmills listed as Bolting Holme Mill and Potter Hill Mill appearing on the extreme left-hand edge of sheet 114 *Lincoln*, for example, are better known as the two mills of North Collingham parish. But the village of North Collingham itself appears on sheet 113 *Ollerton*. Similarly, Streamdike Mill, listed for sheet 73 *Hornsea*, is better known as Catwick Mill but the actual village of Catwick is comfortably inside the border of the adjacent sheet 72 *Beverley*. Also, Hasling Dane Mill, listed as one of the mills on sheet 290 *Dover*, was better known as Shepherdswell Mill based on the evidence of the first edition six-inch, but as Sibertswold Mill if depending on the evidence of the first edition 1:2500. So whatever liberties have been taken in renaming windmills to suit the principle of each sheet and its mills being autonomous, they are no worse than the identity crisis that this last mill must have suffered! The concern over knowing what name to use for a windmill – that of its parish, something more local or one that has since stood the test of time but which was flawed in the first place – is one that the Survey must also have faced. There is an echo of this too in a perennial problem faced by the windmill literature – the precise identification of long-gone windmills. In circumstances where mills were capable of being moved a few yards to better sites, or to be members of a group of mills whose pattern of individual sites could change subtly between map editions, it can be very easy to confuse mills that stood in close proximity to one another. The standard solution to this difficulty that authors have come up with has been to use national grid references to label each mill and, as the evidence of twentieth-century maps is very much included in this second part of the book, this sensible tactic is employed here. This allows us to differentiate between windmills that have been given the same (parish) name in the mill lists. But two caveats do need to be voiced whenever grid references are used: first, their apparent authority is dependent on the accuracy with which they are applied – in the case of conventional 'six-figure' 'rounded-up' references, this is only to the nearest one hundred metres in both the northings and eastings directions. This still provides a level of accuracy adequate for most people's needs and indeed for the need here to distinguish between possibly up to forty mills or so on a map. But it can occasionally lead to situations where neighbouring mills have exactly the same six-figure reference. An example is the two easternmost of the four mills at Bassingbourn on sheet 204 *Biggleswade*, which were in reality just 43 yards apart. It is telling that these two mills later had varied, possibly mixed up, cartographic histories. For those who would wish for greater precision, the degree of work needed in the use of large-scale mapping and aerial photography or on the ground, or all three, can escalate to a point where a forensic touch is needed if accuracy of placement has to be established to the nearest metre, or even ten metres. The need to use judgement never entirely recedes. Whatever the level of accuracy used, windmills can always be found that are sitting astride grid lines and even if only 'working to six figures' this problem does not disappear. If you were to face north with your back to the surviving roundhouse at Earl Soham in Suffolk, the six-figure reference for the point where you would be standing is different to that of the corresponding spot reached by walking just a few feet to stand facing south. Second, and very much aligned with what has just been said, this system of co-ordinates is merely an artificial device overlaying what is already an artificial device – the map – meaning that when we rush to apply rules of measurability and calculative thinking to the placement of mills, something of the nature of their settings gets overlooked.

It would be heartening to think that the variation in the three types of mill recognition on each of the maps reproduced here could be contrasted one sheet against another. This can be done, but only up to a point. Comparing sheet 190 *Eye* and sheet 70 *Leeds* certainly reveals which was the area that supported many windmills acre for acre in the mid-1890s, and which was the area that, with the foothills of the Pennines not too far away, marked an edge to windmill territory. But it is not just raw numbers of mills on sheets that tell a story. The ratio in each case of the mills recorded by a Revised New Series sheet to those it misses – but which are shown on the first edition 1:2500 – is of interest as this offers a clue to the manner of attrition that windmill populations were subject to in separate parts of the country notwithstanding a key factor here being the dates of the larger-scale first editions. Looking at sheet 21 *Sunderland*, the number of active windmills would appear to be minimal if compared to the number of windmills recorded before 1860 by the earliest of the 1:2500 surveys. Was the Revised New Series wrong to record only two windmills and ignore twenty that had been recorded almost forty years beforehand? Certainly not as much as it might seem when one recognises the time difference between the two editions except that, of the large number of mills recorded mid-century but then later ignored by the Revised New Series, seven survived long enough to warrant their recognition by the second edition 1:2500. The suspicion therefore must be that the Revised New Series was remiss here in not recording at least some of the twenty 'mislaid' mills even though it did recognise one of them, but injudiciously so by use of an early New Series style depiction of annotated ground plan – this was the coastal mill at Whitburn. On other sheets, where the first edition 1:2500 is much closer in time to the Revised New Series, the number of mills shown on the larger but not the smaller scale is correspondingly reduced. Even so, across the range of selected sheets, mills recognised by the second edition 1:2500 but not the Revised New Series certainly exist. This endorses the earlier finding of fallibility for the one-inch and gives some measure of confirmation that its lack of complete authority does extend across the country. Similarly, the incidence of windmills whose appearances on an early Old Series sheet are not repeated by any later edition suggests different stories of mill attrition for different areas. For some sheets there will be very few mills or none at all that fit this third category. Sheet 21 *Sunderland* is not alone in not offering any. Old Series sheets above the Preston-Hull line, it will be remembered, were derivatives of the larger scales and were disinclined to show windmills anyway. Sheet 70 *Leeds* and sheet 73 *Hornsea* could, in theory, have offered Old Series depictions derived from the first edition six-inch for mills that then disappeared, but neither actually does. In contrast, sheet 114 *Lincoln* covers a part of the country to which the 1:2500 survey came relatively late – *circa* 1885 – but also whose Old Series treatment dates as far back as 1824. Unsurprisingly, the number of windmills that vanished in the intervening sixty years – so falling into the third category – is much greater than the number in the second category – those recorded on the first edition 1:2500 but not by the Revised New Series just ten years later. Finally, for sheet 114 *Lincoln*, the number of windmills that were recorded by the Revised New Series – those comprising the first category – suggest that mills were still reasonably active here. Other patterns emerge: distinguishing between grinding mills and pumping mills is not as easily done on the Old Series as it is on the Revised New Series and the larger scales. The use of unannotated symbols by the Old Series to record not only corn-grinding mills but what, in the first half of the nineteenth century, may only have been rather insubstantial pumping structures is revealed by lines of mill symbols along irrigation channels and the 'Cuts', 'Drains' and 'Banks' of the Fens. These flimsy structures often did not survive for very long before having to be replaced and the need for high numbers of them lessened with improvements in drainage as the century unfolded, all of which led to cartographers recording far fewer pumping mills by the time of the Revised New Series. Consequently, the numbers of windmills recorded by the Old Series, but which then vanish by the time of the New Series or the larger scales, can be huge in fenland areas. Sheet 158 *Peterborough* demonstrates this very well: large numbers of pumping mills are strung out for example over Great Postland all of which cartographically, and probably actually, disappear by the time of the Revised New Series.

Looking at the patterns of mills shown on each of the facsimile maps and comparing them with one another, it quickly becomes apparent that a consistency in representation across the country is almost impossible to achieve. It has to be remembered for example that, on the maps showing parts of Yorkshire, the chances of any windmill demolished as late as 1890 being depicted are remote. Both the Revised New Series and the first edition 1:2500 would have come too late for such a mill. The greater part of this county is above the Preston-Hull line so the Old Series sheets covering it show windmills in a very partial manner. This compares sadly with both County Durham, where mills demolished as early as *circa* 1860 are shown despite the poverty of the Revised New Series representation, and Lincolnshire where a mill only had to survive until *circa* 1823 in order to be recognised by the Survey. On a final explanatory note, the numbers of mills shown by the sheet listings as surviving to more recent times do not hide any untoward complications. All depictions on Third Edition, Popular Edition and Seventh Series maps of the listed mills have been noted including, for example, where a mill was described as *'Old'* by either of the earlier editions – abbreviated to ✓*Old* for the Third Edition and *OW* for the Popular Edition – or, more recently, whether or not a Seventh Series map depiction had 'sails'. The use of '✓' on its own signifies the combination of a symbol with a *Windmill* annotation for a Third Edition map but on a Popular Edition map it means the use of just a symbol (remembering that unannotated symbols on these two editions imply different things). Otherwise, for the other map editions the use of a '✓' or a '×' means a suitable recognition for a mill, or none at all. A last point to make is that sheets of the Seventh Series had a fairly long life and, as with sheets of the Old Series, windmills could be depicted on their early states but not on later ones. Whitburn Mill, seen on an early state of sheet 78 *Newcastle upon Tyne* but then excluded from later ones is just one example of a windmill to which this happened. For the more windmill-laden countryside of East Suffolk shown on Seventh Series sheets 136 *Bury St. Edmunds* and 137 *Lowestoft* the mill count certainly diminished over the lifetime of the series. States of both sheets, fully revised between 1950 and 1952 and then 'reprinted with minor changes' in 1962 and 1963, understandably depicted a tower mill dismantled in 1962 at Debenham as well as, for example, post mills at Eye, Pettaugh and Thornham Magna all of which had disappeared between 1955 and 1959. Moreover, these depictions made use of 'with sails' symbols (as briefly discussed in Chapter Six). States of these two sheets were later published in 1969 based on a revision of 1966-7, and they corrected the information offered in respect of these four windmills and others, but for someone who had acquired copies of these sheets in 1966 and based their windmill-discovery holidays on them, the later corrected states came as little subsequent consolation! However, given the severely reduced numbers of windmills after the middle of the twentieth century, this difficulty with the Seventh Series is not one that demands full examination. The Revised New Series sheets selected for this part of the book are highlighted in the sheetlines map, showing just part of the country, which appears on the next page. This abridged version of Figure 1.2 (see page 8) only needs to show the central and eastern parts of the country – the region that contained the great majority of English windmills – since the eighteen sheets have, sensibly, all been taken from this area.

Sheets

21	*Sunderland*
70	*Leeds*
73	*Hornsea*
88	*Doncaster*
104	*Alford*
114	*Lincoln*
146	*Fakenham*
155	*Atherstone*
158	*Peterborough*
185	*Northampton*
190	*Eye*
204	*Biggleswade*
224	*Colchester*
237	*Thame*
257	*Romford*
290	*Dover*
304	*Tenterden*
318	*Brighton*

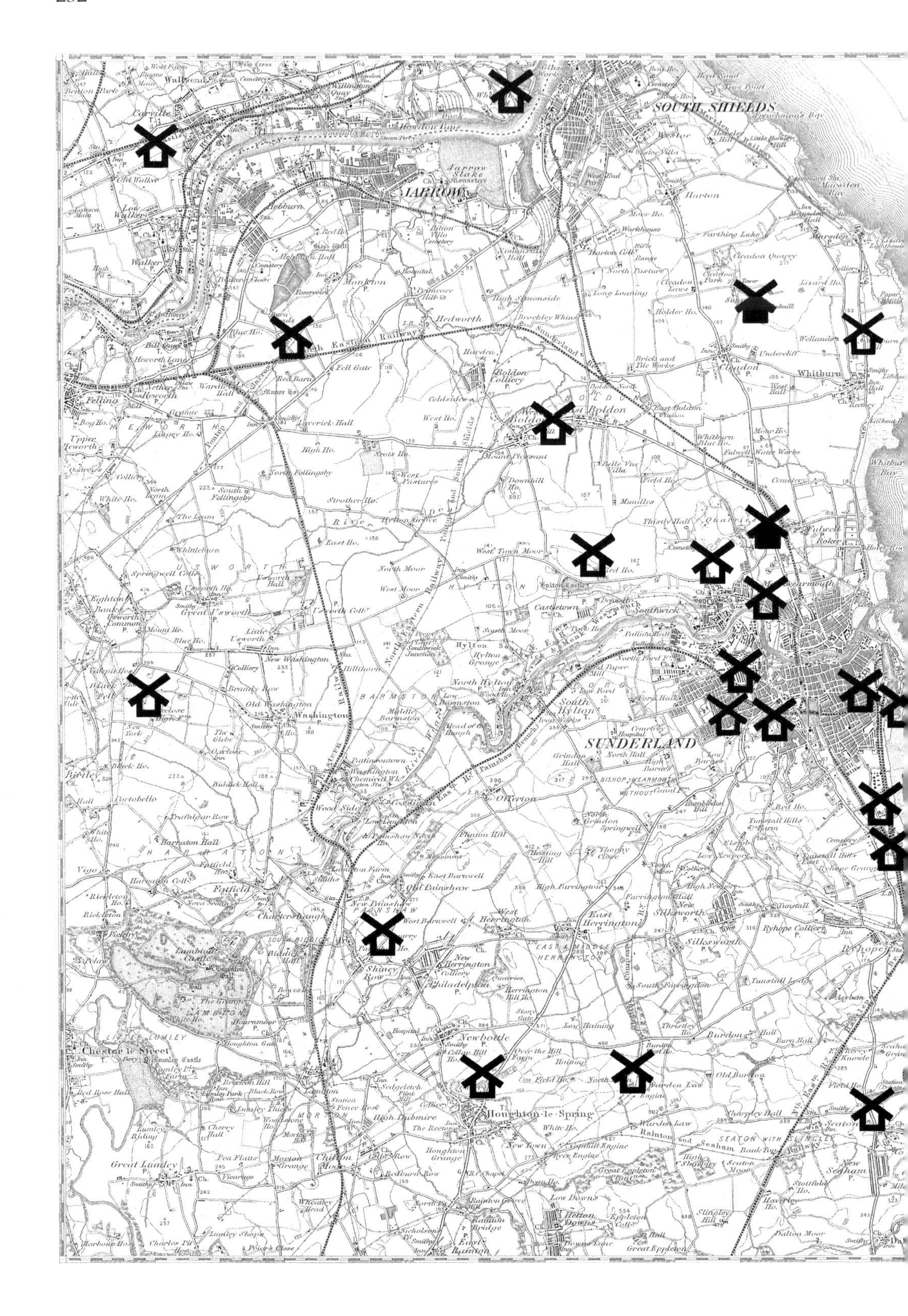
SOUTH SHIELDS
JARROW
SUNDERLAND
Wallsend
Willington Quay
Howdon
Jarrow Slake
Hebburn
Walker
Felling
Monkton
Hedworth
Harton
Westoe
Cleadon
Whitburn
Marsden
Boldon Colliery
East Boldon
Fulwell
Roker
Monkwearmouth
Southwick
Castletown
West Town Moor
Hylton
North Hylton
South Hylton
Great Usworth
Washington
Old Washington
New Washington
Barmston
Offerton
Grindon
Silksworth
Ryhope
Tunstall
East Herrington
West Herrington
New Herrington
Philadelphia
Shiney Row
Penshaw
Fatfield
Harraton Hall
Lambton Castle
Chester le Street
Lumley Castle
Great Lumley
Newbottle
Houghton-le-Spring
Seaham
New Seaham
Burdon
Warden Law
Hetton Downs
East Rainton
Great Eppleton

Sheet 21 *Sunderland*

(see page 232)

Windmill locations as recorded by the Revised New Series *circa* 1895 and/or the first edition 1:2500	19th century recognition			20th century recognition				
	Old Series sheet state 1871	First edition six-inch	Second edition 1:2500	New Series Third Edition	Popular Edition	National Grid Reference	Seventh Series	Current 1:25,000
Cleadon	×	✓	✓	✓*Old*	✓*OW*	NZ389632	✓	✓
Fulwell	×	✓	✓	✓	✓*OW*	NZ392595	✓	✓
Houghton-le-Spring	×	✓	×	×	×	NZ343507	×	×
Hylton	×	✓	×	×	×	NZ359590	×	×
Monkton	×	✓	×	×	×	NZ312626	×	×
Monkwearmouth	×	✓	×	×	×	NZ382588	×	×
Monkwearmouth	×	✓	×	×	×	NZ390585	×	×
Painshaw	×	✓	×	×	×	NZ327529	×	×
Ryhope	×	✓	✓	×	×	NZ409546	×	×
Ryhope	×	✓	✓	×	×	NZ410544	×	×
Seaham	×	✓	✓	×	×	NZ406499	×	×
Sunderland	×	✓	×	×	×	NZ388570	×	×
Sunderland	×	✓	×	×	×	NZ385569	×	×
Sunderland	×	✓	×	×	×	NZ388565	×	×
Sunderland	×	×	×	×	×	NZ404566	×	×
Sunderland	×	✓	×	×	×	NZ409564	×	×
Walker	*Mill*	✓	×	×	×	NZ289658	×	×
Warden Law	×	✓	×	×	×	NZ365503	×	×
Washington	×	✓	✓	×	×	NZ285570	×	×
West Boldon	×	✓	✓	×	✓*OW*	NZ355613	✓	×
Whitburn	*Mill*	✓	✓	×	✓*OW*	NZ407626	✓	✓
Whitehill Point	×	✓	✓	×	×	NZ348669	×	×

This Revised New Series sheet is a good example of one that, while recognising a minimal number of windmills (two), is nonetheless of an area where a much larger number of mills (twenty-two) had been recorded by the first edition 1:2500. Admittedly, the large-scale sheets were issued forty years before the one-inch sheet but even so, nine mills were later recorded by the second edition 1:2500. The Old Series sheet for this area, which would later become the New Series sheet, was dependent on the first 1:2500 sheets so that no third category of windmills exists for this region. In any case, the Old/New Series sheet recognised only two windmills – at Whitburn and Walker – and then only by annotations of *Mill*. The suspicion has to be that the Revised New Series was remiss here in not recognising at least some of the other seven mills seen on the second edition 1:2500 even though one of them *was* recognised, but injudiciously so with its early New Series annotated-ground-plan depiction not being removed for the revised sheet – at Whitburn. Most of the mills recorded by the first edition 1:2500 will have been stone-built tower mills as suggested by the surviving mill shells in this area at Cleadon, East Boldon, Fulwell and Whitburn. But in the 1880s, there were possibly two post mills still working in this area. Monkton Mill was of similar appearance to the post mill seen in Plate 37, and so also was one of the two mills at Ryhope, presumably the one closer to Sunderland since the first and the second 1:2500 editions both annotate this mill as *Stoup Mill*. This, in all likelihood, was a corruption of 'stob mill' – the local description for a post mill. Described in 1899 as probably the last windmill of its kind in the north of England, this post mill was quite derelict by then and only had the remnants of two sails. But it was to survive until 1926 before becoming useful as a source of firewood during that year's General Strike. The list of windmills for Sunderland includes a mill described by the first edition 1:2500 as a *Mortar Mill.* This suggests the presence of a pug mill (see p.69) where edge-stones would have ground lime for mortar. This *could* have been wind powered but the presence of an Engine House implies not. Its inclusion here is very probably inappropriate but the point of interest is the similarity of its ground plan to that of a post mill trestle.

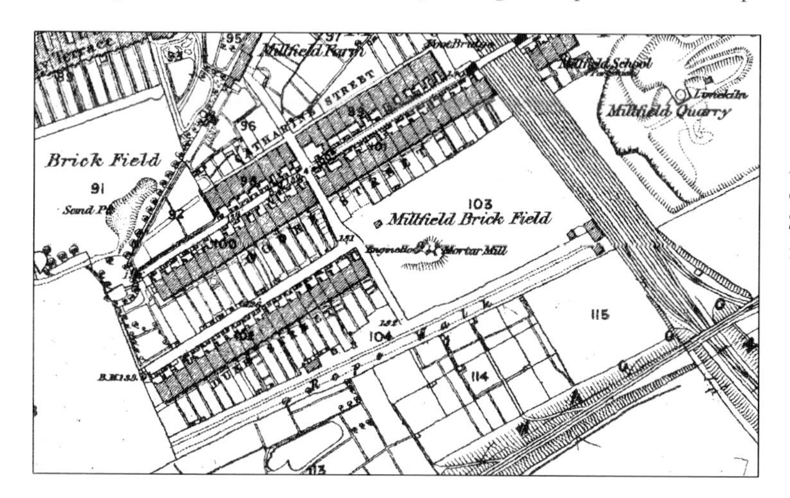

Left: The depiction of a 'mortar mill' at Sunderland on the first edition 1:2500.

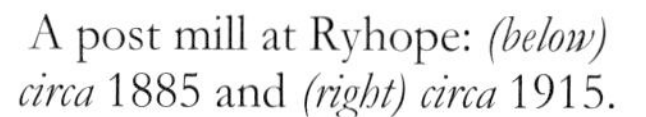

A post mill at Ryhope: *(below) circa* 1885 and *(right) circa* 1915.

Sheet 70 *Leeds*

(see pages 236 & 237)

Windmill locations as recorded by the Revised New Series *circa* 1895 and/or the first edition 1:2500	19th century recognition			20th century recognition				
	Old Series sheet state *ca* 1860	First (derivative) edition six-inch	Second edition 1:2500	New Series Third Edition	Popular Edition	National Grid Reference	Seventh Series	Current 1:25,000
Aberford	×	×	✓	✓*Old*	✓*OW*	SE436369	✓	×
Aberford	×	×	✓	✓*Old*	✓*OW*	SE435359	✓	✓
Askham Richard	✓	✓	✓	×	×	SE542472	×	×
Austhorpe	×	×	✓	✓*Old*	✓*OW*	SE373336	×	×
Barwick in Elmet	×	×	✓	✓*Old*	✓*OW*	SE397370	×	×
Bramham	×	✓	✓	✓*Old*	✓*OW*	SE431433	✓	✓
Church Fenton [1]	×	P	P	×	×	SE506372	×	×
Seacroft	✓	✓	✓	✓*Old*	✓*OW*	SE359363	×	×
Scott Hall	×	×	✓	✓*Old*	×	SE298365	×	×
Sherburn in Elmet	×	✓	✓	✓	✓	SE493336	×	×
Stutton	×	×	✓	✓*Old*	✓*OW*	SE476420	✓	✓
Ulleskelf	✓	✓	✓	✓	✓*OW*	SE519390	✓	✓
Woolas Grange	✓	×	✓	*Old*	✓*OW*	SE543425	✓	✓

[1] The symbol depicting this mill is unannotated. This mill is annotated *Windmill (Pumping)* on the first editions of the 1:2500 and six-inch. It becomes a *Windpump* at a Brick and Tile Works on the second edition 1:2500.

One windmill not included here is Colton Common (Swillington) Mill. This mill is recognised with a circular ground plan on the first edition 1:2500 though there is no accompanying *Windmill* annotation. From the map, it is also not a mill that occupies a hilltop position or is noticeably some distance from a watercourse. As such, on the evidence of the map alone, this is not a windmill that can be listed. In contrast, Barwick in Elmet Mill, though also not explicitly described as a windmill (just like others on these sheets), does occupy a less uncertain topographical position and it does also appear on the early six-inch map. It needs to be remembered that the 1:2500 scale arrived in Yorkshire only after all other counties had been mapped at this scale and it could be that the remnants of other mills, correctly not listed here, had survived until quite late but not then been recorded by the 1:2500 or previously by the Old and New Series sheets which, as discussed in Part One, were particularly unreliable for this area.

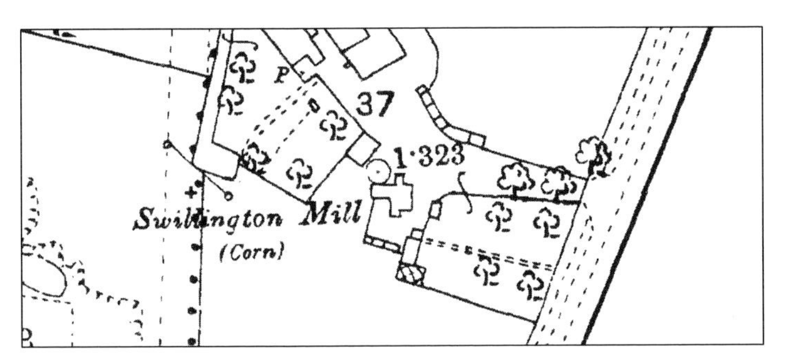

Left: Colton Common (Swillington) Mill as seen on the first edition 1:2500.

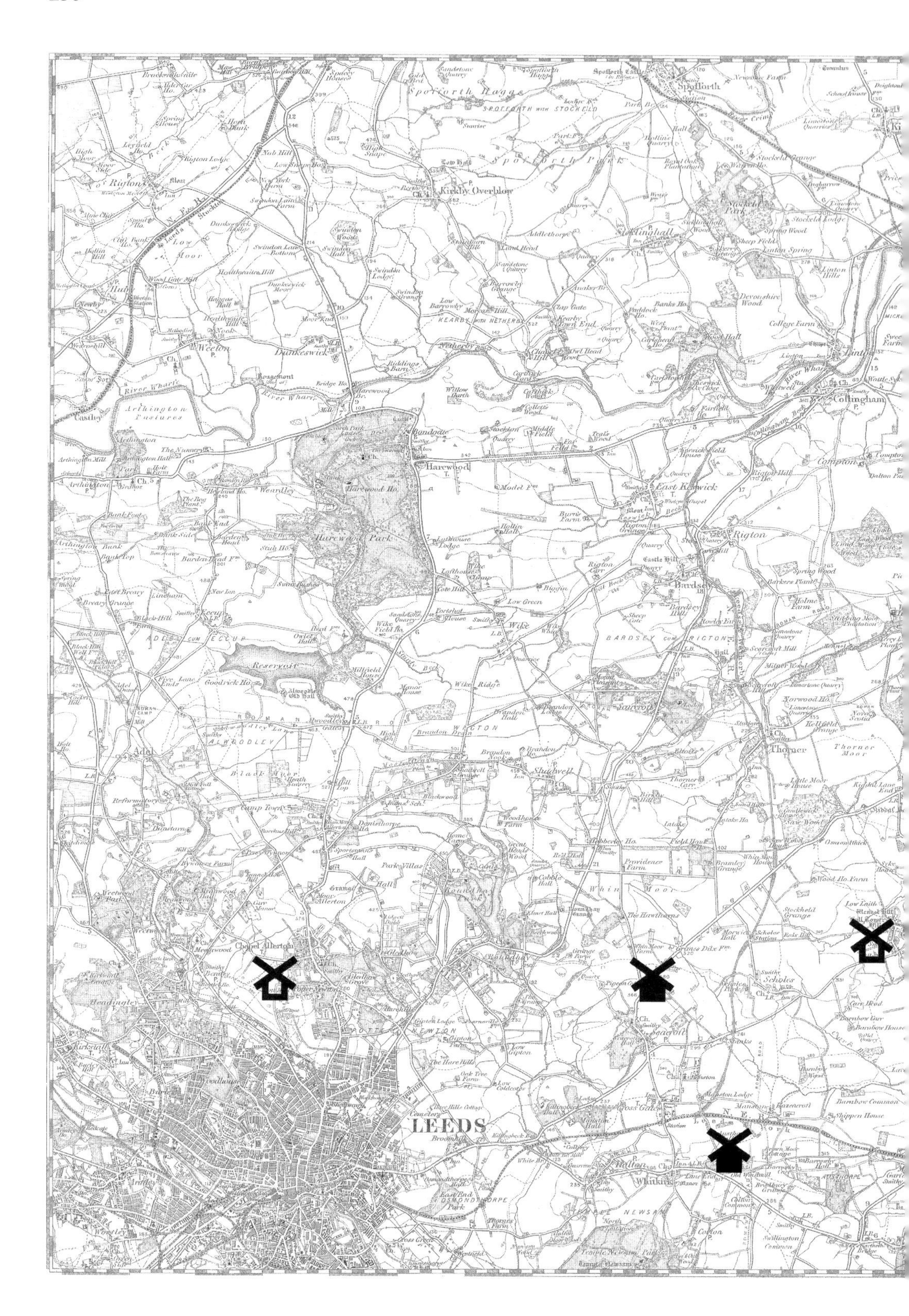
LEEDS
Spofforth
Spofforth Haggs
Kirkby Overblow
Rigton
Weeton
Dunkeswick
Harewood
Harewood Park
Harewood Ho.
River Wharfe
Arthington
Arthington Pastures
Castley
Collingham
Linton
East Keswick
Bardsey
Scarcroft
Thorner
Shadwell
Wike
Eccup
Reservoir
Adel
Alwoodley
Moor Allerton
Chapel Allerton
Headingley
Kirkstall
Roundhay
Seacroft
Whinmoor
Scholes
Halton
Whitkirk
Colton
Temple Newsam
Osmondthorpe
Stockeld Park
Sicklinghall
Weardley
Compton
Thorner Moor
Barwick

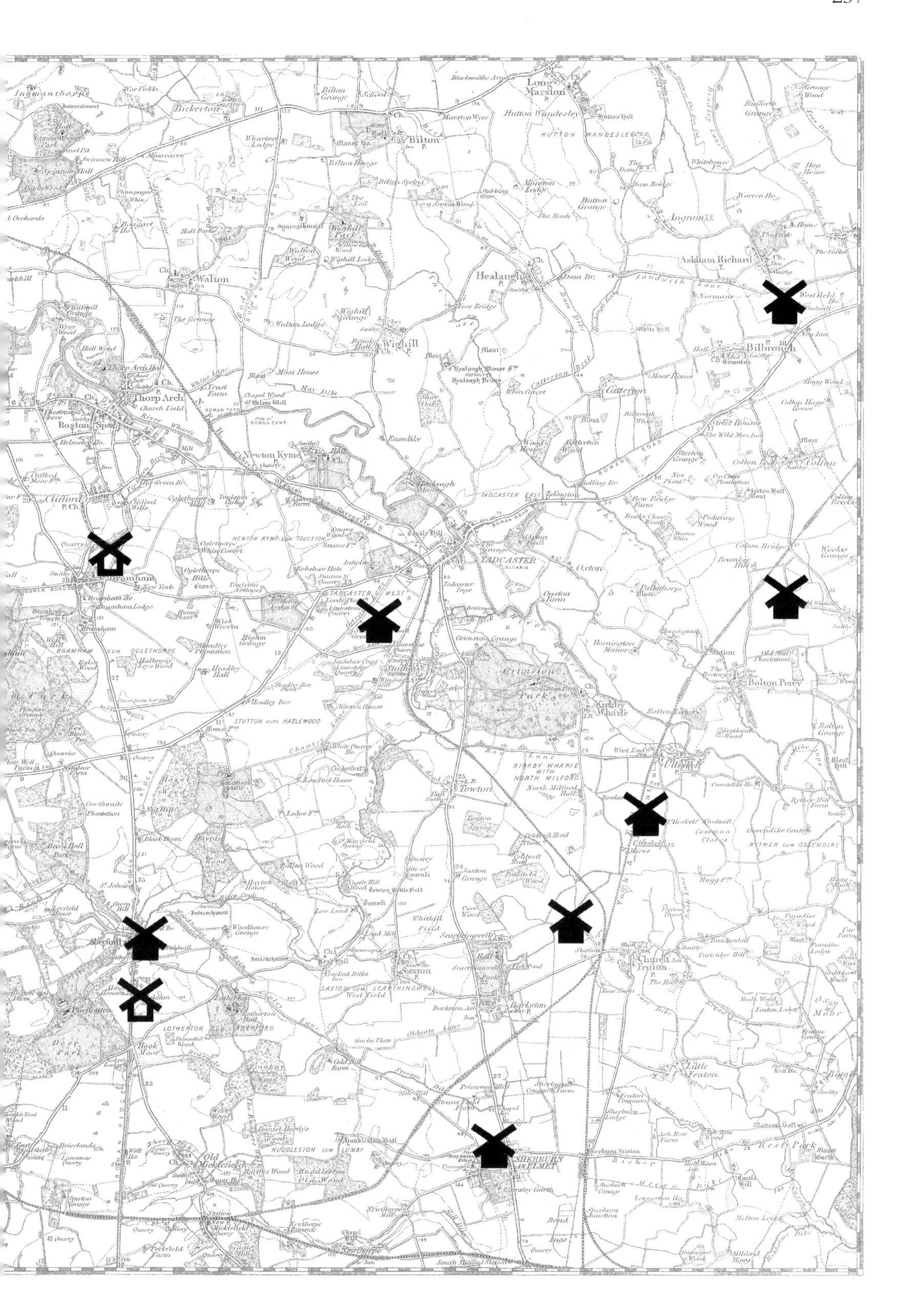
Bickerton
Long Marston
Hutton Wandesley
Bilton
Bilton Grange
Wighill Park
Walton
Healaugh
Askham Richard
Bilbrough
Thorp Arch
Boston Spa
Newton Kyme
Clifford
Bramham
Tadcaster
Oxton
Catterton
Colton
Stutton
Grimston Park
Kirkby Wharfe
Bolton Percy
Ulleskelf
Towton
Hazelwood Castle
Saxton
Scarthingwell
Barkston
Church Fenton
Little Fenton
Aberford
Parlington
Lotherton
Hook Moor
Huddleston
Old Micklefield
Sherburn in Elmet
Sherburn Station
Newthorpe

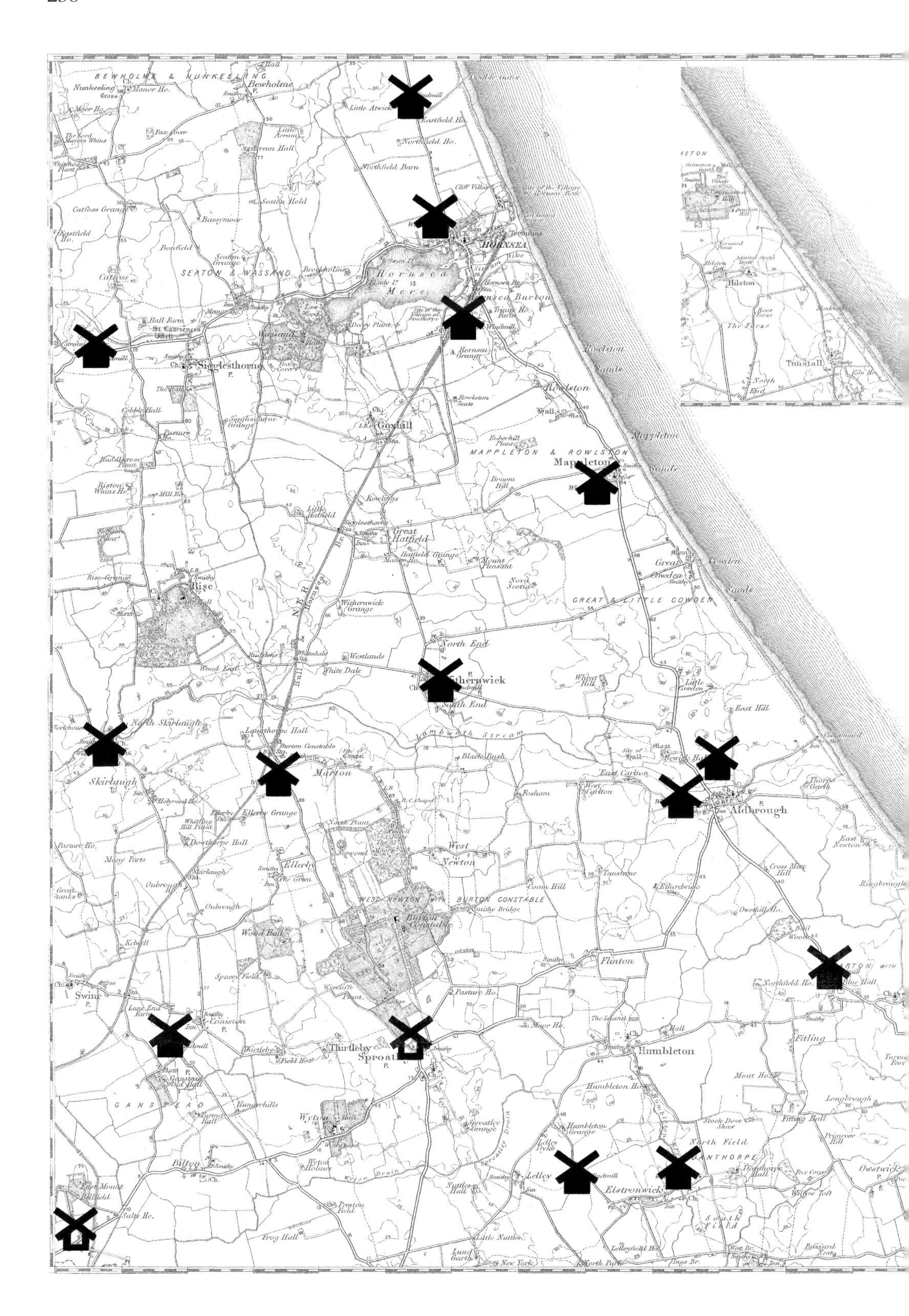
BEWHOLME & NUNKEELING
Nunkeeling
Bewholme
Manor Ho.
Little Atwick
Eastfield Ho.
Northfield Ho.
Northfield Barn
Arram Hall
Catfoss Grange
Seaton Hold
Bassymoor
Cliff Villas
HORNSEA
Hornsea Mere
SEATON & WASSAND
Wassand
Hornsea Burton
Hall Farm
St. Lawrences Well
Sigglesthorne
Rowlston
Goxhill
MAPPLETON & ROWLSTON
Mappleton
Mappleton Sands
Cobble Hall
Pasture Ho.
Huddlecross Plant
Riston Whins Ho.
Mill Ho.
Little Hatfield
Great Hatfield
Hatfield Grange
Mount Pleasant
Nova Scotia
Great Cowden
Cowden Sands
GREAT & LITTLE COWDEN
Rise Grange
Rise
Witherwick Grange
Westlands
White Dale
North End
Withernwick
South End
Whitedale
Wood End
North Skirlaugh
Langthorpe Hall
Burton Constable
Marton
Black Bush
East Hill
East Carlton
Aldbrough
Skirlaugh
Fosham
Ellerby Grange
Dowthorpe Hall
Pasture Ho.
Many Parts
Ellerby
West Newton
WEST NEWTON WITH BURTON CONSTABLE
Coom Hill
Cross Mere Hill
Ringbrough
Owsthill Ho.
Burton Constable
Wood Hall
Flinton
Smithy Bridge
Swine
Coniston
Spacey Field
Pasture Ho.
Northfield Ho.
Thirtleby
Sproatley
Humbleton
Humbleton Ho.
Fitling
Moat Ho.
Longbrough
Fitling Hall
GANSTEAD
Hungerhills
Wyton Hall
Sproatley Grange
Humbleton Grange
Lelley Dyke
Stock Dove Shaw
North Field
DANTHORPE
Danthorpe Hall
Primrose Hill
Owstwick
Bilton
Wyton Holmes
Lelley
Elstronwick
Nuttles Hall
South Field
East Mount
Bellfield
Salts Ho.
Preston Field
Frog Hall
Little Nuttles
Lund Garth
New York
North Park
Lelleyfield Ho.
Ings Br.
West Br.
Blizzard Nest
Hilston
Tunstall
North End
The Furze

Sheet 73 *Hornsea*

(see page 238)

Windmill locations as recorded by the Revised New Series *circa* 1895 and/or the first edition 1:2500	19th century recognition			20th century recognition				
	Old Series sheet state *ca* 1860	First (derivative) edition six-inch	Second edition 1:2500	New Series Third Edition	Popular Edition	National Grid Reference	Seventh Series	Current 1:25,000
Aldbrough	×	✓	✓	✓	✓*OW*	TA237386	×	×
Aldbrough	*Mill*	✓	✓	✓	×	TA243391	×	×
Danthorpe	✓	✓	×	×	×	TA236325	×	×
Ellerby	✓	✓	✓	✓	✓	TA170390	×	×
Ganstead	×	✓	✓	✓*Old*	×	TA152348	×	×
Garton	×	✓	✓	✓*Old*	✓*OW*	TA261357	✓	✓
Hornsea	×	✓	✓	✓	✓	TA198479	✓	×
Hornsea	×	✓	×	Pump	×	TA201463	×	×
Lelley	×	✓	✓	✓	✓	TA219326	✓	✓
Little Atwick	*Mill*	×	×	×	×	TA193498	×	×
Mappleton	✓	✓	✓	✓	✓*OW*	TA224437	✓	✓
Salts House	×	✓	×	×	×	TA135318	×	×
South Skirlaugh	×	✓	✓	✓	✓*OW*	TA141395	×	×
Sproatley	×	✓	×	✓*Old*	×	TA192347	×	×
Streamdike Bridge	×	✓	✓	✓	×	TA140459	×	×
Withernwick	×	✓	×	×	×	TA197405	×	×

This area is typical of many now largely denuded of windmills. Those left are all tower mill shells with post mills now unknown this far north, although the roundhouse belonging to nearby Wetwang Mill was still visible in the late 1970s.

Above: Atwick Road Mill, Hornsea *circa* 1907.

Right: South Skirlaugh Mill in 1907.

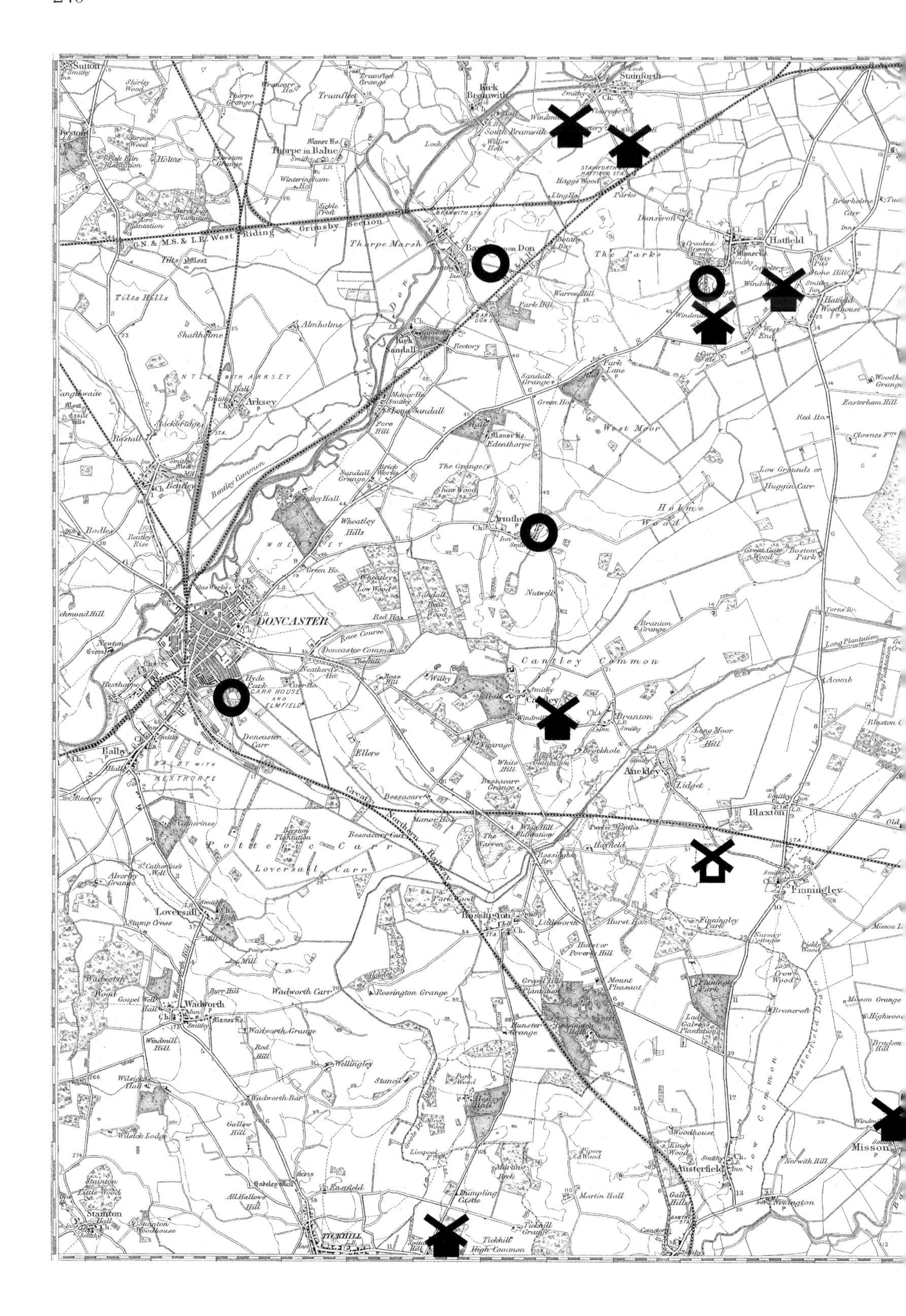

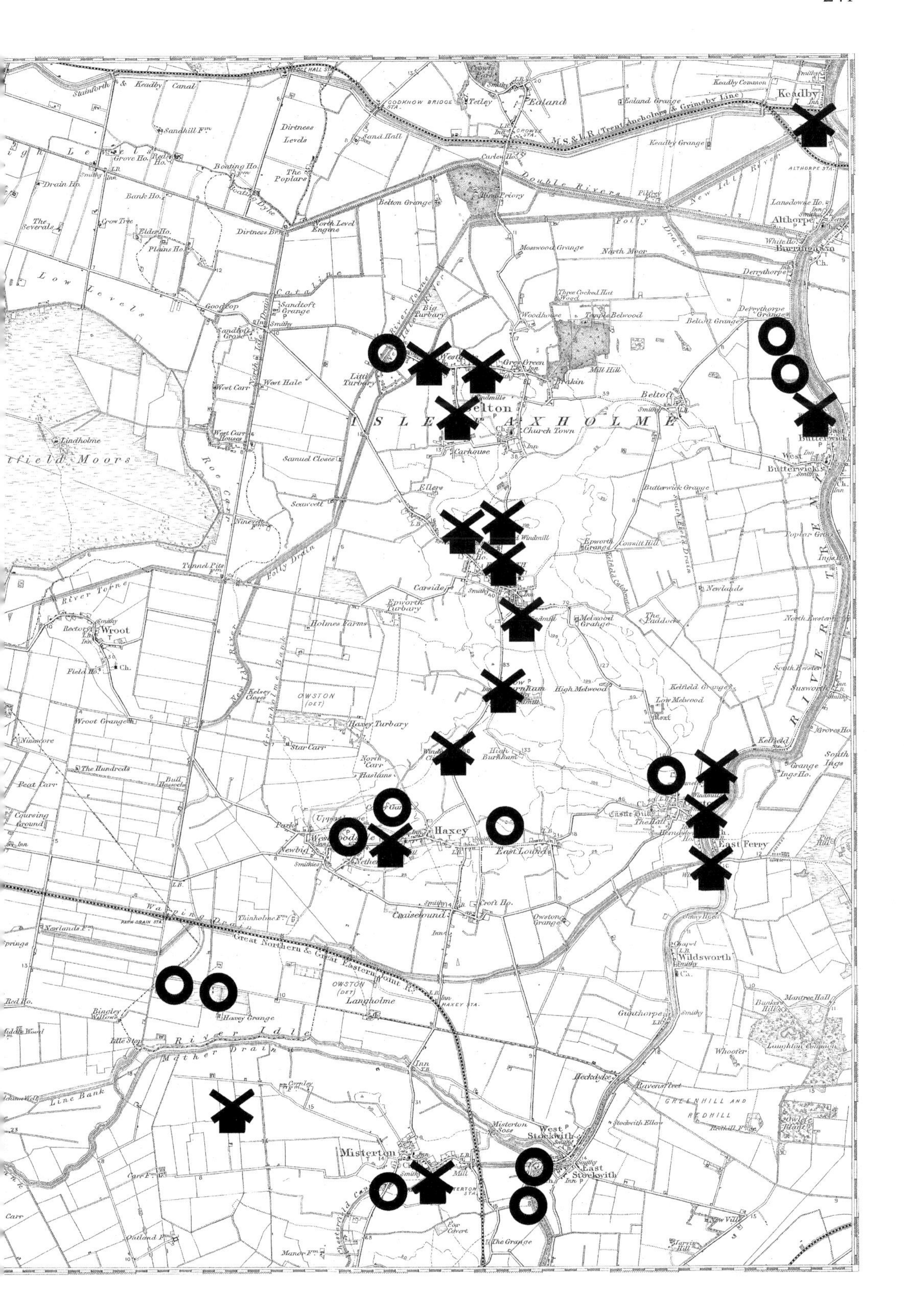
Stainforth & Keadby Canal
Sandhill Fm.
Dirtness Levels
The Poplars
Sand Hall
Godnow Bridge
Tetley
Crowle
Ealand
Ealand Grange
Keadby Common
Keadby
Keadby Grange
M.S.& L.R. (Trent Ancholme & Grimsby Line)
Double Rivers
New Idle River
Althorpe Sta.
Lansdowne Ho.
Althorpe
Burringham
Derrythorpe
Grove Ho.
Red Ho.
Boating Ho.
Drain Ho.
Bank Ho.
The Severals
Crow Tree
Elder Ho.
Plains Ho.
Dirtness Br.
North Level Engine
Belton Grange
Mosswood Grange
North Moor
Low Levels
Goodcop
Sandtoft Grange
Big Turbary
Little Turbary
Woodhouse
Temple Belwood
Three Cocked Hat Wood
Mill Hill
Beltoft Grange
Derrythorpe Grange
Beltoft
West Hale
West Carr
Belton
Church Town
Carhouse
ISLE OF AXHOLME
West Carr Houses
Lindholme
Hatfield Moors
Samuel Closes
Scawcett
Ellers
East Butterwick
West Butterwick
Butterwick Grange
Ninevah Fm.
Folly Drain
Tunnel Pits Fm.
Epworth Grange
Carside
Epworth Turbary
Holmes Farms
Melwood Grange
The Paddocks
Newlands
Wroot
River Torne
Field Ho.
Kelsey Closes
Owston (Det)
High Melwood
Low Melwood
Moat
Kelfield Grange
Susworth
Wroot Grange
Haxey Turbary
Star Carr
The Hundreds
Bull Hassocks
Peat Carr
Coursing Ground
North Carr
High Burnham
Kelfield
Grange Ings Ho.
River Trent
Haxey
Newbig
Westwoodside
East Lound
Castle Hill
The Hall
East Ferry
Craiselound
Croft Ho.
Owston Grange
Thinholme Fm.
Newlands Fm.
Great Northern & Great Eastern Joint Ry.
Wildsworth
Owston (Det)
Langholme
Haxey Sta.
Haxey Grange
Bingley Willows
River Idle
Mother Drain
Gunthorpe
Manor Hall
Whooter
Line Bank
Heckdyke
Ravensfleet
Greenhill and Redhill
Misterton Soss
West Stockwith
East Stockwith
Misterton
Misterton Sta.
Fox Covert
The Grange
Manor Fm.
Carr Fm.
Oatland Fm.

Sheet 88 *Doncaster*

(see pages 240 & 241)

Windmill locations as recorded by the Revised New Series *circa* 1895 and/or the first edition 1:2500	19th century recognition			20th century recognition				
	Old Series sheet states *ca* 1862	First (derivative) edition six-inch	Second edition 1:2500	New Series Third Edition	Popular Edition	National Grid Reference	Seventh Series	Current 1:25,000
Belton	✓	✓	✓	✓	✓	SE770075	✓	✓
Belton	✓	✓	✓	✓	×	SE778070	×	×
Belton	✓	✓	✓	×	×	SE777069	×	×
Cantley	✓	✓	✓	✓	✓*OW*	SE632018	✓	✓
Epworth	✓	✓	✓	✓	✓	SE777047	✓	✓
Epworth	✓	✓	✓	✓*Old*	✓*OW*	SE782046	×	×
Epworth	✓	✓	✓	✓	✓	SE781045	✓	✓
Epworth	✓	✓	✓	✓	✓	SE785034	✓	✓
Finningley	✓	✓	×	×	×	SK657995	×	×
Hatfield	✓	✓	✓	✓	✓*OW*	SE660083	✓	✓
Hatfield	✓	✓	✓	✓*Old*	✓*OW*	SE671088	✓	✓
Haxey	✓	✓	✓	✓*Old*	✓*OW*	SK761997	✓	✓
Haxey	✓	✓	✓	✓	✓	SE773012	✓S	✓
Keadby	×	✓	✓	✓	✓	SE836111	×	×
Low Burnham	×	×	✓	✓	×	SE781021	×	×
Misson	✓	✓	✓	✓	✓	SK688952	×	×
Misterton [1]	×	×	✓	×	×	SK735954	×	×
Misterton	✓	✓	✓	✓	×	SK769943	×	×
Owston	×	✓	✓	✓	✓	SE817005	×	×
Owston	✓	✓	✓	✓	✓	SE817004	×	×
Owston	✓	✓	✓	✓	✓	SK815992	✓	✓
Spital Hill	×	×	✓	✓	✓	SK612933	×	×
Stainforth	✓	✓	✓	✓	✓	SE636115	×	×
Stainforth	✓	✓	✓	✓*Old*	✓*OW*	SE645113	×	×
West Butterwick [2]	✓	✓	✓	✓	✓*OW*	SE836066	✓	✓

[1] The symbol depicting this mill is unannotated. This mill was not recorded by the first editions of the larger scales but was later recorded by the second edition 1:2500.

[2] The circular ground plan for this windmill appears in a loose complex of buildings, all of which are collectively annotated as *Flour Mill* so making this one of more than a few instances where one might think that the windmill status of this mill is better expressed by the one-inch scale rather than the 1:2500.

As the first of the selected sheets to lie south of the line between Preston and Hull, sheet 88 *Doncaster* offers examples for each of the three categories of windmill depiction that the facsimile maps of Part Two have been designed to distinguish between. This is the first sheet from Part Two, in other words, that includes mills recorded by the Old Series (in this case partly by a sheet from the Lincolnshire Map of 1824) that then vanished before the first 1:2500 survey took place. The numbers of mills shown for each of the three categories are fairly small for this sheet compared with those seen on sheets for areas further south, and the distributions of the mills in each are uneven too. Unlike the mills recorded only by the relevant sheets of the Old Series (whose initial states offer a large date range), a clear majority of the windmills recorded by the Revised New Series, for example, are located close together on the Isle of Axholme or on the nearby bank of the River Trent.

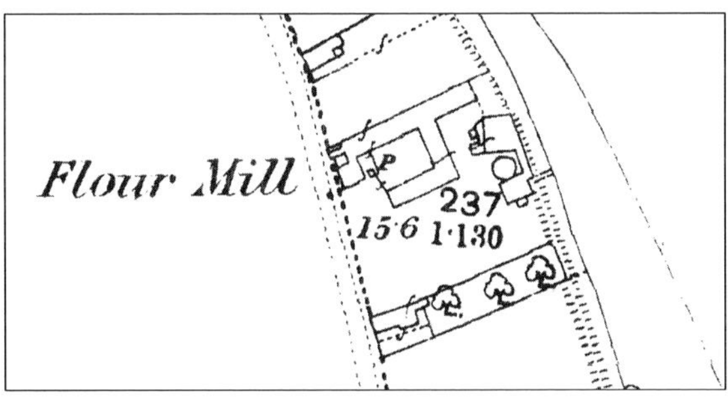

Above: West Butterwick Mill very close to the west bank of the River Trent *circa* 1905.

Left: The milling complex at West Butterwick as seen on the first edition 1:2500.

Two views of the post mill at Stainforth taken in different seasons and probably within three years of one another in the period just before the First World War. The mill is seen at rest *circa* 1911 *(left)* when it is carrying a pair of roller sails and a pair of shuttered spring sails. By *circa* 1913 *(right)*, the roller sails have gone – the straps on this sort of sail were not hard-wearing – and an old-fashioned set of cloth sails put in their place. These were cheap and sensitive to light winds, albeit more awkward to handle. Even so, at this time of adapt-and-make-do, these cloth sails have kept their shuttered leading edges.

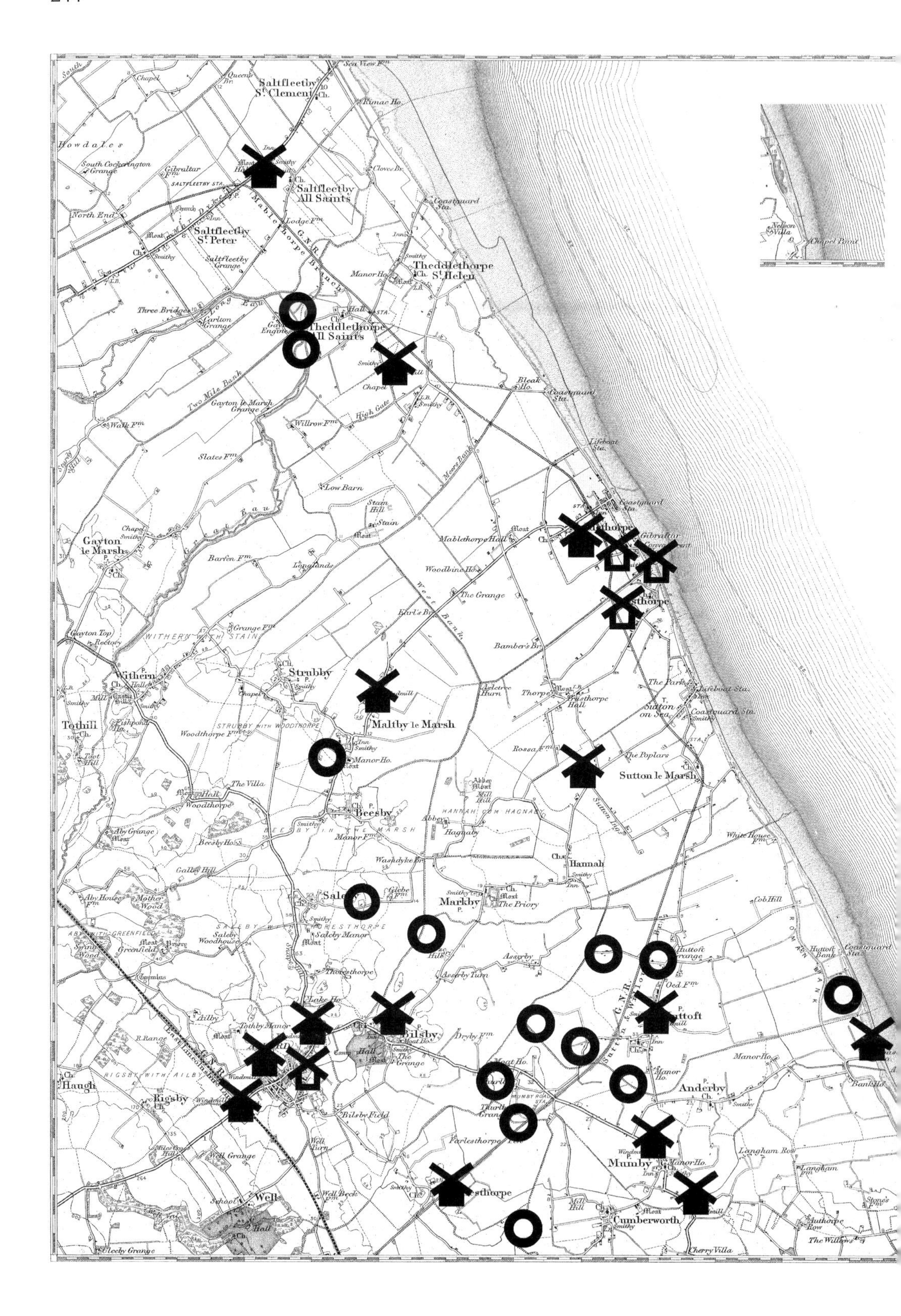

Sheet 104 *Alford*

(see page 244)

Windmill locations as recorded by the Revised New Series *circa* 1895 and/or the first edition 1:2500	19th century recognition			20th century recognition				
	Old Series sheet state *ca* 1858	First edition six-inch	Second edition 1:2500	New Series Third Edition	Popular Edition	National Grid Reference	Seventh Series	Current 1:25,000
Alford	✓	✓	✓	✓	✓	TF445755	✓	×
Alford	✓	✓	✓	✓	✓	TF449759	×	×
Alford	✓	×	✓	×	×	TF454759	✓S	×
Alford	✓	✓	✓	✓	✓	TF457765	✓S	✓
Anderby	×	×	P	Pump	Pump	TF550763	WP	×
Bilsby	✓	✓	✓	✓	✓	TF470766	✓	✓
Farlesthorpe	×	×	P	Pump	×	TF480739	×	×
Huttoft	×	✓	✓	✓	✓	TF514767	✓	✓
Mablethorpe [1]	×	P	P	Pump	Pump	TF502845	WP	×
Mablethorpe	×	P	P	Pump	×	TF505844	×	×
Maltby le Marsh	×	✓	✓	✓	✓	TF469820	✓	✓
Mumby	✓	✓	✓	✓	✓	TF514746	×	×
Mumby	✓	✓	×	×	×	TF520738	×	×
Saltfleetby All Saints	✓	✓	✓	✓	×	TF451906	×	×
Sutton le Marsh [2]	×	P	P	Pump	×	TF503808	×	×
Th'thorpe St Helen	✓	✓	✓	✓	✓	TF471873	✓	×
Trusthorpe	✓	×	✓	×	✓	TF513840	×	×
Trusthorpe	×	×	×	×	×	TF513840	×	×

[1] The symbol depicting this mill is unannotated. The annotation used by the large scales for this mill, and for its near neighbour, is *Windmill (Pumping)*.

[2] The symbol depicting this mill is unannotated. Again, the large scales use an annotation of *Windmill (Pumping)* here. At Anderby Creek and Farlesthorpe, unannotated windmill symbols are used to depict windpumps that are recorded for the first time by the second editions of the large scales. Many of the earlier Old Series symbols are annotated with *Water Engine* or just *Engine*.

Left: Bilsby Mill *circa* 1904. Distinctly visible in the brickwork of the tower is the original level of the curb. This shows the extent to which the tower had been raised at some point in the past to improve the flow of wind to the sails.

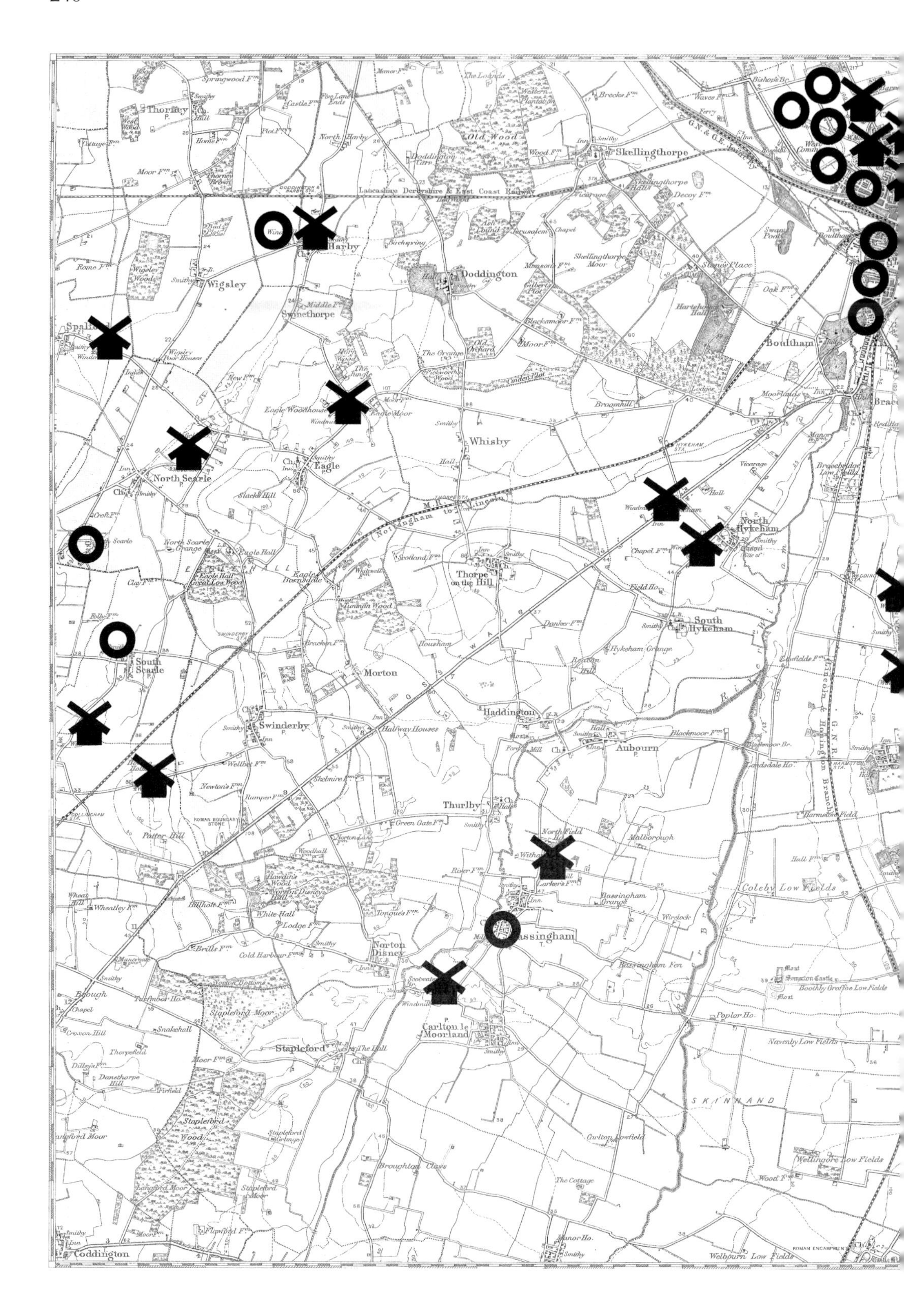
Thorney
Springwood Fm
Skellingthorpe
Old Wood
North Harby
Harby
Doddington
Wigsley
Swinethorpe
Whisby
Eagle
Eagle Moor
North Scarle
Boultham
North Hykeham
Thorpe on the Hill
South Hykeham
South Scarle
Morton
Swinderby
Haddington
Aubourn
Thurlby
Bassingham
Norton Disney
Carlton le Moorland
Stapleford
Stapleford Moor
Stapleford Wood
Coddington
Coleby Low Fields
Wellingore Low Fields
Welbourn Low Fields
Nottingham to Lincoln
Lancashire Derbyshire & East Coast Railway

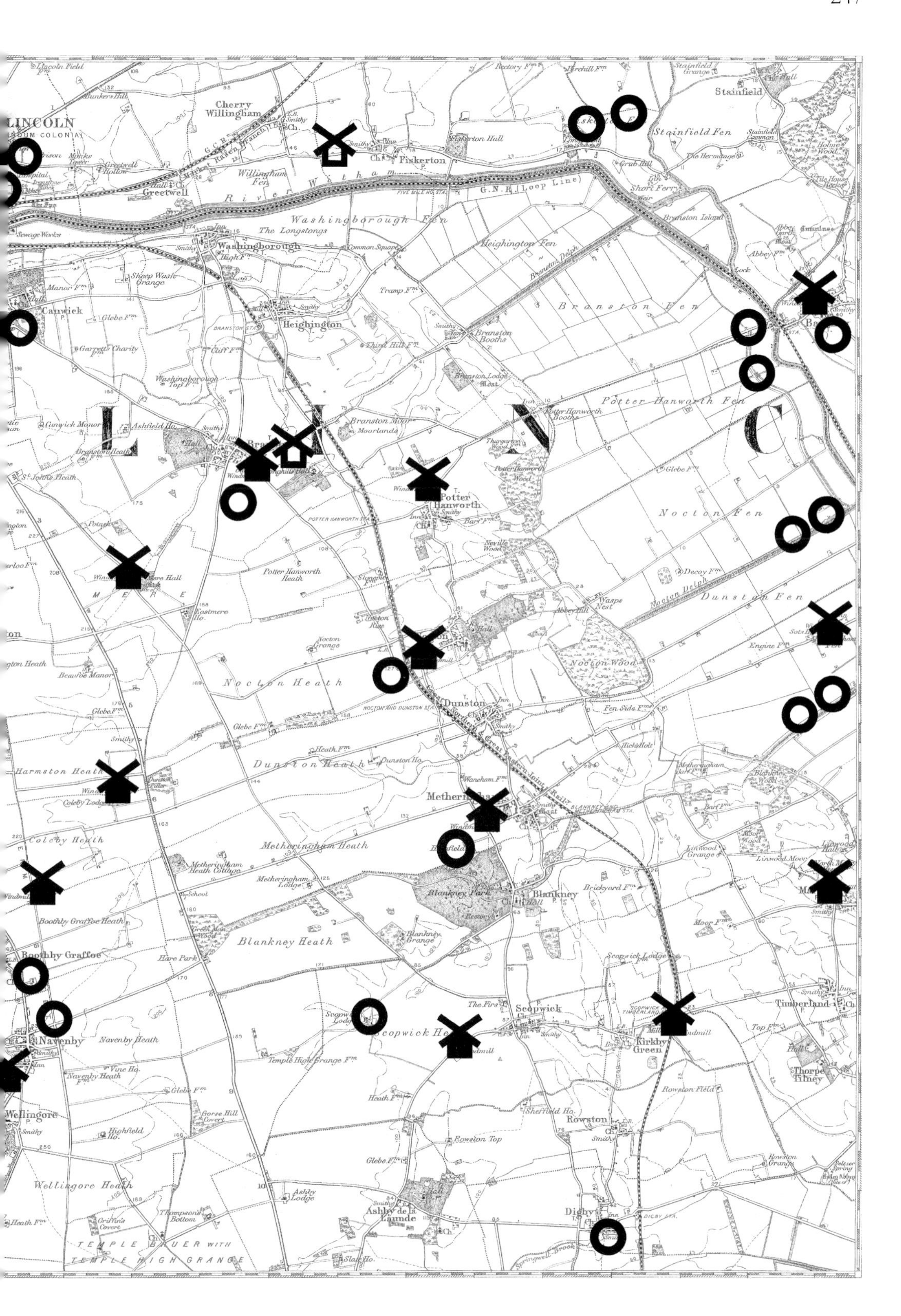
LINCOLN
Cherry Willingham
Fiskerton
Fiskerton Hall
Stainfield
Stainfield Fen
Greetwell
River Witham
G.N.R. (Loop Line)
Washingborough Fen
The Longstongs
Washingborough
Heighington Fen
Canwick
Heighington
Branston Fen
Branston Booths
Potter Hanworth Fen
Potter Hanworth
Nocton Fen
Dunston Fen
Nocton Heath
Nocton Wood
Dunston
Dunston Heath
Harmston Heath
Metheringham Heath
Coleby Heath
Blankney Park
Blankney
Blankney Heath
Boothby Graffoe
Navenby
Navenby Heath
Scopwick
Kirkby Green
Timberland
Thorpe Tilney
Rowston
Wellingore
Wellingore Heath
Ashby de la Launde
Digby
TEMPLE BRUER WITH TEMPLE HIGH GRANGE

Sheet 114 *Lincoln*

(see pages 246 & 247)

Windmill locations as recorded by the Revised New Series *circa* 1895 and/or the first edition 1:2500	19th century recognition			20th century recognition				
	Old Series sheet states *ca* 1830	First edition six-inch	Second edition 1:2500	New Series Third Edition	Popular Edition	National Grid Reference	Seventh Series	Current 1:25,000
Bardney [1]	✓	×	✓	✓	✓	TF120696	×	×
Bassingham	✓	✓	✓	✓	×	SK915604	×	×
Bolting Holme	✓	✓	✓	✓	✓*OW*	SK837627	×	×
Branston	✓	✓	×	✓*Old*	✓*OW*	TF026670	×	×
Branston	✓	✓	×	×	×	TF028672	×	×
Carlton le Moorland	✓	✓	✓	✓	✓	SK897585	×	×
Coleby	×	✓	✓	✓	✓	SK990601	WP	×
Coleby [2]	×	×	×	Pump	Pump	TF003618	×	×
Eagle	✓	✓	✓	✓*Old*	×	SK883679	×	×
Fiskerton	✓	✓	×	×	×	TF040720	×	×
Harby	×	✓	✓	✓	✓*OW*	SK877708	✓	✓
Kirkby Green	×	✓	✓	✓	✓*OW*	TF094578	×	×
Lincoln	✓	✓	✓	✓	✓	SK970726	×	×
Lincoln	✓	✓	×	✓*Old*	✓*OW*	SK971723	✓	×
Lincoln	✓	✓	✓	✓	✓	SK971722	×	✓
Lincoln	×	×	×	×	×	SK972725	×	×
Martin	×	✓	✓	✓	✓	TF123600	×	×
Mere Hall [3]	×	×	×	×	×	TF006652	×	×
Metheringham	×	✓	✓	✓	✓	TF064612	✓	✓
Metheringham Fen	×	✓	✓	✓*Old*	✓*OW*	TF122643	×	×
Nocton	×	✓	×	×	×	TF053639	×	×
North Hykeham	✓	✓	✓	✓	✓*OW*	SK934663	×	×
North Hykeham	×	✓	✓	✓	✓	SK940656	✓	×
North Scarle	×	✓	✓	✓	✓*OW*	SK855671	×	×
Potter Hanworth	×	✓	✓	✓	✓*OW*	TF055666	×	×
Potter Hill	✓	✓	✓	✓	✓	SK849620	×	×
Scopwick [4]	×	✓	✓	✓	×	TF058576	✓	✓
Spalford	✓	✓	✓	✓	×	SK842690	×	×
Waddington	✓	✓	×	×	×	SK974647	×	×
Waddington	✓	✓	✓	✓	✓*OW*	SK975635	✓	✓
Wellingore	✓	✓	✓	✓	✓	SK984570	×	✓

[1] This mill is annotated *Tower Mill (Flour)* on the first and second editions 1:2500 but ignored by the six-inch. The annotation *Flour* is used for other mills on this sheet, for example at Coleby, but *Corn* appears far more frequently.

[2] The use of an annotated mill symbol here is unsupported by either the first and the second editions of the 1:2500 or by the earlier one-inch New Series. None of them provides any trace of a ground plan that might account for this Revised New Series depiction. The solution is that the structure recognised here was a barn-top mill similar to the one at West Ashling in Sussex. It survived for a while: the sail-less remnants of it were still in

place *circa* 1935.

[3] The use of an annotated mill symbol here, just two miles from Coleby (Lodge) Mill discussed above, is again unsupported by either the New Series or the large scales. It is highly probable that these two structures, recorded only by the one-inch Revised New Series with annotated symbols, either had short operational lives of little consequence or were sufficiently small domestic windpumps for later surveyors to feel that they could be overlooked.

[4] An example of a windmill whose status was not made explicit on the first edition 1:2500 either by the annotation provided – *Scopwick Mill (Corn)* – or by any discernible circular ground plan. This situation arose from the number of different buildings contributing to the complex here, of which the windmill was only a small part.

The greatest concentration of windmills shown on this sheet appears in the area of the City of Lincoln. When the Old Series was first issued, fourteen windmills were recorded here of which eight formed a line on the ridge that runs north-west from the city centre. Another three were sited to the south of the centre and a further three to the east of it. On later states of Old Series sheet 83, one of the eastern mills disappears. Only one of these fourteen mills appears on the one-inch New Series but then two more are resurrected, leading to the Revised New Series showing three windmills in Lincoln. Even so, a relatively new tower mill close to the north-western ridge is missed though it had been shown by the large-scale first editions.

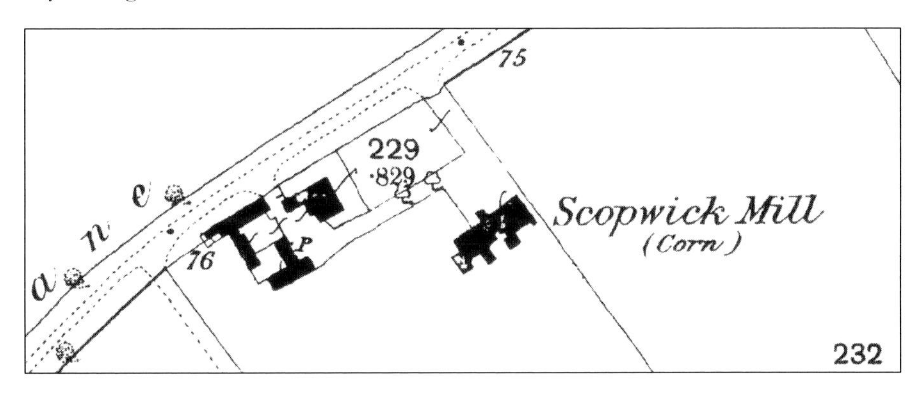

Above: The windmill at Scopwick can not be considered as easily discernible here on the first edition 1:2500, sited as it is within a complex of associated buildings.

Left: The charismatic old post mill at North Hykeham in Lincolnshire *circa* 1904 when it still carried four sails. In spite of a general look of dilapidation, enhanced it would seem by everything being 'out of plumb', this mill carried on working against the odds for some years. It eventually collapsed at some point during the 1930s.

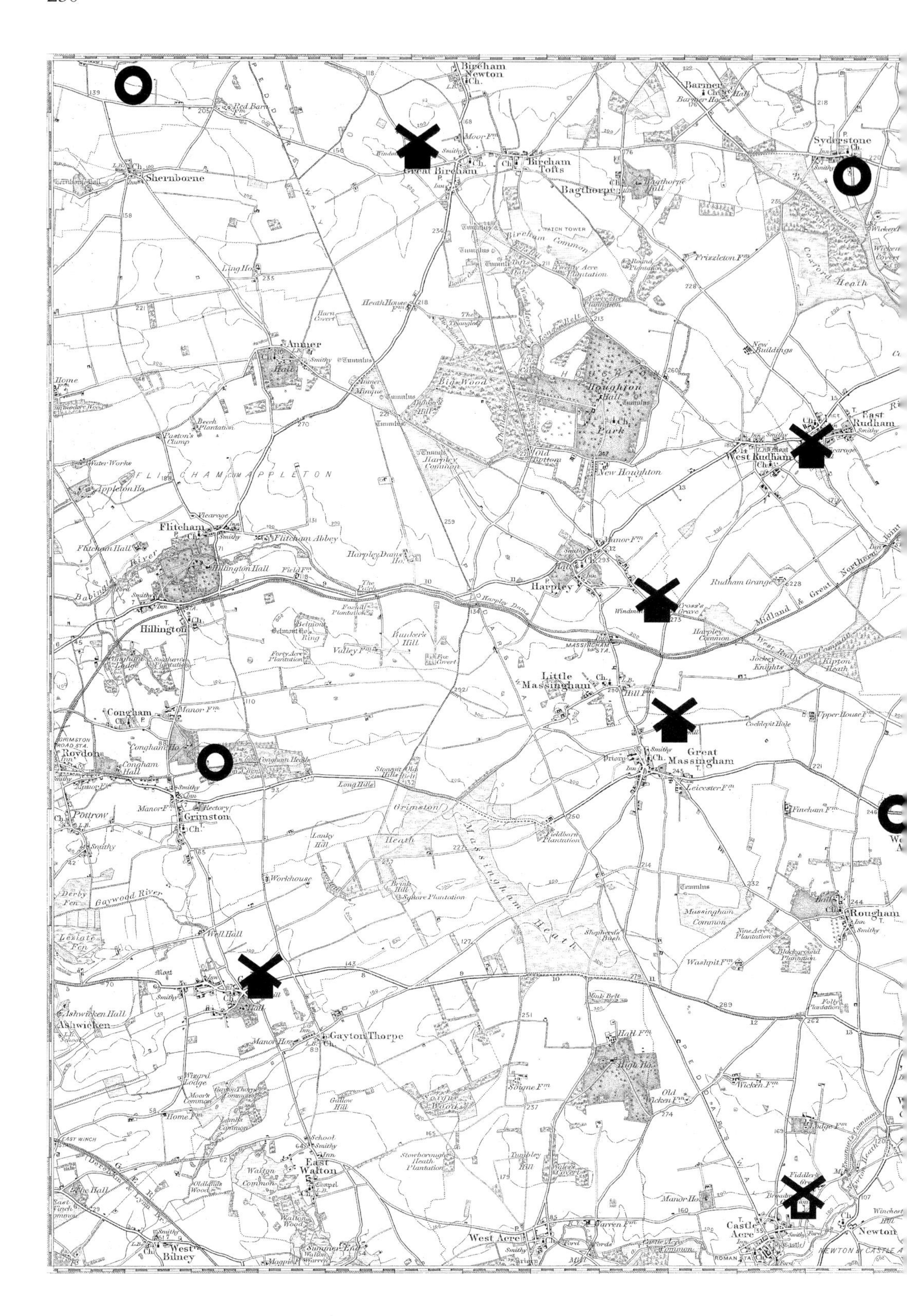
Bircham Newton
Barmer
Syderstone
Great Bircham
Bircham Tofts
Bagthorpe
Shernborne
Red Barn Fm
Moor Fm
Bircham Common
Frizzleton Fm
Coxford Heath
Ling Ho.
Heath House Fm
Anmer
Anmer Hall
Big Wood
Houghton Hall
Park
New Buildings
East Rudham
West Rudham
New Houghton
Beech Plantation
Paston's Clump
Water Works
Appleton Ho.
FLITCHAM CUM APPLETON
Flitcham
Flitcham Abbey
Flitcham Hall
Hillington Hall
Babingley River
Hillington
Harpley Dams
Harpley
Manor Fm
Rudham Grange
Midland & Great Northern Joint
Massingham Sta.
Little Massingham
Harpley Common
Kipton Heath
Congham
Congham Lodge
Manor Fm
Upper House Fm
Cocklepit Hole
Grimston Road Sta.
Roydon
Congham Hall
Congham Ho.
Congham Heath
Long Hills
Great Massingham
Leicester Fm
Grimston
Pottrow
Rectory
Fieldbarn Plantation
Fincham Fm
Lanky Hill
Heath
Massingham Heath
Workhouse
Square Plantation
Tumulus
Massingham Common
Roughton
Gaywood River
Derby Fen
Leziate Fen
Well Hall
Shepherd's Bush
Nine Acre Plantation
Washpit Fm
Ashwicken Hall
Ashwicken
Gayton Thorpe
Mink Belt
Hall Fm
High Ho.
Wizard Lodge
Home Fm
Soigne Fm
Wicken Fm
Old Wicken Fm
East Winch
East Walton
Walton Common
Stowborough Heath Plantation
Tumbley Hill
Manor Ho.
Castle Acre
Newton
West Acre
Warren Fm
West Bilney
The Hall
Summer End
Roman Sta.
Newton by Castle Acre

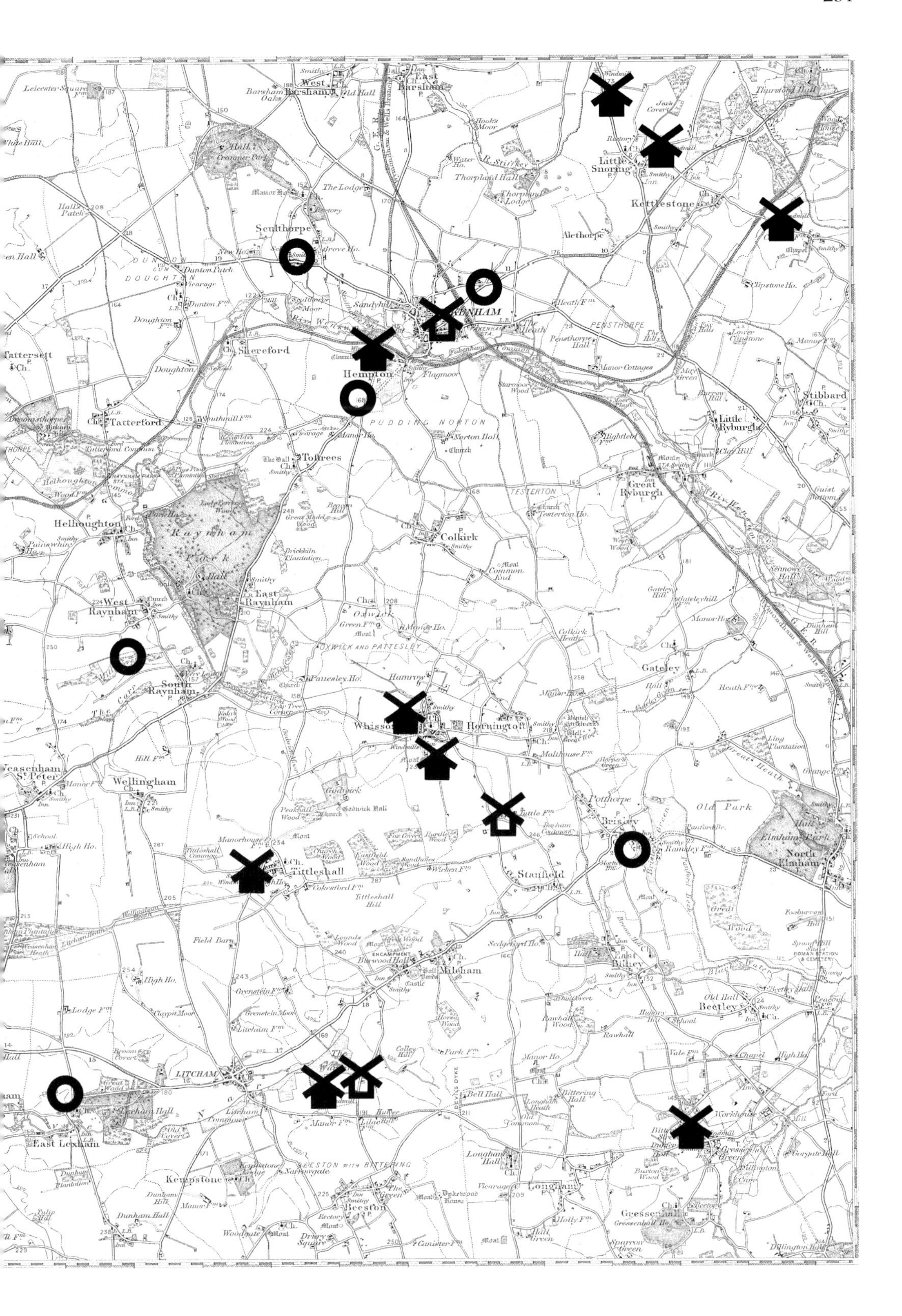
West Barsham
East Barsham
Barsham Oaks
Leicester Square Fm
Cranmer Park
Hook's Moor
R. Stiffkey
Thorpland Hall
Thorpland Lodge
Little Snoring
Kettlestone
Thursford Hall
Sculthorpe
Alethorpe
Clipstone Ho.
Dunton Patch
Doughton Fm
Tattersett
Shereford
Hempton
Flagmoor
Pensthorpe Hall
Manor Cottages
Stibbard
Little Ryburgh
Tatterford
Toftrees
Pudding Norton
Norton Hall
Great Ryburgh
Testerton
Helhoughton
Raynham Park
Colkirk
East Raynham
West Raynham
Oxwick
Gateley
South Raynham
Pattesley Ho.
Hamrow
Horningtoft
Weasenham St. Peter
Wellingham
Godwick
Old Park
North Elmham
Potthorpe
Brisley
Tittleshall
Stanfield
Manorhouse Fm
Great Wood
Mileham
East Bilney
Burwood Hall
Old Hall
Beetley
LITCHAM
Bittering Hall
Longham Hall
East Lexham
Kempstone
Beeston
Longham
Gressenhall
Drury Square
Dunham Hall
Rawhall
Sedgeford Ho.

Sheet 146 *Fakenham*

(see pages 250 & 251)

Windmill locations as recorded by the Revised New Series *circa* 1895 and/or the first edition 1:2500	19th century recognition			20th century recognition				
	Old Series mixed sheet states [1]	First edition six-inch	Second edition 1:2500	New Series Third Edition	Popular Edition	National Grid Reference	Seventh Series	Current 1:25,000
Castle Acre	✓	✓	✓	×	×	TF824157	×	×
Croxton Farm	✓	✓	✓	✓	✓	TF979313	×	×
East Rudham	✓	✓	✓	✓	×	TF825278	×	×
Fakenham	✓	✓	✓	×	×	TF922299	×	×
Gayton [2]	✓	✓	✓	✓	✓	TF732193	✓	✓
Great Bircham	✓	✓	✓	✓	✓	TF760327	×	✓
Great Massingham	✓	✓	✓	✓	×	TF804234	×	×
Gressenhall	✓	✓	✓	×	×	TF963167	×	×
Harpley [3]	✓	×	×	✓	✓	TF799253	✓	✓
Hempton	✓	✓	✓	✓	✓ *OW*	TF912293	×	×
Little Snoring [4]	✓	✓	✓	✓	✓	TF949338	×	×
Little Snoring	✓	✓	✓	×	×	TF959327	×	×
Mileham [5]	✓	✓	✓	✓	✓	TF901174	✓	✓
Mileham	×	✓	×	×	×	TF901174	×	×
Stanfield [6]	✓	✓	×	×	×	TF932219	×	×
Tittleshall	×	✓	✓	✓	×	TF890209	×	×
Whissonsett	✓	✓	✓	×	×	TF917232	×	×
Whissonsett	✓	✓	✓	×	×	TF918231	×	×

[1] Old Series sheets 65, 66NW, 68SW and 69 are needed here. Both quarter-sheets used were Margary state 2 (of 7) *circa* 1844. The copy of sheet 65 used was state 1 (of 14) *circa* 1825 and that of sheet 69 was state 3 (of 16) *circa* 1830.

[2] Annotated only as *Gayton Mill (Corn)* on the large-scale first editions. Nevertheless, a circular ground plan is visible within the composite ground plan for the collection of buildings belonging to this mill.

[3] It is as well that the Revised New Series records this mill since neither the first nor the second large-scale editions offer a discernible circular ground plan within the composite plan of this milling business. This is a notable triumph for the one-inch scale and the Revised New Series in particular.

[4] This mill offers a good illustration of the difficulties that can be encountered when proximity to a parish border becomes a key determinant in whether or not a mill is depicted effectively. Quite simply, this mill was so near the border between the parishes of Little Snoring and Great Snoring (estimated at less than fifty feet away) that careful use of the 1:2500 is needed to assess the circumstances of this windmill. Unhelpfully, this scrutiny is made difficult if not impossible on the first edition 1:2500 because the ground plan of this open-trestle post mill – four individual brick pier plans – is obliterated by other annotation relating to acreage of the small plot of land on which the mill was sited. The second edition is much better and reveals the truth of the situation. It is also not particularly helpful that the annotation at 1:2500 is *Snoring Windmill (Corn)*!

[5] An example of a mill whose Old Series annotation incorporated the parish name. The possibility here is that this was done to avoid confusion, since this mill is closer to the village of Litcham than it is to Mileham.

[6] The large-scale depiction here is unquestionably one of a windmill despite an annotation of just *Tuttle Mill (Corn)* that serves only to emphasise its association with the adjacent Tuttle Farm.

Sheet 155 *Atherstone*

(see pages 254 & 255)

Windmill locations as recorded by the Revised New Series *circa* 1895 and/or the first edition 1:2500	19th century recognition			20th century recognition				
	Old Series sheet states *ca* 1852	First edition six-inch	Second edition 1:2500	New Series Third Edition	Popular Edition	National Grid Reference	Seventh Series	Current 1:25,000
Austrey	✓	✓	✓	×	×	SK302072	×	×
Austrey	✓	✓	×	×	×	SK300063	×	×
Desford	✓	✓	✓	✓*Old*	×	SK475030	×	×
Earl Shilton	✓	✓	×	×	×	SP458974	×	×
Earl Shilton	✓	✓	✓	✓	✓	SP487972	×	×
Groby	×	✓	×	×	×	SK526069	×	×
Kirby Muxloe	×	✓	✓	×	✓*OW*	SK519055	×	×
Markfield	✓	✓	×	×	×	SK483109	×	×
Netherseal	✓	✓	✓	✓	×	SK278136	×	×
Newbold Verdon	✓	✓	✓	✓*Old*	×	SK444030	×	×
Sheepy Magna	✓	✓	×	×	×	SP308996	×	×
Warton	✓	✓	✓	✓	✓*OW*	SK286032	×	×
Woodhouse Eaves	✓	✓	✓	✓*Old*	✓*OW*	SK526143	✓	×

The mix of the three types of cartographic depiction for the mills of this border area of Leicestershire with Derbyshire, Staffordshire and Warwickshire conforms to the pattern of how most people would assume windmills went gently into decline from the mid-nineteenth century onwards. To the west of this area windmills were never thick on the ground other than in isolated pockets, while to the east, in the remainder of Leicestershire and beyond, some of the densest concentrations of windmills in the country were to be found. On this sheet of 1899 the six recorded windmills are more than matched by seven unrecorded mills that had appeared on the first edition 1:2500 of *circa* 1884. But, even combined, these thirteen mills are eclipsed by the sixteen recorded by the Old Series (on states of *circa* 1852), but which then disappeared before 1880. However, as seen on other sheets, the timetable for the decay in a windmill population and its associated large-scale mapping could be very different in other areas.

Left: The post mill at Netherseal *circa* 1908.

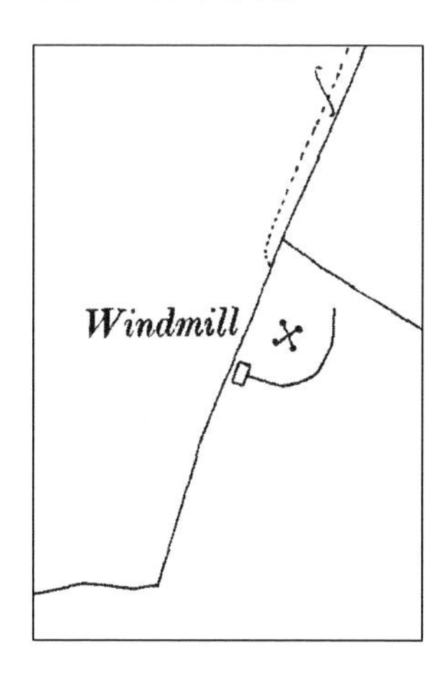

Right: The open-trestle post mill at Groby on the first edition 1:2500.

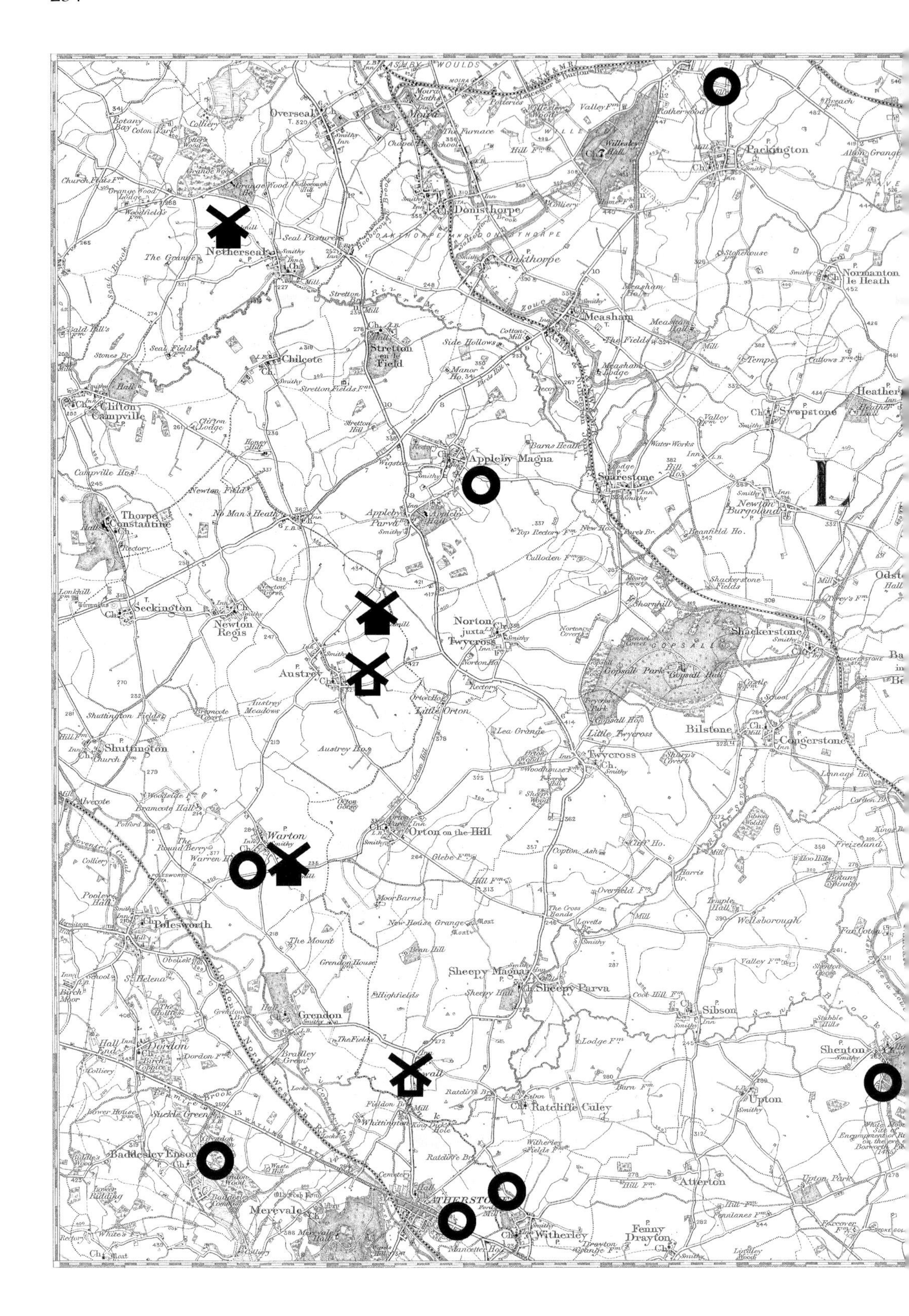
Overseal
Moira
Donisthorpe
Oakthorpe
Packington
Normanton le Heath
Netherseal
Measham
Chilcote
Stretton en le Field
Clifton Campville
Swepstone
Heather
Appleby Magna
Snarestone
Newton Burgoland
Thorpe Constantine
Appleby Parva
No Man's Heath
Seckington
Newton Regis
Norton juxta Twycross
Shackerstone
Austrey
Gopsall Park
Little Orton
Bilstone
Congerstone
Shuttington
Twycross
Little Twycross
Orton on the Hill
Warton
Alvecote
Polesworth
Wellsborough
Sheepy Magna
Sheepy Parva
Grendon
Sibson
Dordon
Shenton
Ratcliffe Culey
Upton
Baddesley Ensor
Merevale
Atterton
Witherley
Fenny Drayton
Atherstone

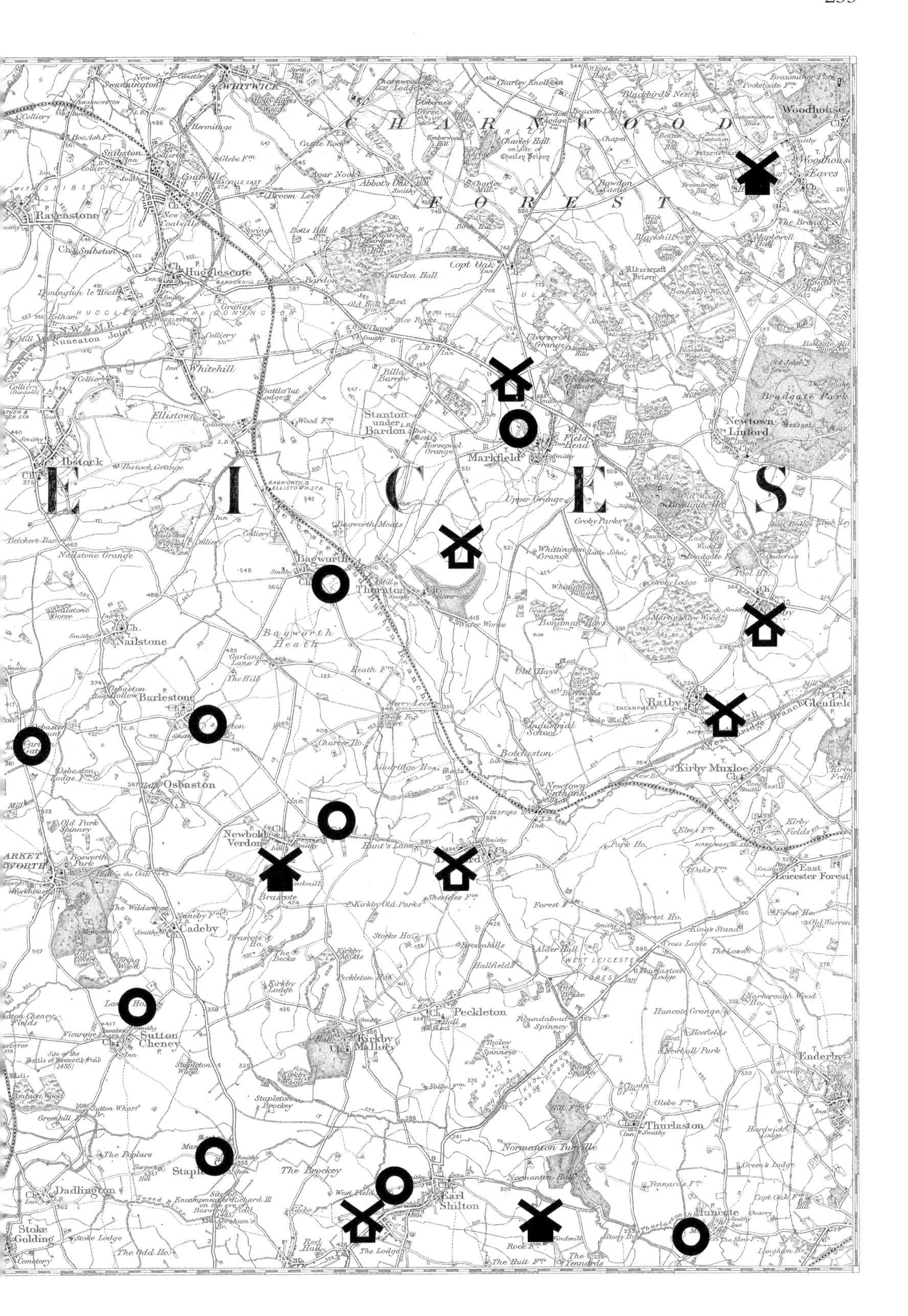
C H A R N W O O D
F O R E S T
L E I C E S
Whitwick
Coalville
Ravenstone
Hugglescote
Ibstock
Nailstone
Barlestone
Osbaston
Newbold Verdon
Cadeby
Sutton Cheney
Dadlington
Stoke Golding
Bagworth
Thornton
Stanton under Bardon
Markfield
Copt Oak
Bardon Hall
Ulverscroft Priory
Woodhouse Eaves
Newtown Linford
Bradgate Park
Groby
Ratby
Glenfield
Kirby Muxloe
East Leicester Forest
Peckleton
Kirkby Mallory
Earl Shilton
Normanton Turville
Thurlaston
Huncote
Enderby
Stapleton
Brascote
Botcheston
Desford

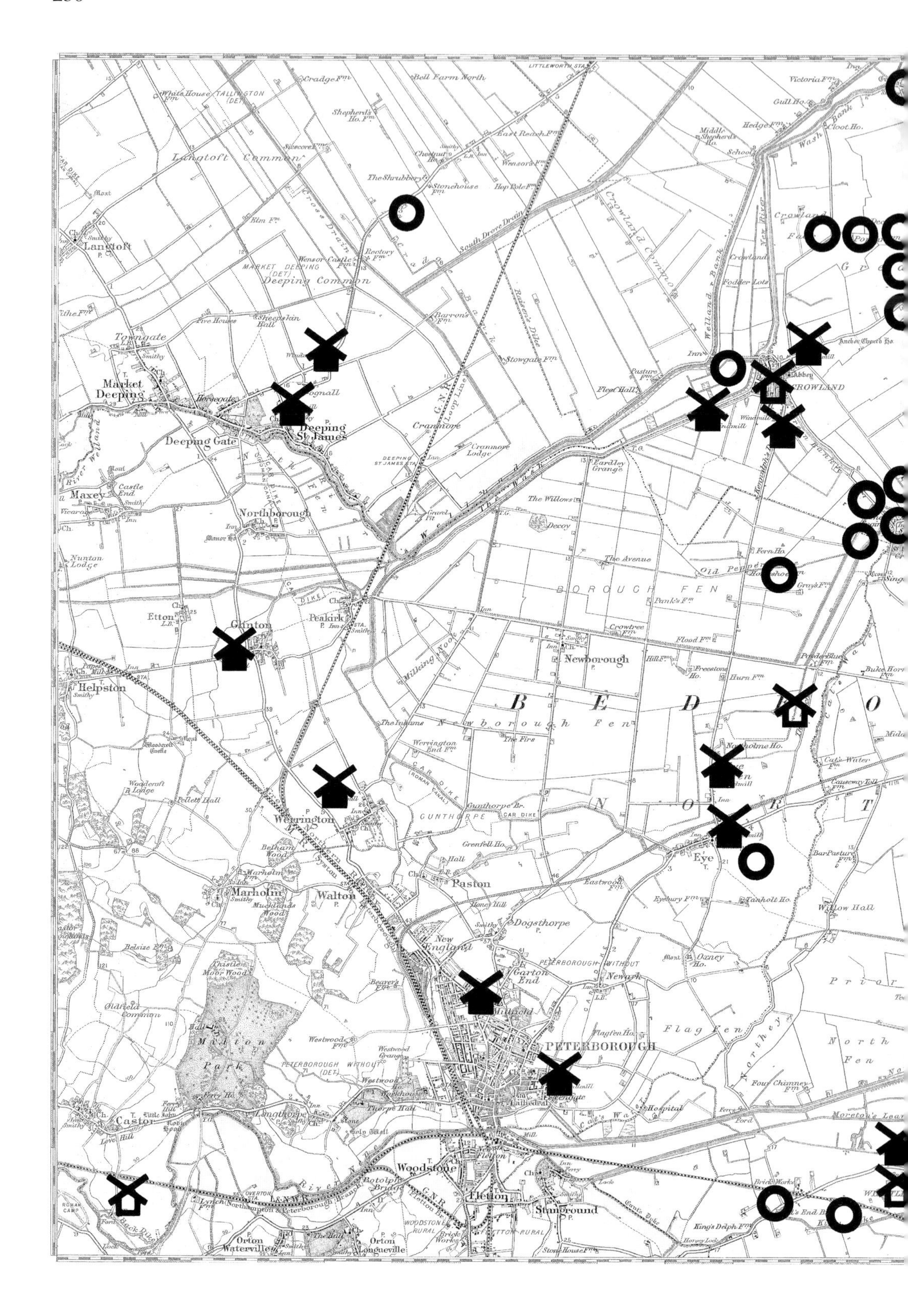
Langtoft Common
Langtoft
Deeping Common
Crowland Common
Market Deeping
Deeping Gate
Deeping St James
Maxey
Northborough
Etton
Glinton
Peakirk
Helpston
Borough Fen
Newborough
Newborough Fen
Werrington
Marholm
Walton
Paston
Dogsthorpe
Eye
New England
Garton End
Newark
Flag Fen
North Fen
Milton Park
PETERBOROUGH
Castor
Longthorpe
Woodstone
Fletton
Stanground
Orton Waterville
Orton Longueville
CROWLAND

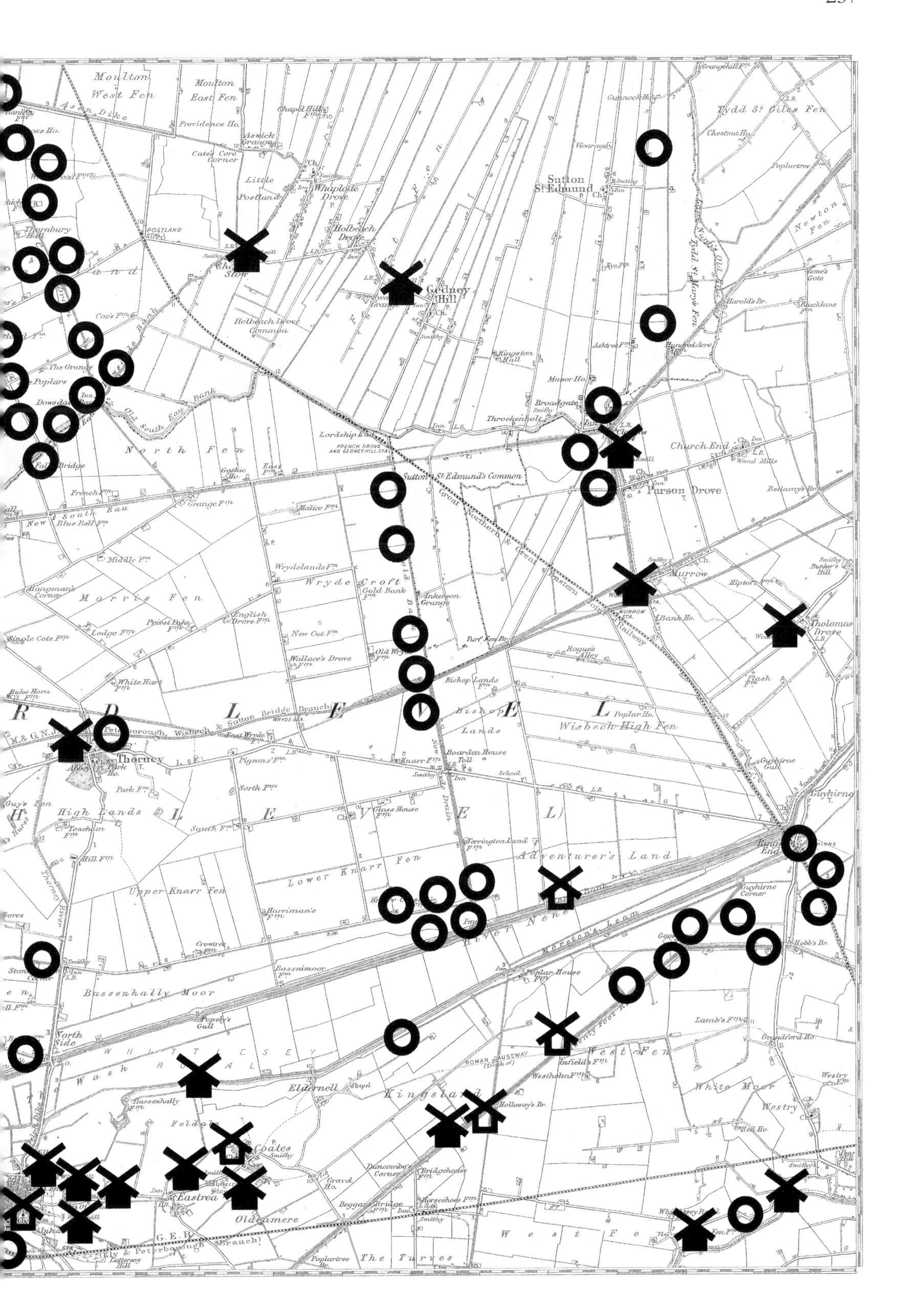
Moulton West Fen
Moulton East Fen
Providence Ho.
Asgarby Grange
Cates' Cove Corner
Little Postland
Whaplode Drove
Holbeach Drove
Gedney Hill
Holbeach Drove Common
Sutton St. Edmund
Tydd St. Giles Fen
Newton Fen
Kingston Hall
Manor Ho.
Broadgate
Throckenholt
Lordship End
North Fen
Church End
Parson Drove
Sutton St. Edmund's Common
Great Northern & Great Eastern Railway
Morris Fen
Murrow
Tholomas Drove
Wryde Croft
Bishop Lands
Wisbech High Fen
Thorney
Peterborough, Wisbech & Sutton Bridge Branch
High Lands
Guyhirne
Adventurer's Land
Lower Knarr Fen
Upper Knarr Fen
River Nene
Morton's Leam
Bassenhally Moor
North Side
Whittlesey
Eldernell
Kingsland
Feldale
Coates
Eastrea
Oldeamere
West Fen
White Moor
Westry
The Turves
Ely & Peterborough (Branch)

Sheet 158 *Peterborough*

(see pages 256 & 257)

Windmill locations as recorded by the Revised New Series *circa* 1895 and/or the first edition 1:2500	19th century recognition			20th century recognition				
	Old Series sheet states *ca* 1866	First edition six-inch	Second edition 1:2500	New Series Third Edition	Popular Edition	National Grid Reference	Seventh Series	Current 1:25,000
Adventurer's Land	✓	✓	×	×	×	TF358019	×	×
Castor	✓	✓	✓	✓*Old*	✓*OW*	TL129968	✓	✓
Coates [1]	×	×	✓	✓	✓	TL303975	×	×
Coates	✓	✓	×	×	×	TL304976	×	×
Coates	✓	P	P	×	×	TL299989	×	×
Crowland	✓	✓	✓	✓*Old*	×	TF227096	×	×
Crowland	✓	✓	✓	✓	×	TF240096	×	×
Crowland [2]	✓	✓	✓	×	×	TF238100	×	×
Crowland [3]	✓	✓	✓	✓	×	TF244106	×	×
Deeping St James	✓	✓	✓	✓	✓	TF157099	✓	×
Deeping St James	✓	✓	✓	✓	✓*OW*	TF164107	×	×
Eastrea	✓	✓	✓	×	×	TL297974	×	×
Eye	✓	✓	✓	✓	✓*OW*	TF229037	×	×
Eye	✓	✓	✓	✓	✓	TF230030	✓S	×
Eye	✓	✓	✓	✓*Old*	×	TF243048	×	×
Gedney Hill	✓	✓	✓	✓	✓*OW*	TF334116	✓	✓
Glinton	✓	✓	✓	✓	×	TF148058	×	×
Kingsland	✓	P	P	Pump	×	TL341980	×	×
Kingsland	✓	P	P	×	×	TL343983	×	×
Murrow	×	✓	✓	×	×	TF373067	×	×
Parson Drove	✓	✓	✓	✓	✓*OW*	TF369089	×	×
Peterborough	×	✓	✓	✓	×	TF189002	×	×
Peterborough	✓	✓	✓	✓*Old*	✓	TL203988	×	×
Shepeau Stow	✓	✓	✓	✓	✓*OW*	TF308122	✓	✓
Tholomas Drove [4]	✓	×	✓	✓	✓*OW*	TF398061	×	×
Thorney	✓	✓	✓	✓	✓	TF278043	✓	✓
Werrington	✓	✓	✓	✓	✓*OW*	TF165034	✓	×
West Fen	✓	P	P	×	×	TL357994	×	×
West Fen Farm	✓	P	P	×	×	TL382962	×	×
West Fen Farm	✓	P	P	×	×	TL397969	×	×
Whittlesey	✓	✓	✓	✓	×	TL264975	×	✓
Whittlesey	✓	✓	×	×	×	TL263973	×	×
Whittlesey	×	✓	×	×	×	TL263972	×	×
Whittlesey	×	✓	✓	✓	✓	TL276975	×	×
Whittlesey	✓	✓	✓	×	×	TL277971	×	×
Whittlesey	✓	✓	✓	×	×	TL281971	×	×
Whittlesey	×	✓	✓	✓*Old*	×	TL276969	×	×

[1] This mill appears on the second edition 1:2500, but not on the first edition. In contrast, another mill, just a few yards away on the north side of the Eastrea to Coates road, was recorded by the first but not by the second edition 1:2500. Given that the annotation for the first mill was *Windmill (Corn)* and thus unlikely to be signifying a brand new structure, this, without resorting to any further knowledge of the mills here, suggests that a windmill may have simply been moved from one side of the road to the other. However, the one-inch Third Edition, which postdates the second edition 1:2500 by some years, unfortunately does not corroborate this speculation.

[2] This mill is annotated simply as *Corn Mill* on the first large-scale editions and then as *Windmill (Disused)* on the second edition 1:2500.

[3] This second mill at Crowland is treated in exactly the same way as the above mill.

[4] The circular ground plans for this mill on the first and second editions of the 1:2500 are clear if rather cramped. Both are accompanied by the somewhat odd annotation of *Unexpected Mill (Corn)*, though this is not repeated for either six-inch edition. Moreover, the annotation is separated from the plan to an extent that is unusual.

Where an unannotated mill symbol appears on this sheet, reference to the larger scales reveals it to be signifying a *Draining Pump*. Examples of this can be seen at Coates, Kingsland (twice), West Fen and West Fen Farm (twice). In the case of the mill at Coates, the annotation alters to *Engine House (Pumping)* for the second edition 1:2500. In addition, the great majority of the large number of Old Series mill depictions that are unsupported by later large-scale depictions were recording pumping mills, as the nature of their sites suggests. In all probability, these would have been structures of no great substance.

As on other sheets, the annotations for some windmills include a place name and so often this can preclude the use of *Windmill* in favour of just *Mill*. Usually, however, the topography of a mill site and an obvious circular ground plan – open in the case of the first edition 1:2500 and shaded for the second – safely point to the presence of a windmill. Examples on this sheet include the use of *Gedney Hill Mill (Corn)*, *Murrow Mill (Corn)*, *Parson Drove Mill (Corn)* and *Shepeau Stow Mill (Corn)*. A further point of note here is that the three windmills at Gedney Hill, Parson Drove and Shepeau Stow – all appearing in the top right hand corner of the sheet – were recorded using annotations of just *Mill* on the one-inch New Series. Other windmills on the sheet were accorded unannotated symbols, as had become usual for New Series sheets this far north. The situation was rationalised for the Revised New Series when these three mills were treated in the normal way for that edition.

Left: The smock mill at Parson Drove *circa* 1908.

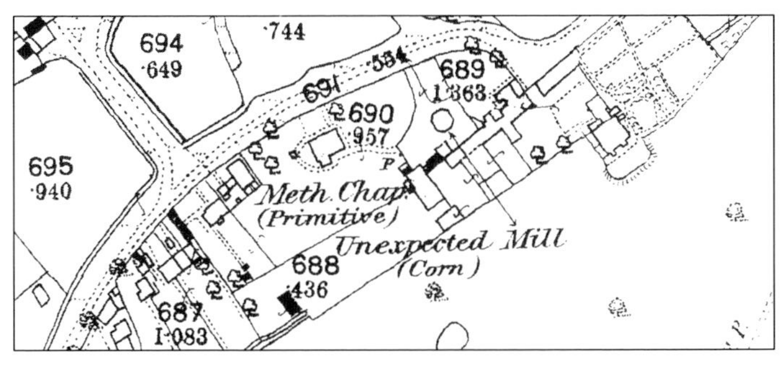

Left: Tholomas Drove Mill shown on the first edition 1:2500.

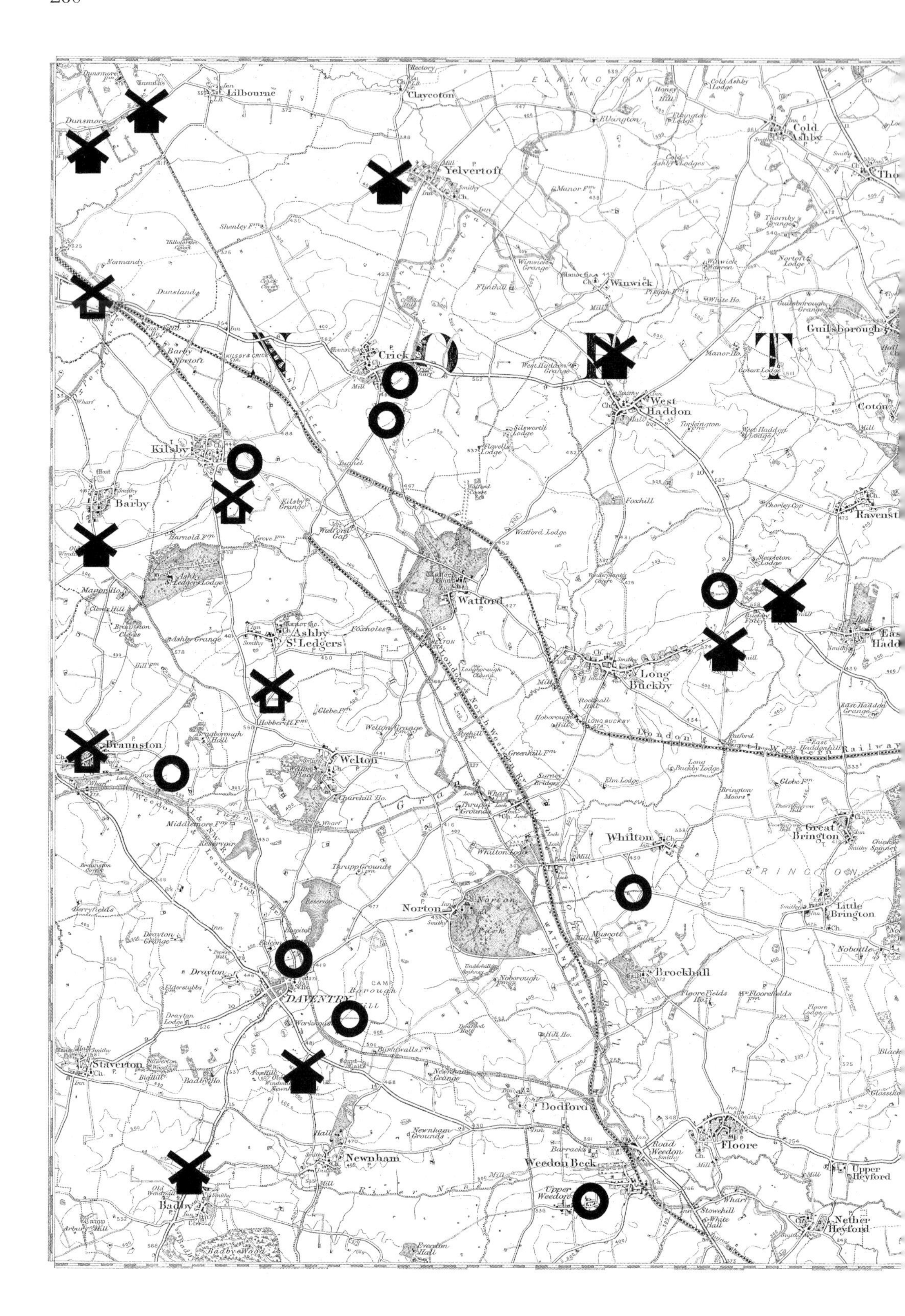
Lilbourne
Claycoton
Yelvertoft
Cold Ashby
Winwick
Guilsborough
Crick
West Haddon
Kilsby
Barby
Watford
Ashby St. Ledgers
Long Buckby
Braunston
Welton
Whilton
Great Brington
Little Brington
Norton
Brockhall
Daventry
Staverton
Dodford
Newnham
Weedon Beck
Floore
Upper Heyford
Nether Heyford
Badby
Badby Wood
Upper Weedon
Road Weedon
Ravensthorpe
Dunsmore
Watling Street
London & North Western Railway

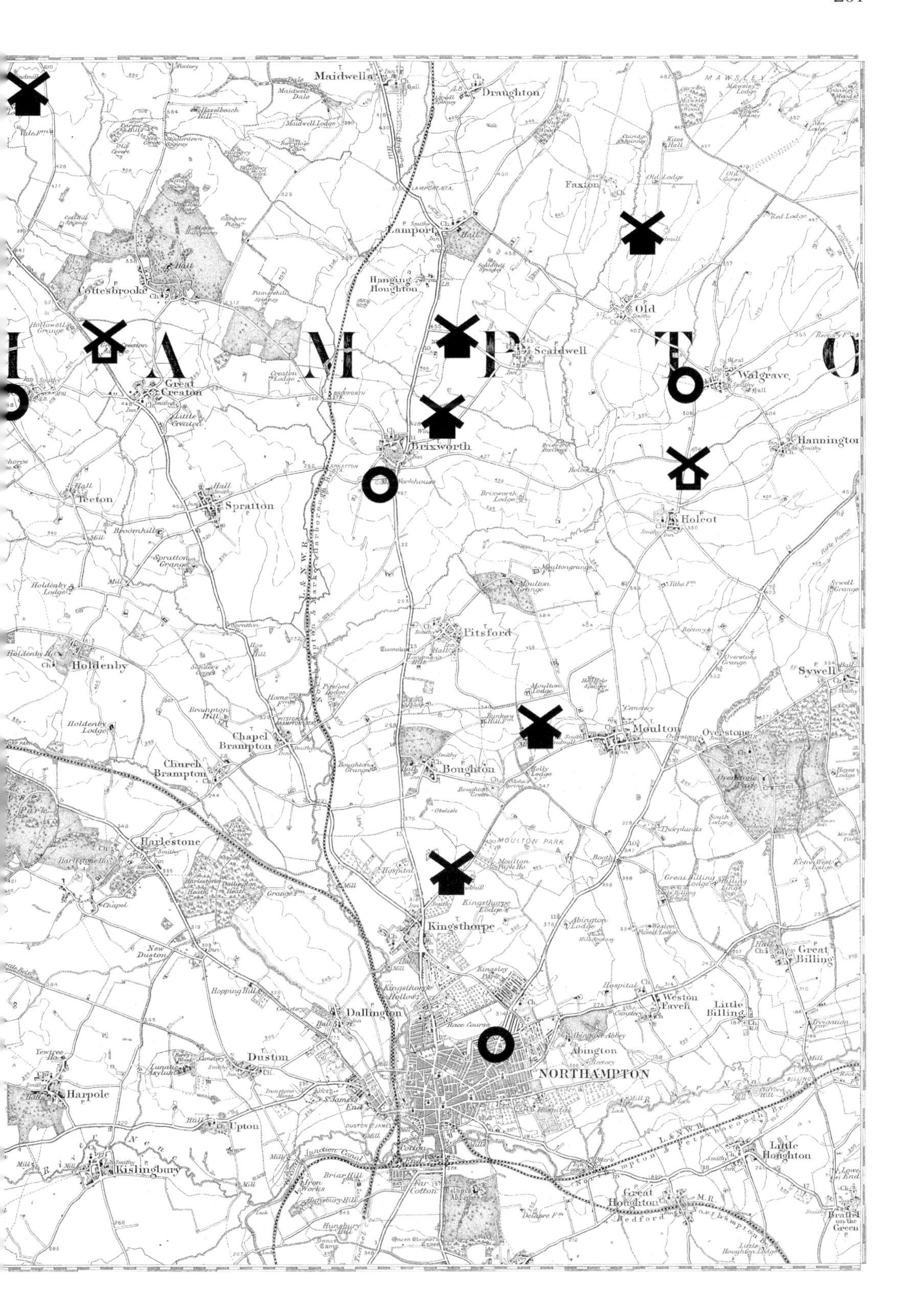
Maidwell
Draughton
MAWSLEY
Faxton
Lamport
Hanging Houghton
Cottesbrooke
Great Creaton
Old
Scaldwell
Walgrave
Brixworth
Hannington
Teeton
Spratton
Holcot
Pitsford
Holdenby
Sywell
Moulton
Overstone
Chapel Brampton
Church Brampton
Boughton
Harlestone
Kingsthorpe
Great Billing
Dallington
Weston Favell
Little Billing
Abington
Duston
NORTHAMPTON
Harpole
Upton
Kislingbury
Little Houghton
Great Houghton
Far Cotton

Sheet 185 *Northampton*

(see pages 260 & 261)

Windmill locations as recorded by the Revised New Series *circa* 1895 and/or the first edition 1:2500	19th century recognition			20th century recognition				
	Old Series mixed sheet states [1]	First edition six-inch	Second edition 1:2500	New Series Third Edition	Popular Edition	National Grid Reference	Seventh Series	Current 1:25,000
Ashby St Ledgers	×	✓	×	×	×	SP571670	×	×
Badby	✓	✓	✓	✓*Old*	✓	SP555590	×	×
Barby [2]	✓	✓	✓	✓*Old*	✓*OW*	SP541696	✓	✓
Braunston [3]	✓	×	×	×	✓*OW*	SP538662	✓	✓
Brixworth	✓	✓	×	×	×	SP757714	×	×
Dunsmore	×	×	×	×	×	SP537760	×	×
Dunsmore [4]	×	×	P	Pump	Pump	SP549767	×	×
East Haddon	✓	✓	✓	✓*Old*	✓*OW*	SP656686	✓	×
Great Creaton	✓	✓	×	×	×	SP699729	×	×
Holcot	✓	✓	×	×	×	SP797704	×	×
Kilsby	✓	✓	×	×	×	SP564703	×	×
Kingsthorpe	✓	✓	✓	✓*Old*	×	SP757640	×	×
Long Buckby	✓	✓	✓	✓*Old*	×	SP647677	×	×
Moulton	✓	✓	✓	✓*Old*	×	SP772662	×	×
Naseby Lodge	×	✓	✓	✓	✓*OW*	SP686771	×	×
Newnham	✓	✓	✓	✓*Old*	✓*OW*	SP575609	×	✓
Old	✓	✓	✓	✓*Old*	×	SP790743	×	×
Scaldwell	×	✓	✓	✓	✓*OW*	SP760727	×	×
West Haddon [5]	✓	✓	×	×	×	SP628725	×	×
Wharf Inn	✓	✓	×	×	×	SP541735	×	×
Yelvertoft	✓	✓	×	×	×	SP590754	×	×

[1] Old Series quarter-sheets 52NW, 52SW, 53NE and 53SE are needed here. All states used here are dated between 1844 and 1848. The copy of quarter-sheet 52SW used, by way of example, was Margary state 4 (of 13) *circa* 1846.

[2] Annotated as *Barby Corn Mill* on the first edition 1:2500, this mill was annotated more conventionally on the first edition six-inch as *Barby Mill (Corn)*. On the second edition 1:2500 it reverts to *Barby Corn Mill (Disused)*.

[3] Though shown conventionally on the first edition 1:2500, this mill was not shown by the six-inch.

[4] The unannotated symbol used here is recording a pumping mill that made its large-scale debut on the second edition 1:2500 as *Windpump*. More oddly, the annotated symbol that appears a mile to the south-west was also only recording a pumping mill though this was not recorded by the 1:2500 until the third edition, with nothing in place on either of the two earlier editions.

[5] The position of this mill happened to coincide with where the Survey needed to superimpose the first letter 'R' from the county name NORTHAMPTONSHIRE. This label was written in sufficiently large and bold a font size that it spread across two Revised New Series sheets. The discretion that the Survey had over where it placed such necessarily prominent county labels was limited and, inevitably, a lot of detail was obscured. Careful inspection of the way that the ten large letters from the county name on this Revised New Series sheet were written and the way in which some detail was indeed obscured, yet some was not, is interesting. The ability to read the name at a glance has not been compromised yet, for example, the symbol for the church at Scaldwell can be seen since part of the letter 'P' has been removed to allow this to be the case. So it is with West Haddon Mill (except that this has been made impossible to see since the symbol used for the facsimile map has had to be superimposed over it).

Sheet 190 *Eye*

(see pages 266 & 267)

Windmill locations as recorded by the Revised New Series *circa* 1895 and/or the first edition 1:2500	19th century recognition			20th century recognition				
	Old Series sheet states *ca* 1842	First edition six-inch	Second edition 1:2500	New Series Third Edition	Popular Edition	National Grid Reference	Seventh Series	Current 1:25,000
Bacton	✓	✓	×	×	×	TM045663	×	×
Badwell Ash	✓	✓	✓	✓	✓	TL985687	×	×
Bedfield	×	✓	✓	×	×	TM222663	×	×
Bedingfield	✓	✓	✓	✓	✓*OW*	TM176683	×	×
Botesdale	✓	✓	✓	✓	✓	TM049756	×	×
Botesdale	✓	✓	✓	✓	✓*OW*	TM054757	×	×
Botesdale	×	✓	✓	✓	×	TM058749	×	×
Brandeston	✓	✓	×	×	×	TM246611	×	×
Brundish	✓	✓	✓	✓	✓	TM257701	×	×
Brundish	✓	✓	✓	✓	✓	TM263699	×	×
Buxhall	✓	✓	✓	✓	✓	TL996577	✓	✓
Charsfield	✓	✓	✓	✓	✓	TM256570	×	×
Cotton	✓	✓	✓	✓	×	TM068673	×	×
Crowfield	×	✓	✓	✓	✓	TM151571	✓	×
Debenham	✓	✓	✓	✓	✓	TM165630	✓	×
Debenham [1]	×	✓	×	×	×	TM171638	×	×
Debenham	×	✓	✓	✓	✓	TM182638	×	×
Denham	×	✓	✓	✓*Old*	×	TM195740	×	×
Earl Soham	✓	✓	✓	✓	✓	TM219634	×	×
Earl Soham	✓	✓	✓	✓	✓*OW*	TM227629	×	×
Earl Soham	×	✓	✓	×	×	TM246645	×	×
Earl Stonham [2]	✓	✓	×	×	×	TM089598	×	×
Earl Stonham	✓	✓	×	×	×	TM090598	×	×
Eye	✓	✓	✓	✓	✓	TM139743	×	×
Eye	✓	✓	✓	✓	✓	TM160735	×	×
Eye [3]	×	✓	✓	✓	✓	TM160727	×	×
Framsden	✓	✓	✓	✓	✓	TM192598	✓S	✓
Framsden	✓	✓	✓	✓	✓*OW*	TM200614	×	×
Gislingham	✓	✓	×	✓*Old*	×	TM048717	×	×
Gislingham	✓	✓	✓	✓	✓	TM067720	×	×
Great Ashfield	✓	✓	✓	✓	✓*OW*	TL990665	×	×
Haughley	✓	✓	×	×	×	TM031644	×	×
Haughley [4]	✓	✓	✓	✓	×	TM028626	×	×
Haughley	✓	✓	✓	✓	✓	TM031621	×	×
Hepworth	✓	✓	×	×	×	TL996742	×	×
Horham	✓	✓	✓	✓	✓	TM209724	×	×
Kettleburgh	×	✓	×	×	×	TM267598	×	×

Little Stonham	×	✓	✓	✓	×	TM114601	×	×
Mellis	✓	✓	×	×	×	TM085738	×	×
Mendlesham [5]	✓	✓	✓	✓	×	TM102657	×	×
Mendlesham	×	✓	✓	✓	✓	TM092639	×	×
Mendlesham	✓	✓	✓	✓	✓ *OW*	TM090634	×	×
Monewdon	✓	✓	✓	✓	✓	TM240585	×	×
Monk Soham	✓	✓	✓	✓	✓ *OW*	TM207663	×	×
Occold	✓	✓	✓	✓	✓	TM156705	✓	×
Old Newton	✓	✓	×	×	×	TM055623	×	×
Onehouse	✓	✓	×	×	×	TM032591	×	×
Pettaugh	✓	✓	✓	✓	✓	TM167595	✓	×
Rickinghall Inferior	✓	✓	×	×	×	TM039759	×	×
Rishangles	✓	✓	✓	×	×	TM164674	×	×
Saxtead Green	✓	✓	✓	✓	✓	TM253644	✓S	✓
Saxtead Lodge [6]	×	×	✓	✓	×	TM265643	×	×
Stonham Aspall	✓	✓	✓	✓	✓	TM137604	×	×
Stowmarket	✓	✓	✓	×	×	TM045594	×	×
Stowmarket	✓	✓	✓	×	×	TM046594	×	×
Stowmarket	✓	✓	×	×	×	TM042585	×	×
Stowupland	✓	✓	✓	✓	×	TM067598	×	×
Stradbroke [7]	×	✓	×	×	×	TM223754	×	×
Stradbroke	✓	✓	✓	✓	✓	TM229743	×	×
Stradbroke	✓	✓	✓	✓	✓	TM240740	×	×
Tannington [8]	×	✓	✓	×	×	TM251681	×	×
Thorndon	✓	✓	✓	✓	✓	TM139697	×	×
Thornham Magna	✓	✓	✓	✓	✓	TM110711	×	×
Walsham le Willows	✓	✓	✓	✓	✓	TM005718	×	×
Walsham le Willows	✓	✓	×	×	×	TM006704	×	×
Ward Green	✓	✓	×	×	×	TM048643	×	×
Wattisfield	✓	✓	✓	✓	×	TM013745	×	×
Wetheringsett	✓	✓	✓	✓	✓	TM131651	×	×
Wetheringsett	✓	✓	✓	✓	✓	TM146648	×	✓
Wetherden	✓	✓	✓	✓	×	TL996629	×	×
Wetherden	✓	✓	✓	✓	✓	TM006630	×	×
Worlingworth	×	✓	✓	✓	✓	TM213689	×	×
Worlingworth	✓	✓	✓	✓	✓	TM213688	×	×
Wyverstone	×	✓	×	×	×	TM031677	×	×
Yaxley	✓	✓	×	×	×	TM120737	×	×

[1] This mill was an open-trestle post mill whose first edition 1:2500 ground plan depiction consists of each of the four piers being diminutively shown. There were other open-trestle post mills working in Suffolk at the end of the nineteenth century. Another example shown on this sheet was the mill at Horham.

[2] This mill is shown on the first edition six-inch but, rather surprisingly, not on the preceding first edition 1:2500 yet the same annotation of *Earl Stonham Windmills* was applied to each. It was an open-trestle post mill.

[3] This mill had *Windmill* as part of its annotation on the first edition 1:2500, but then becomes cartographically less visible as a windmill when the annotation on the second edition was reduced to simply *Cranley Mills (Corn)*.

[4] It could be thought that this windmill was lucky to be recorded here on the one-inch Revised New Series since the second edition 1:2500 depiction for this mill is simply an unannotated open circle, though the first edition had shown it as a presumably working mill. The simple explanation here is that this was one of many windmills whose status changed suddenly and dramatically on burning down – in this case of this mill *circa* 1900.

[5] This mill, in contrast to the above mill, cartographically came into its own on the second edition 1:2500 when it was annotated quite normally as *Windmill (Corn)*, even though the corresponding six-inch depiction is just

Windmill. On the first editions, this mill, as part of a broader milling enterprise, had been subsumed within an annotation of *Corn Mills* though at least a circular ground plan identified exactly the mill's position.

[6] This mill was new to the large-scale second editions. Apparently it was a small much-travelled smock mill that only came to this site very late in the nineteenth century and then only worked until *circa* 1914 before disappearing soon after.

[7] This open-trestle mill sat on a pronounced mound that is still marked on the modern 1:25,000 as **Mill Mound**.

[8] The depiction for this windmill on the second edition 1:2500 is an unannotated shaded circle. The fact that this ground plan is shaded – as is normal for this edition – implies that more substantial 'covered-in' remains survived here than had done so at Haughley after a mill there burned down (see Note 4). This surmise is not contradicted by the appearance of the Haughley Mill on the Revised New Series as it would still have been at work in the 1890s, but Tannington Mill is not shown on the Revised New Series because the body of the mill here is thought to have blown down in 1879 leaving just a roundhouse.

The relatively high survival pattern for windmills in Suffolk is made very clear by this sheet. This part of the country was cited in Part One as an area notable for the large proportion of its windmills still at work when the First World War began. Of such mills, there were even some that continued working for several more years, albeit often under straitened circumstances. The corollary of this is that a big majority of all the windmills noted by the facsimile of sheet 190 come within the category of mills that the one-inch Revised New Series felt moved to record – 58 mills out of a total of 86. Furthermore, the number of these 58 mills that were annotated as *Old Windmill* rather than as *Windmill* is a mere two – a mill at Gislingham and, rather disturbingly, the mill at Haughley, discussed in the notes above when it was conveniently assumed that this mill worked up until its destruction by fire. In an age before one thinks of windmills as being targets for vandals and arsonists, the most likely cause of fire in a mill was a failure to ensure a supply of grain to the stones when the mill was working and the stones were in motion. Without grain passing between a stationary bed stone and a revolving top stone, the two were likely to come into contact and sparks would ensue. Since so many of the already exceptionally large number of windmills noted on this sheet were recorded by the Revised New Series, it does show that the mills of this area were generally still capable of working relatively late in the day. This is in contrast to the position on most other sheets considered in Part Two, where the decline of the mill can be seen either to have all but run its course, or at least to have been well established by the 1890s.

Left: The mill at Barley Green, Stradbroke in pristine working order *circa* 1905.

Below: Horham open-trestle post mill as depicted on the first edition 1:2500.

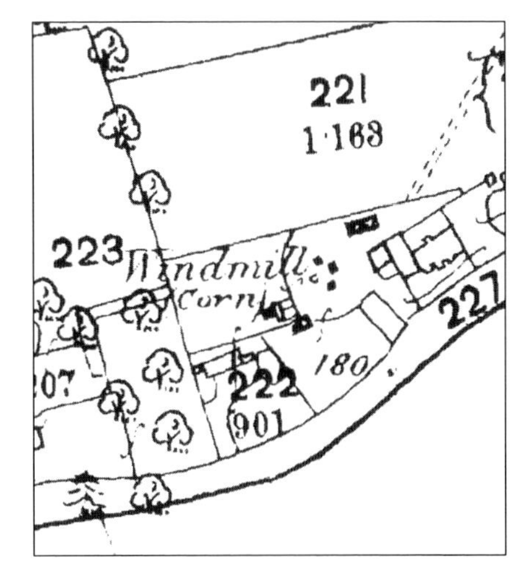

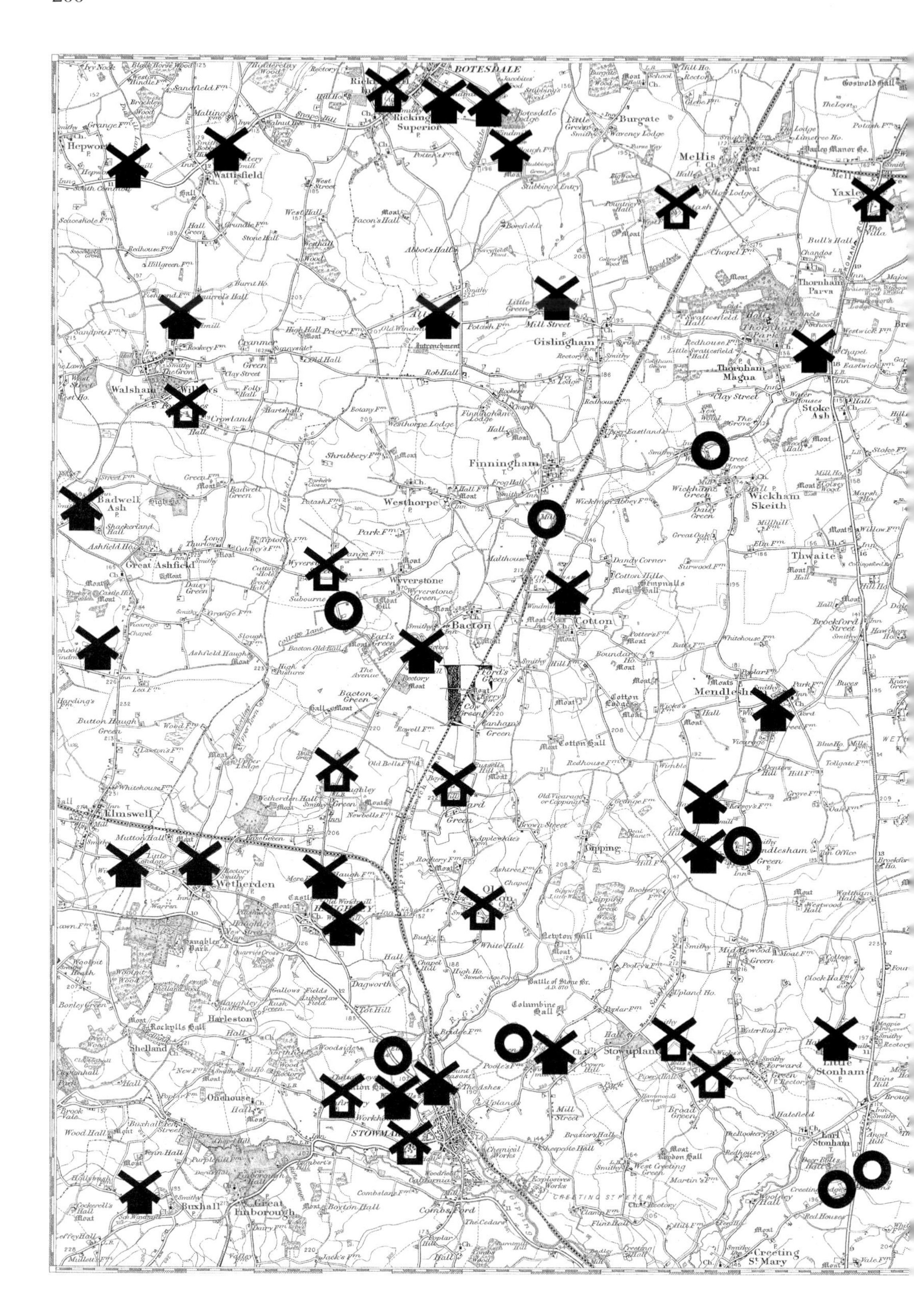
BOTESDALE
Rickinghall Superior
Wattisfield
Burgate
Mellis
Gislingham
Finningham
Westhorpe
Wyverstone
Bacton
Cotton
Badwell Ash
Great Ashfield
Walsham
Elmswell
Wetherden
Mendlesham
Wickham Skeith
Thwaite
Thornham Magna
Thornham Parva
Stoke Ash
Gipping
Old Newton
Haughley
Harleston
Shelland
Onehouse
Stowupland
Little Stonham
Earl Stonham
Combs Ford
Great Finborough
Buxhall
Creeting St Mary
Dagworth
Columbine Hall
Brockford Street
Yaxley

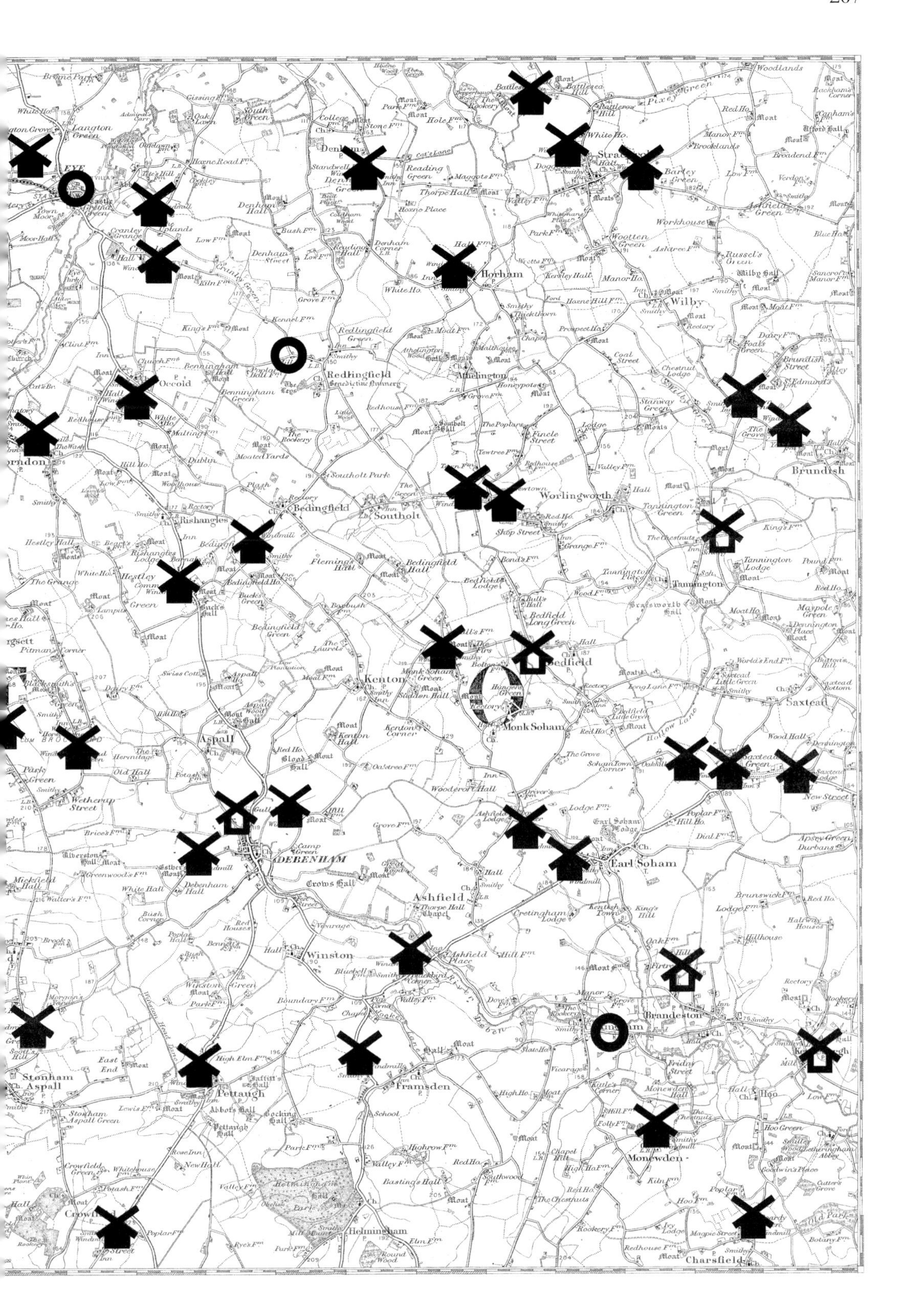

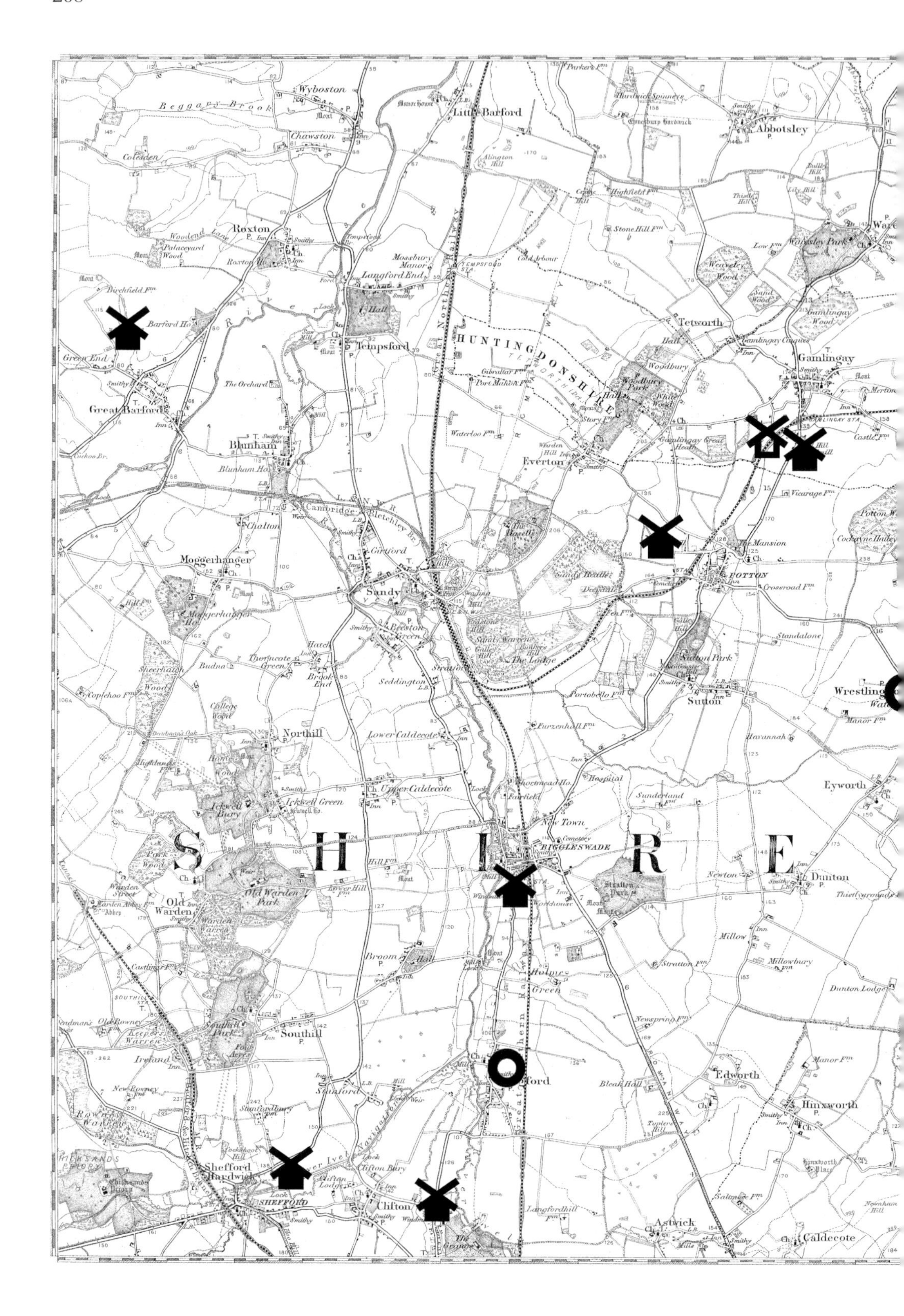
Wyboston
Beggary Brook
Little Barford
Abbotsley
Chawston
Colesden
Roxton
Palaceyard Wood
Roxton Ho.
Birchfield Fm.
Barford Ho.
Green End
Great Barford
The Orchard
Blunham
Blunham Ho.
Tempsford
Langford End
Mossbury Manor
Tempsford Sta.
Great Northern Railway
HUNTINGDONSHIRE
Gibraltar Fm.
Port Mahon Fm.
Waterloo Fm.
Cold Arbour
Stone Hill Fm.
Highfield Fm.
Tetworth
Woodbury
Woodbury Park
Everton
Gamlingay
Gamlingay Great Heath
Waresley Park
Weaveley Wood
Sand Wood
Gamlingay Wood
Low Fm.
Castle Fm.
Vicarage Fm.
Cambridge & Bletchley Br.
Chalton
Girtford
Moggerhanger
Moggerhanger Ho.
Sandy
Beeston Green
The Hasells
Sandy Heath
Sandy Warren
The Lodge
Potton
The Mansion
Crossroad Fm.
Sutton Park
Sutton
Standalone
Wrestlingworth
Manor Fm.
Potton Wood
Cockayne Hatley
Sheerhatch Wood
Copleheo Fm.
Budna
Thorncote Green
Hatch
Brook End
Seddington
Portobello Fm.
Furzenhall Fm.
College Wood
Northill
Lower Caldecote
Upper Caldecote
Ickwell Green
Ickwell Bury
Home Wood
Highlands Fm.
Havannah
Eyworth
Hospital
Sunderland
New Town
BIGGLESWADE
S H I R E
Park Wood
Old Warden Park
Old Warden
Warden Abbey Fm.
Warden Warren
Hill Fm.
Lower Hill Fm.
Stratton Park
Newton
Dunton
Thistlyground Fm.
Millow
Millowbury Fm.
Dunton Lodge
Broom
Holme Green
Stratton Fm.
Castlings Fm.
Southill Sta.
Old Rowney
Keeper's Warren
Southill Park
Southill
Ireland
New Rowney
Rowney Warren
Newspring Fm.
Bleak Hall
Edworth
Manor Fm.
Hinxworth
Stanford
Stanfordbury
Shefford Hardwick
Chicksands Priory
Shefford
River Ivel Navigation
Clifton Bury
Clifton Lodge
Clifton
The Grange
Langfordhill Fm.
Astwick
Topler's Hill
Salamber Fm.
Hinxworth Place
Newnham Hill
Caldecote

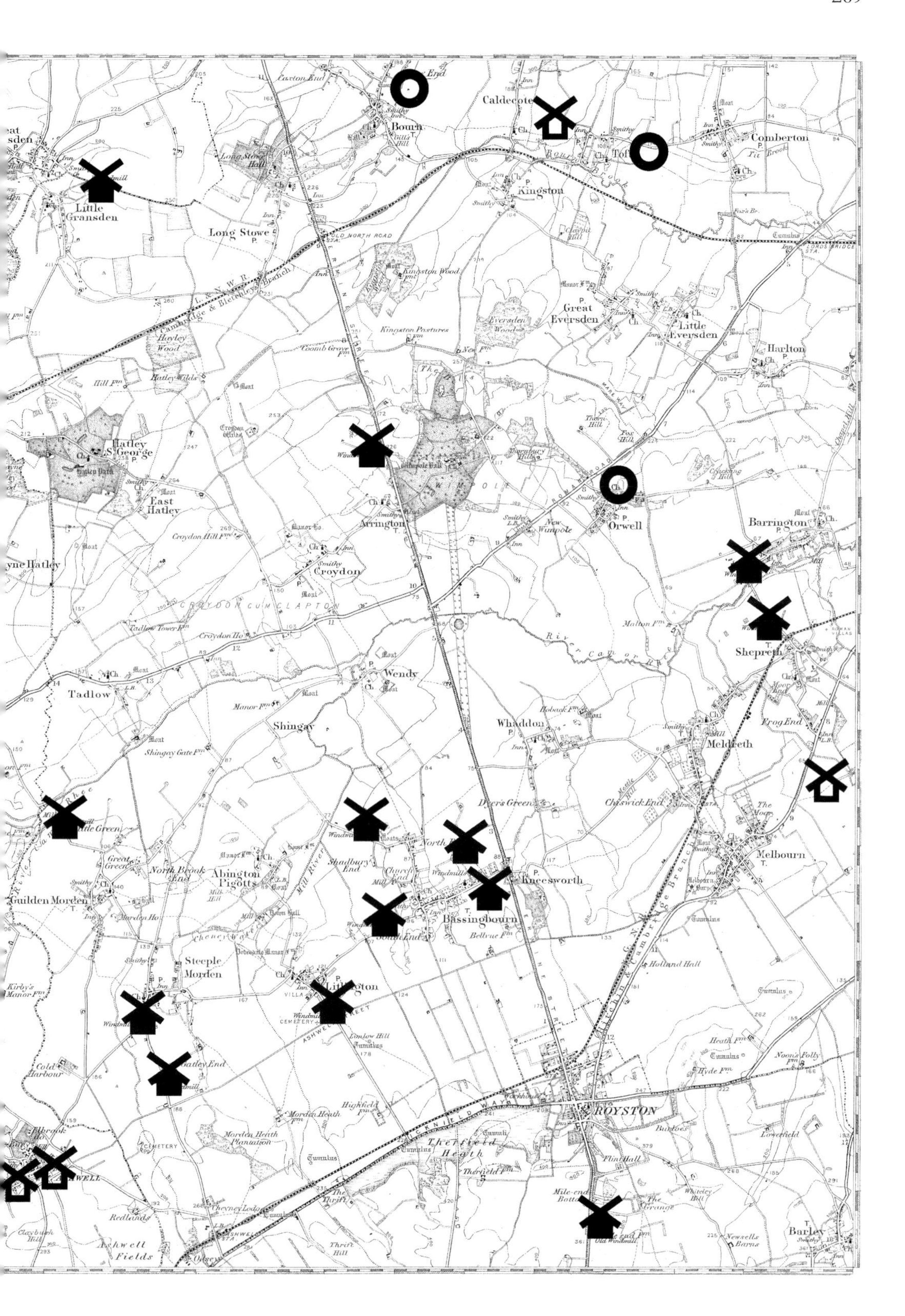
Caldecote
Bourn
Comberton
Toft
Kingston
Little
Gransden
Long Stowe
Great
Eversden
Little
Eversden
Harlton
Hatley
S. George
East
Hatley
Arrington
Orwell
Barrington
Croydon
Shepreth
Wendy
Tadlow
Shingay
Whaddon
Meldreth
Melbourn
Abington
Pigotts
Guilden Morden
Kneesworth
Bassingbourn
Steeple
Morden
ROYSTON
Barley

Sheet 204 *Biggleswade*

(see pages 268 & 269)

Windmill locations as recorded by the Revised New Series *circa* 1895 and/or the first edition 1:2500	19th century recognition			20th century recognition				
	Old Series mixed sheet states [1]	First edition six-inch	Second edition 1:2500	New Series Third Edition	Popular Edition	National Grid Reference	Seventh Series	Current 1:25,000
Arrington [2]	✓	✓	✓	✓	✓	TL323511	×	×
Ashwell	✓	✓	×	×	×	TL268392	×	×
Ashwell	✓	✓	×	×	×	TL269394	×	×
Barrington	✓	✓	✓	✓ *Old*	✓ *OW*	TL386492	×	×
Bassingbourn	×	✓	×	×	×	TL322450	×	×
Bassingbourn [3]	✓	✓	×	Pump	×	TL325436	×	×
Bassingbourn	×	✓	✓	×	✓ *OW*	TL339444	×	×
Bassingbourn	×	✓	✓	×	×	TL339444	✓	×
Biggleswade	×	✓	✓	✓	✓	TL189440	✓	×
Gamlingay	✓	✓	×	×	×	TL237513	×	×
Gamlingay	✓	✓	✓	✓	✓ *OW*	TL237512	✓	×
Great Barford	×	✓	✓	✓ *Old*	×	TL123532	×	×
Great Gransden [4]	✓	✓	✓	✓	✓	TL277555	✓S	✓
Guilden Morden	×	✓	✓	✓ *Old*	✓ *OW*	TL271452	✓	✓
Litlington	✓	✓	✓	✓	×	TL316421	×	×
Melbourn	✓	✓	×	×	×	TL400458	×	×
Mile End Farm	✓	✓	✓	✓ *Old*	✓ *OW*	TL360386	✓	✓
Potton	✓	✓	✓	✓	✓	TL213498	✓	✓
Shepreth	✓	✓	✓	✓ *Old*	×	TL390483	×	×
Southill	×	✓	✓	✓ *Old*	✓ *OW*	TL150395	✓	✓
Steeple Morden	✓	✓	✓	✓	✓	TL284420	✓	✓
Steeple Morden	×	✓	✓	✓	✓	TL289410	×	×
The Grange	✓	✓	✓	✓	✓	TL175390	×	×
Toft [5]	✓	✓	✓	×	×	TL353564	×	×

[1] Old Series quarter-sheets 46NE, 51SW, 52SE together with full sheet 47 are needed here. All states used here are *circa* 1846 with the exception of sheet 52SE whose Margary state 2 (of 12) is *circa* 1840.

[2] This mill is one of those examples where a 1:2500 annotation, in this case *Arrington Mill (Corn)*, on both first and second editions is not really helpful for the map user looking for windmills. But, in this instance, the clear circular ground plan in an open setting provides some compensation.

[3] This mill cannot have made a good impression on the large-scale revisers when they visited it prior to the second edition since the first edition annotation of *Windmill (Corn)* disappears, though the circular ground plan remains on the new edition but it also stays open rather than shaded. Due to these limitations, this mill effectively disappears from the map, though we cannot yet say it has become the windpump that the one-inch Third Edition depicts.

[4] This mill was (and survives as) an open-trestle post mill. The ground plan depictions of this mill on the first and second editions of the 1:2500 are very different to one another, but both are effective in conveying the nature of this mill to the map user.

[5] This mill was also an open-trestle mill like the surviving mill at Great Gransden (see Note 4). Another surviving

example of this type of windmill is even closer to the site of Toft Mill, but the mill at Bourn is shown on another sheet. As with other open-trestle mills, the first edition 1:2500 depiction is of four brick piers diminutively shown and this style of depiction is retained on the second edition, unlike the equivalent pair of depictions used at Great Gransden. Both depictions of Toft Mill are annotated as *Windmill Disused.*

Left: Great Gransden open-trestle post mill *circa* 1908.

Great Gransden Mill seen at 1:2500: *(below left)* on the first edition and *(below right)* on the second edition.

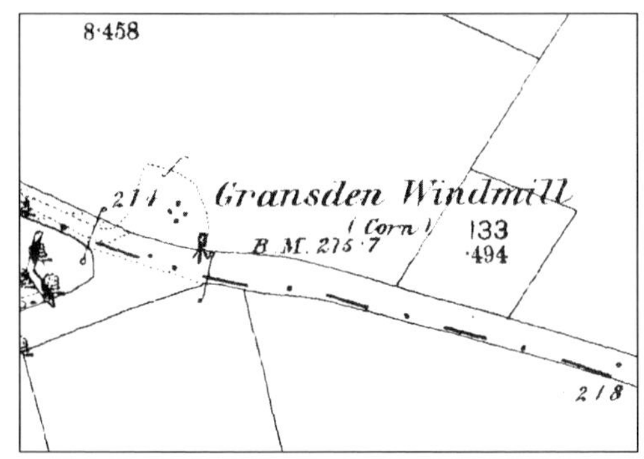

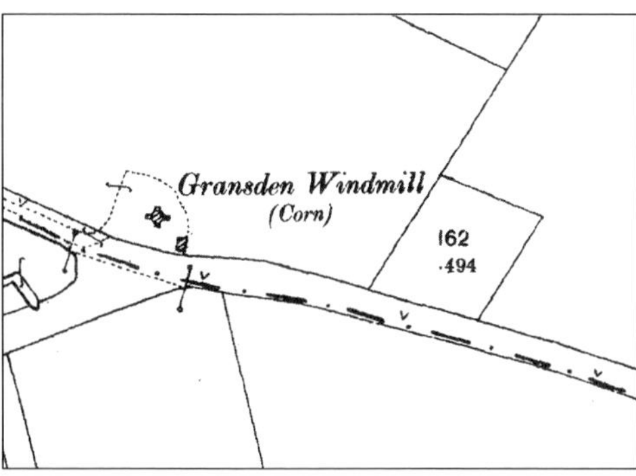

Left: The tower mill that stands outside the small town of Potton close to the road leading to the village of Everton, seen here *circa* 1920.

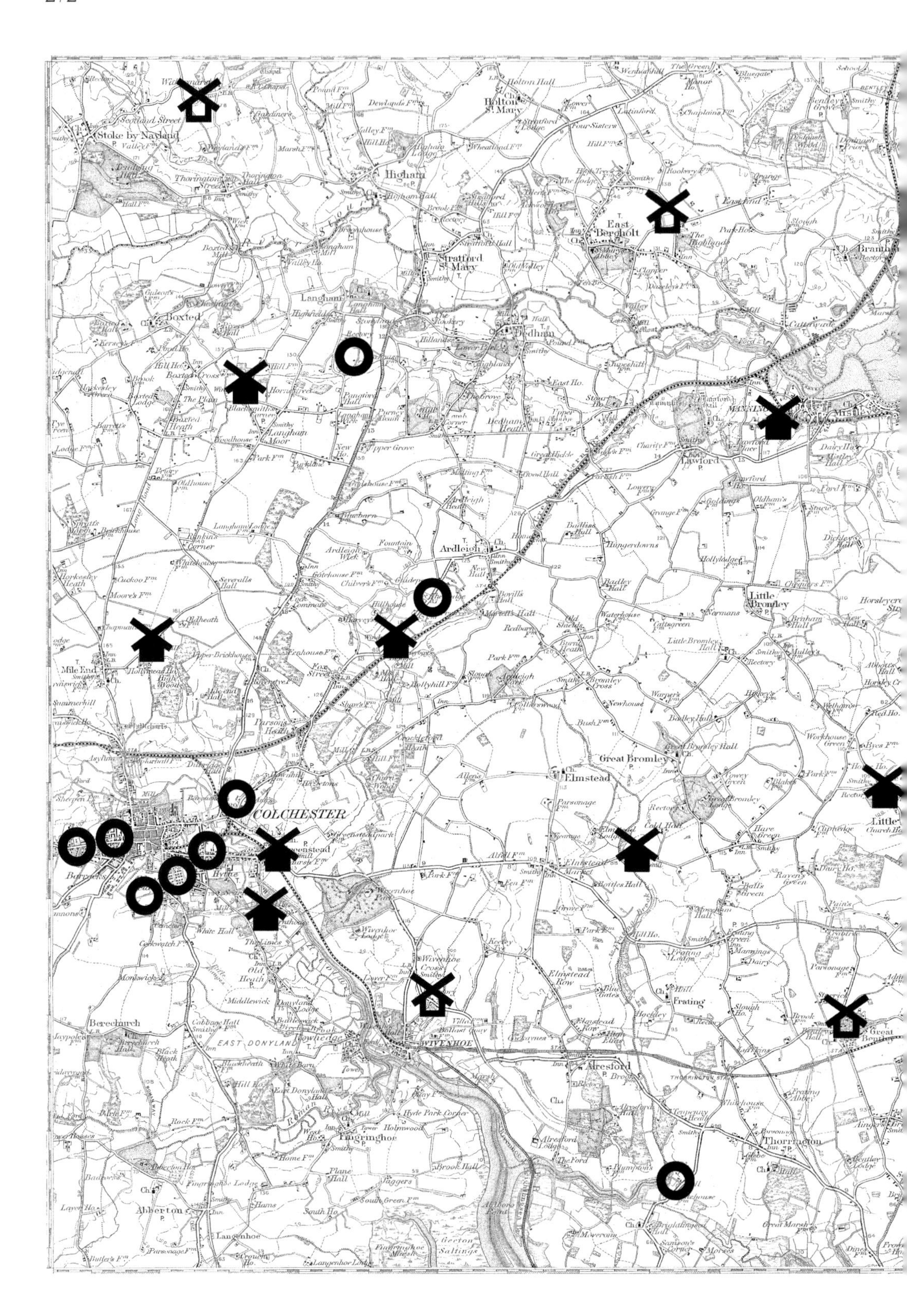

Stoke by Nayland
Scotland Street
Holton Hall
Holton St. Mary
Higham
East Bergholt
Stratford St. Mary
Langham
Boxted
Dedham
Langham Moor
Lawford
Ardleigh
Little Bromley
Mile End
Great Bromley
Elmstead
COLCHESTER
Elmstead Market
Wivenhoe Cross
WIVENHOE
Berechurch
EAST DONYLAND
Rowhedge
Alresford
Frating
Fingringhoe
Thorrington
Abberton
Langenhoe

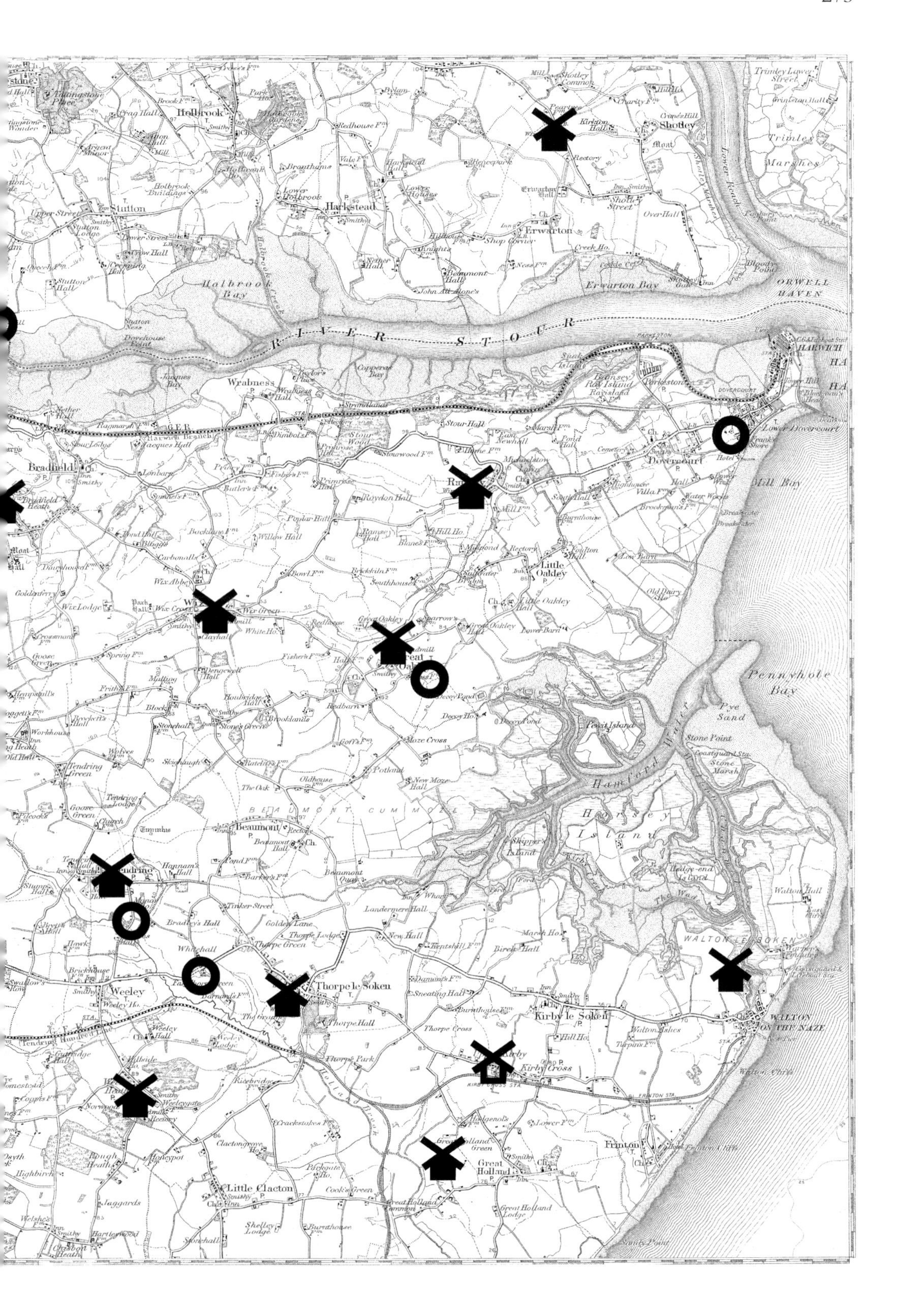
Holbrook
Stutton
Harkstead
Erwarton
Shotley
Holbrook Bay
Erwarton Bay
ORWELL HAVEN
RIVER STOUR
HARWICH
Wrabness
Bradfield
Dovercourt
Mill Bay
Little Oakley
Pennyhole Bay
Pye Sand
Stone Point
Pewit Island
Hamford Water
Horsey Island
Skipper's Island
Beaumont
Tendring Green
Landermere Hall
Thorpe le Soken
Thorpe Hall
Thorpe Park
Weeley
Kirby le Soken
Kirby Cross
Walton Hall
WALTON ON THE NAZE
Great Holland
Frinton
Little Clacton
Rough Heath

Sheet 224 *Colchester*

(see pages 272 & 273)

Windmill locations as recorded by the Revised New Series *circa* 1895 and/or the first edition 1:2500	19th century recognition			20th century recognition				
	Old Series [1] sheet state *ca* 1810	First edition six-inch	Second edition 1:2500	New Series Third Edition	Popular Edition	National Grid Reference	Seventh Series	Current 1:25,000
Ardleigh	✓	✓	✓	×	×	TM038281	×	×
Bradfield	×	✓	✓	✓	✓	TM133302	×	×
Colchester	✓	✓	✓	×	×	TL998281	×	×
Colchester	✓	✓	✓	✓	×	TM018247	×	×
Colchester	✓	✓	×	×	×	TM016237	×	×
East Bergholt	×	✓	×	×	×	TM084350	×	×
Great Bentley	×	✓	×	×	×	TM112219	×	×
Great Bromley [2]	✓	✓	×	×	×	TM077247	×	×
Great Holland [3]	✓	×	×	✓	✓	TM204193	✓	×
Great Oakley [4]	✓	✓	✓	×	×	TM196277	×	×
Langham	×	✓	✓	✓ *Old*	×	TM014322	×	×
Little Bentley	×	✓	✓	✓ *Old*	×	TM120257	×	×
Manningtree	×	✓	✓	✓ *Old*	✓ *OW*	TM102316	×	×
Ramsey	×	✓	✓	✓	✓	TM209304	✓	✓
Shotley	×	✓	✓	✓ *Old*	✓ *OW*	TM224361	×	×
Stoke by Nayland [5]	×	✓	✓	×	×	TM006368	×	×
Tendring [6]	×	×	×	✓	✓ *OW*	TM150239	×	×
Thorpe le Soken	×	✓	✓	✓ *Old*	×	TM178222	×	×
Upper Kirby	×	✓	×	×	×	TM210210	×	×
Walton on the Naze	×	✓	✓	✓	✓	TM252223	×	×
Weeley Heath	×	✓	✓	✓	✓	TM153205	×	×
Wivenhoe [7]	×	✓	×	×	×	TM043223	×	×
Wix	×	✓	✓	×	×	TM167283	×	×

[1] This New Series sheet falls entirely within the coverage of Old Series sheet 48. The recording of windmills by this very early sheet state illustrates the paucity of OS mill recognition in the earliest decades of the nineteenth century. This would later drastically change for this area when sheet 48 became one of the few Old Series sheets to undergo a total overhaul *circa* 1840.

[2] A mill whose depiction on the second edition 1:2500 is meagre – simply an unannotated shaded circle.

[3] The annotation for this mill on the first and second editions 1:2500 as well as on the first edition six-inch is just *Great Holland Mill (Corn)*. Also, crucially, no appropriate ground plan is easily discernible on the first edition 1:2500 but this changes on the second edition to a very obvious shaded circular ground plan forming part of a composite plan.

[4] The one-inch Third Edition is fairly lax here. No symbol is found on the map but a minimal annotation of *Mill* does appear. This may or not be referring to the windmill, or, as is much more likely, this annotation is recording the working steam mill that had been built alongside the windmill. To make matters worse, sheet 224 *Colchester* of the Third Edition had been surveyed in 1904, at exactly the same time that Farries considers this post mill to have been reduced to just a roundhouse or, as he put it '[its] body [was] removed shortly before 1905'.

[5] The farm here has appeared on maps as *Round House Farm* ever since the third edition 1:2500 and it still does so today on the 1:25,000 map. The windmill once here had indeed been a post mill with a roundhouse.

[6] Another windmill whose depictions at 1:2500 on both first and second editions, as well as that of the first edition six-inch, give nothing away. The annotation for all three is *Tendring Mills (Corn)* and no appropriate ground plan is apparent on any edition. The clue lies in the use of the plural expression of 'mills', implying maybe that a windmill has been subsumed within a grander milling enterprise, but without any clear ground plans the map user is forced to conclude that the one-inch scale has been of more use for this windmill than the larger scales.

[7] Just one of many mills where the cartographic evidence is corroborated by the known fate of a mill. The circular ground plan of this mill on the first edition 1:2500 disappears from the second edition and the annotation changes from *Wivenhoe Windmill (Corn)* to become *Wivenhoe Mill (Corn)*. This matches the scenario of an old post mill here that burnt down in 1882 leaving an adjoining steam mill, which then continued working for many years.

Great Holland Mill at 1:2500: *(above)* on the first edition and *(above right)* on the second edition.

Right: Great Holland Mill *circa* 1910.

Walton on the Naze Mill: *(below left)* as it appeared *circa* 1910 and *(below right)* as depicted on the first edition 1:2500 amid watery surroundings.

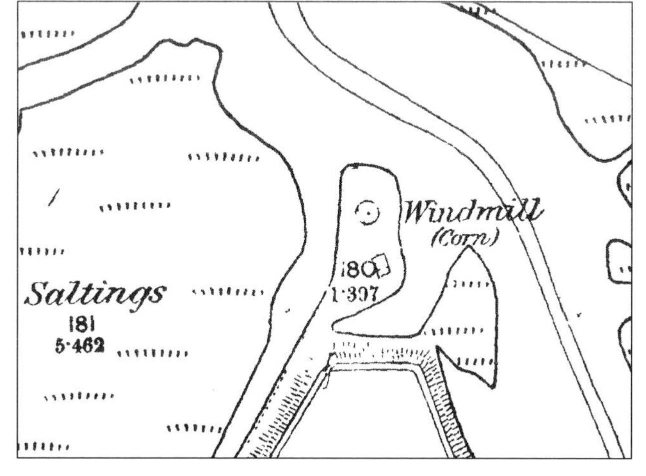

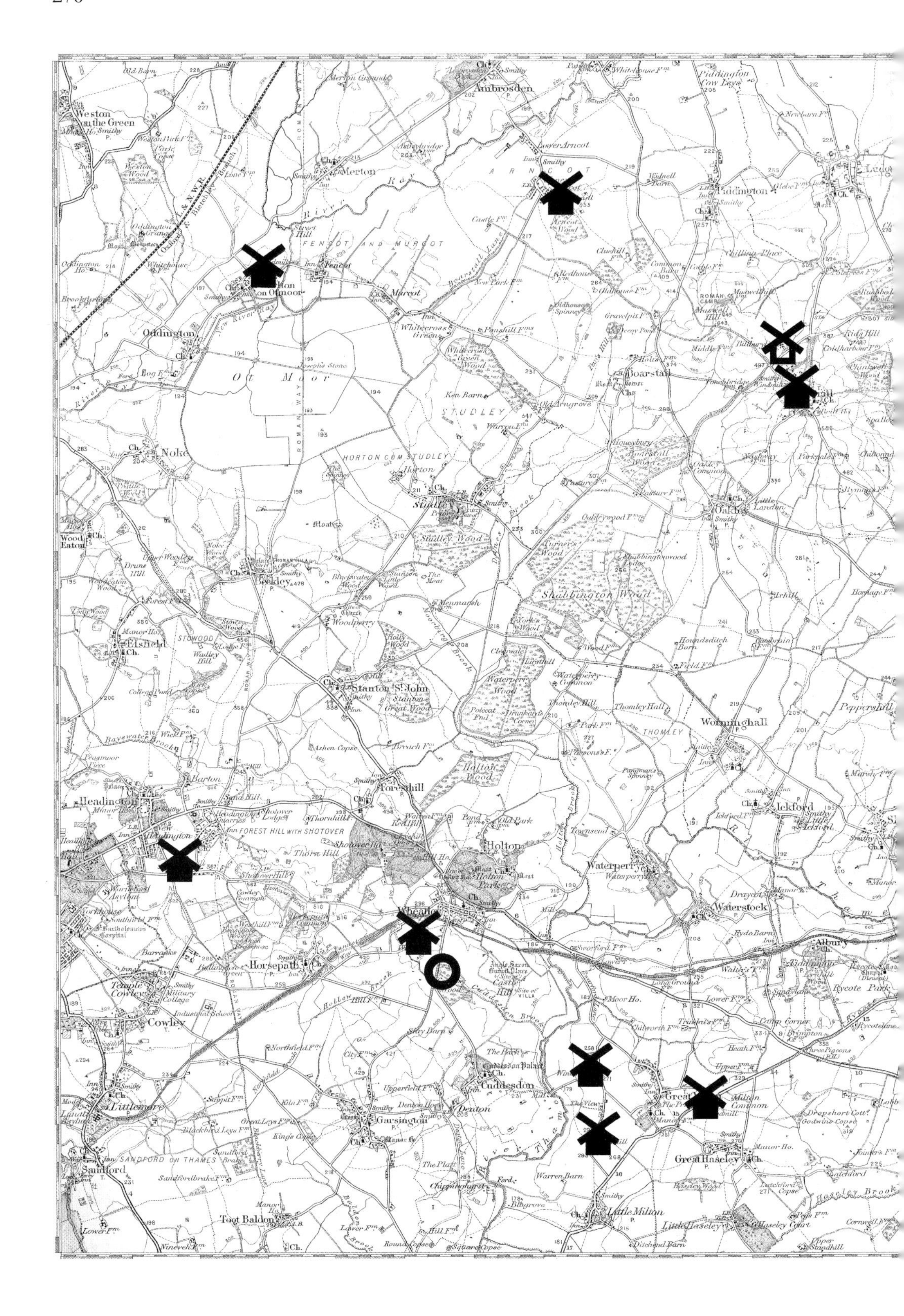
Ambrosden
Weston on the Green
Merton
Lower Arncot
Piddington
Fencot
Murcot
Oddington
Ot Moor
Boarstall
Noke
Studley
Horton
Wood Eaton
Beckley
Elsfield
Woodperry
Stanton St. John
Shabbington Wood
Worminghall
Forest Hill
Headington
Ickford
Holton
Waterperry
Waterstock
Wheatley
Horsepath
Albury
Temple Cowley
Cowley
Cuddesdon
Denton
Garsington
Littlemore
Great Haseley
Sandford
Toot Baldon
Little Milton
Little Haseley

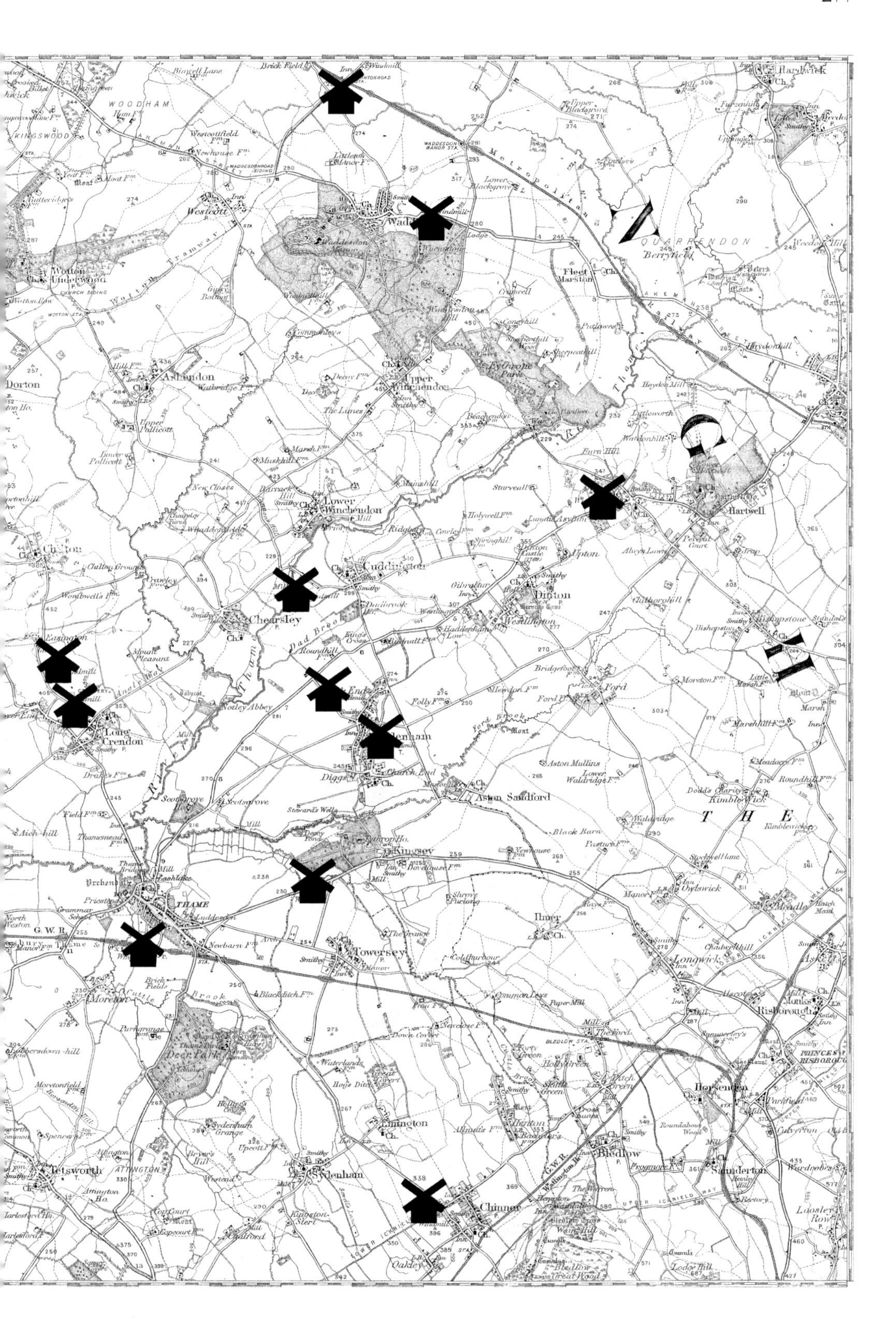

Brick Field
Windmill
Hardwick
WOODHAM
KINGSWOOD
Westcottfield Fm.
Newhouse Fm.
Westcott
Metropolitan
Lower Blackgrove
QUARRENDON
Berryfield
Wotton Underwood
Fleet Marston
Waddesdon Manor
Ashendon
Dorton
Upper Winchendon
Eythrope Park
Haydon Mill
Hartwell
Lower Winchendon
Cuddington
Upton
Dinton
Westlington
Chearsley
Chilton
Easington
Long Crendon
Notley Abbey
Bishopstone
Ford
Haddenham
Church End
Aston Sandford
Aston Mullins
Kimble Wick
THE
Scotsgrove
Kingsey
Thame
THAME
North Weston
G. W. R.
Towersey
Ilmer
Owlswick
Longwick
Moreton
Thame Park
Deer Park
Mortonfield Fm.
Sydenham Grange
Emmington
Henton
Bledlow
Horsenden
Saunderton
Tetsworth
Attington
Sydenham
Chinnor
Kingston Stert
Oakley
Princes Risborough
Lodge Hill
Loosley Row

Sheet 237 *Thame*

(see pages 276 & 277)

Windmill locations as recorded by the Revised New Series *circa* 1895 and/or the first edition 1:2500	19th century recognition			20th century recognition				
	Old Series sheet states *ca* 1834	First edition six-inch	Second edition 1:2500	New Series Third Edition	Popular Edition	National Grid Reference	Seventh Series	Current 1:25,000
Brick Field	✓	✓	✓	✓*Old*	×	SP741192	×	×
Brill	✓	✓	✓	×	×	SP652143	×	×
Brill [1]	✓	✓	✓	✓	✓*W*	SP652141	✓S	✓
Charlton on Otmoor	✓	✓	✓	✓	×	SP564161	×	×
Chinnor	✓	✓	✓	✓	✓	SP749010	✓	×
Cuddington	✓	✓	✓	✓	✓*OW*	SP731109	×	×
Great Haseley	✓	✓	✓	✓	✓*OW*	SP638024	✓	✓
Great Milton	✓	✓	✓	×	×	SP618030	×	×
Haddenham [2]	✓	×	✓	✓	✓	SP734093	×	×
Haddenham	✓	✓	✓	✓*Old*	✓*OW*	SP744085	×	×
Headington	✓	✓	×	×	×	SP549065	×	×
Little Milton	✓	✓	✓	✓*Old*	✓*OW*	SP619020	×	×
Long Crendon	✓	✓	✓	✓	×	SP691098	×	×
Long Crendon	✓	✓	✓	✓*Old*	✓*OW*	SP693093	×	×
Stone	✓	✓	✓	✓	✓*OW*	SP781123	×	×
Thame	✓	✓	✓	×	×	SP705052	×	×
Towersey	✓	✓	✓	✓*Old*	×	SP733062	×	×
Upper Arncot	×	✓	✓	✓	✓*OW*	SP614173	✓	×
Waddesdon	✓	✓	✓	✓*Old*	✓*OW*	SP753168	×	×
Wheatley [3]	✓	✓	✓	✓	✓	SP589052	✓	✓

[1] Very unusually, this mill has an annotation of *Windmill* rather than *Old Windmill* on the one-inch Popular Edition.

[2] This mill does not appear on the first editions of the large scales but makes its debut on the second editions. This is strange as this was a normal tower mill that photographs show to have been in a disused state as early as *circa* 1910.

[3] The 1:2500 annotations for this mill changed from *Windmill (Ochre)* on the first edition to *Windmill (Corn)* on the second edition.

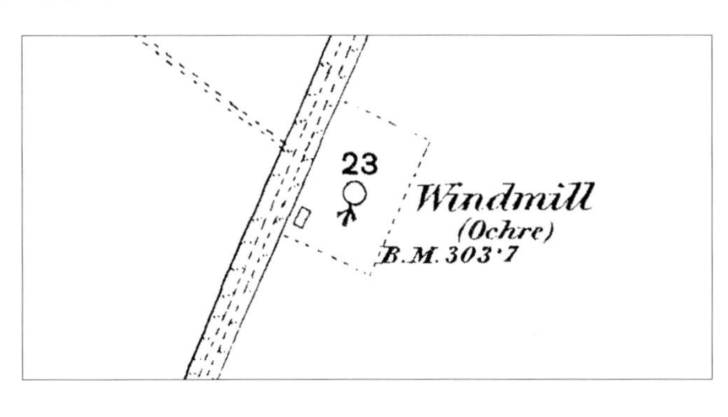

Left: Little Milton Mill seen on the first edition 1:2500. A feature of this mill is its use as a benchmark by the Survey, hence *B.M.* and the War Department 'arrow' on the map showing where on the mill this height above sea level had been measured to.

Perhaps rather surprisingly for a part of the country that was not, relatively speaking, isolated and that would have been influenced by its proximity to the capital more so than most other areas discussed in Part Two, the sophistication of the windmills here was not one might have expected it to be. The mix between post mills and tower mills was more in favour of post mills than for almost every other sheet discussed. Moreover, although the post mills of, say, Suffolk and Kent could be more than a match, technically speaking, for their tower and smock mill counterparts, the post mills here in Oxfordshire and Buckinghamshire were small and crude by comparison. Many were open-trestle post mills which, as we have seen, epitomises the most basic form of a windmill on the assumption that if a simple brick roundhouse never got built for a mill, then it was unlikely to have had costly internal machinery vested in it.

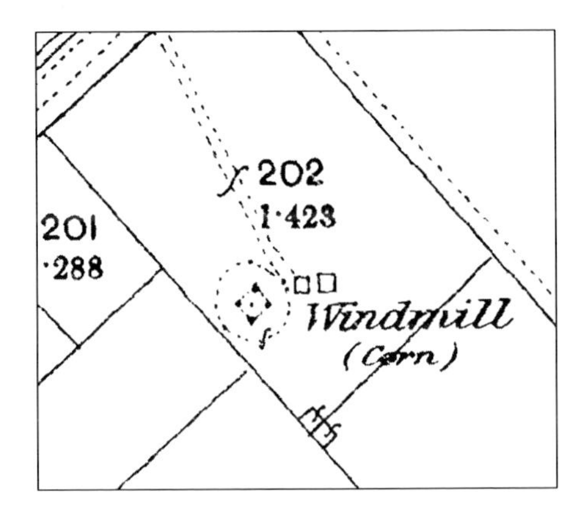

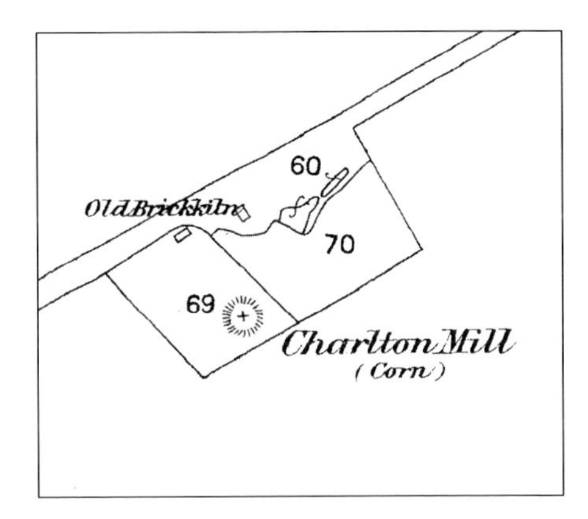

Typical of the idiosyncratic post mill depictions seen on the first edition 1:2500 are: *(above left)* the mill at Brick Field and *(above right)* the mill at Charlton on Otmoor.

The post mill that still survives at Brill *(above) circa* 1908 and *(right)* on the first edition 1:2500 with a second post mill that was pulled down shortly after the start of the twentieth century. Both of these windmills can be seen in Plate 2.

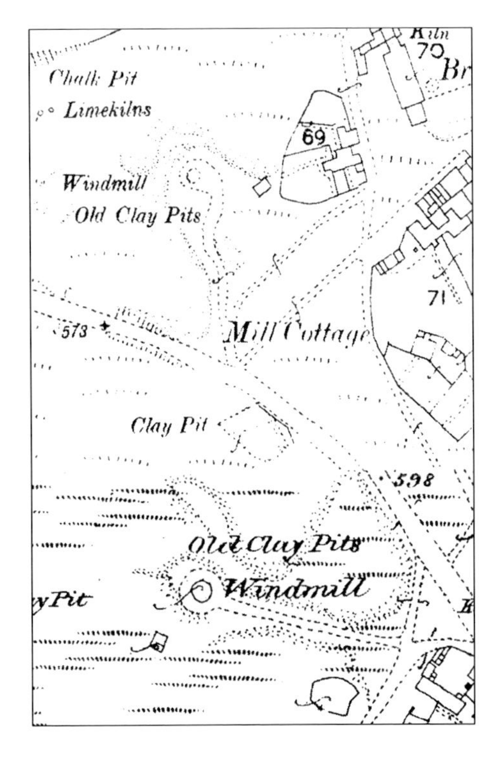

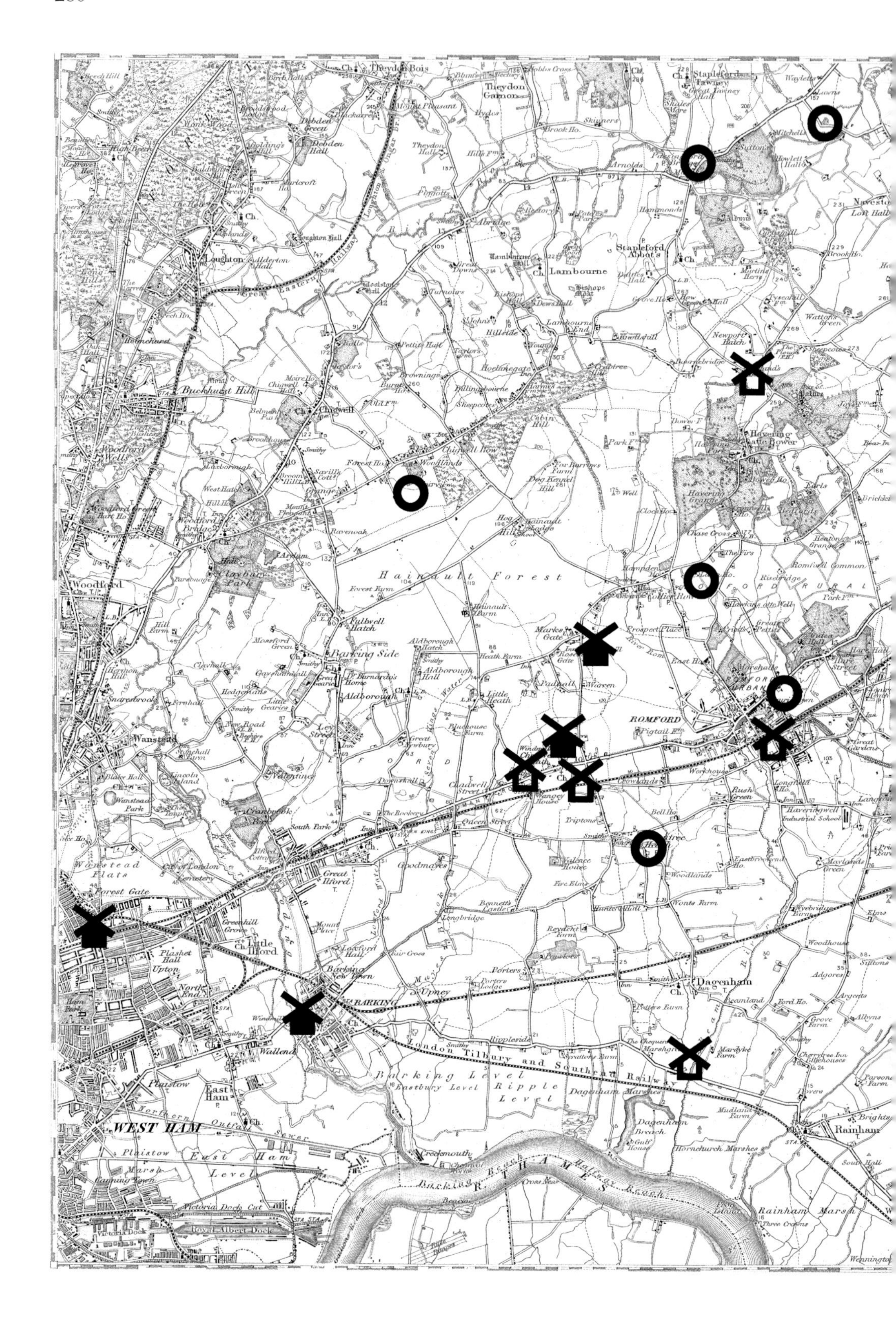
Theydon Bois
Theydon Garnon
Stapleford Tawney
Loughton
Buckhurst Hill
Chigwell
Lambourne
Stapleford Abbot's
Havering atte Bower
Woodford
Wanstead
Hainault Forest
Barking Side
Aldborough
ROMFORD
Chadwell Street
Great Ilford
Little Ilford
Forest Gate
Barking
Dagenham
WEST HAM
East Ham
Plaistow
Wallend
Rainham
Barking Level
London Tilbury and Southend Railway
Royal Albert Dock
Victoria Dock
R. THAMES

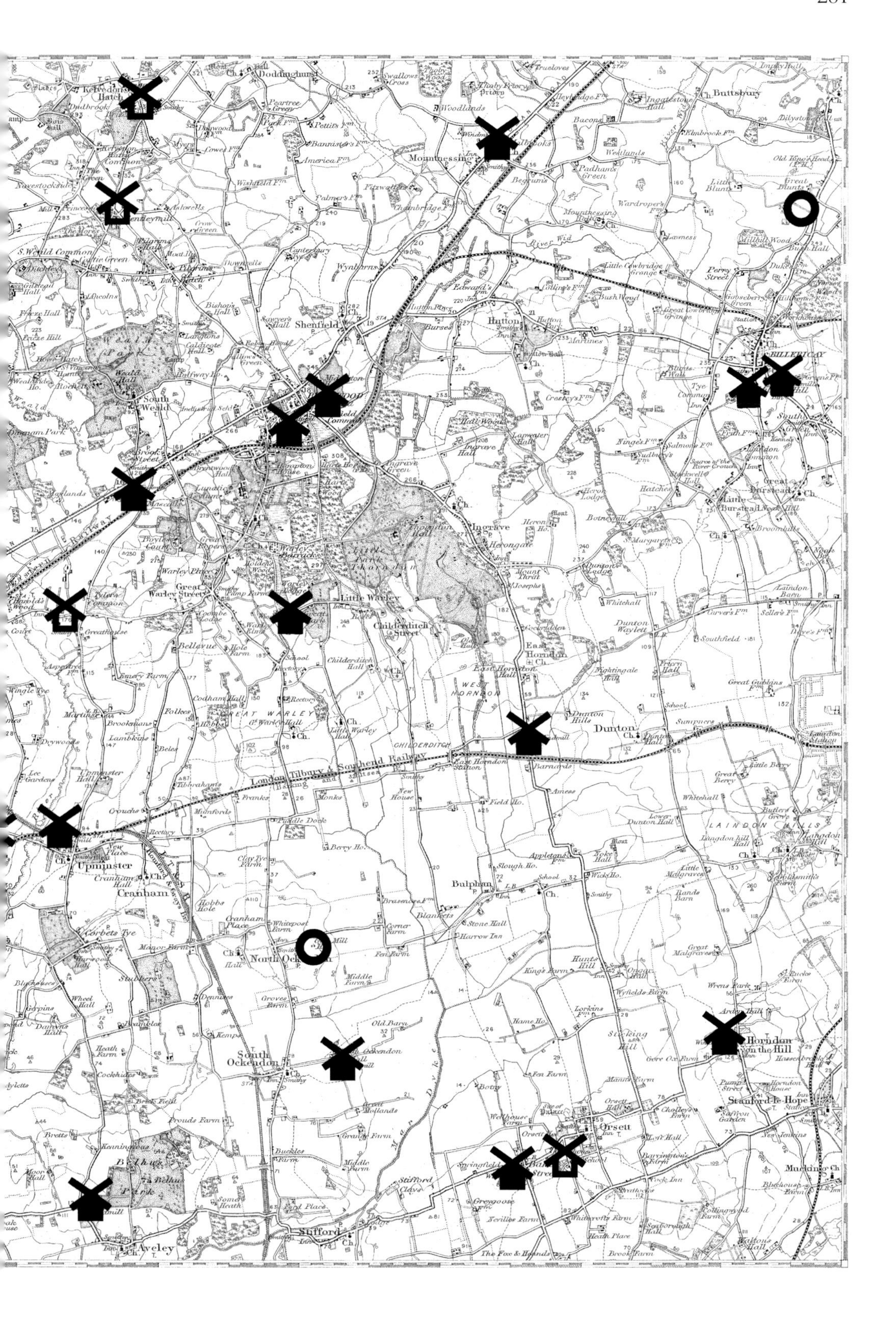

Sheet 257 *Romford*

(see pages 280 & 281)

Windmill locations as recorded by the Revised New Series *circa* 1895 and/or the first edition 1:2500	19th century recognition			20th century recognition				
	Old Series [1] sheet states *ca* 1845	First edition six-inch	Second edition 1:2500	New Series Third Edition	Popular Edition	National Grid Reference	Seventh Series	Current 1:25,000
Aveley	✓	✓	✓	✓*Old*	×	TQ562807	×	×
Barking [2]	✓	×	✓	✓	✓*OW*	TQ436839	×	×
Bentleymill [3]	✓	✓	×	×	×	TQ568968	×	×
Billericay	✓	✓	✓	✓	×	TQ677939	×	×
Billericay	✓	✓	✓	✓	✓	TQ678939	×	×
Boylands Oak	✓	✓	✓	✓*Old*	✓*OW*	TQ511943	×	×
Brentwood	✓	✓	×	×	×	TQ597934	×	×
Brentwood	✓	✓	×	×	×	TQ601937	×	×
Chadwell Heath	✓	✓	✓	✓*Old*	×	TQ480883	×	×
Chadwell Heath	✓	✓	×	×	×	TQ480883	×	×
Chadwell Heath	✓	✓	×	×	×	TQ480882	×	×
Dagenham [4]	✓	✓	✓	×	×	TQ502832	×	×
East Horndon [5]	✓	✓	✓	✓*Old*	×	TQ636883	×	×
Forest Gate	×	✓	×	×	×	TQ399854	×	×
Hornchurch	✓	✓	✓	✓	✓	TQ544867	×	×
Horndon on the Hill	✓	✓	✓	✓*Old*	✓*OW*	TQ669833	×	×
Kelvedon Hatch	✓	✓	✓	✓	✓	TQ572987	×	×
Little Warley	✓	✓	×	×	×	TQ596903	×	×
Marks Gate	✓	✓	✓	✓*Old*	✓*OW*	TQ485899	×	×
Mountnessing	✓	✓	✓	✓	✓	TQ631980	✓	✓
Orsett	✓	✓	✓	✓*Old*	✓*OW*	TQ633813	✓	✓
Orsett	✓	✓	✓	×	×	TQ641814	×	×
Romford	✓	✓	×	×	×	TQ515885	×	×
South Ockendon	✓	✓	✓	✓	✓	TQ604831	✓	×
South Weald	✓	✓	✓	✓	×	TQ571924	×	×
Tylers Common	✓	✓	×	×	×	TQ558904	×	×
Upminster	✓	✓	✓	✓	✓	TQ557867	✓	✓

[1] This New Series sheet falls entirely within the coverage of Old Series sheet 1. Sheet 1 (as discussed in Part One) was extensively revised *circa* 1840 from the full sheet that had first been published in 1805. Subsequently issued in quarter-sheets, the states used here are the first for these new sheets published in 1843-4. They differ significantly in their windmill counts from the last known state of the full sheet that was still current in 1840. The list of 27 mills above shows that all but one of them was recorded by the revised sheet 1, whereas the unrevised sheet of just three years beforehand had only recorded sixteen of them.

[2] The decline in importance of this windmill so close to London is shown by the familiar change from first edition 1:2500 depiction to second edition depiction. This mill appears as *Wellington Mill (Windmill)* on the first edition, but becomes *Wellington Mill (Corn)* for the second edition. The windmill's polygonal ground plan remains prominent on the second edition though now it is shaded as is normal for this edition. This was a large smock mill and, even after losing its sails, it remained an imposing structure until it was demolished in 1926.

[3] This mill, not too far from London, as well as others from the area had turned to the supplementary use of steam power. This option open to a miller was raised in Part One when discussing Upminster Mill. Eventually this type of power source became more convenient and dependable to use and it ousted the use of wind power. Consequently, many steam plants worked well into the twentieth century, but not here at Bentleymill. The first large-scale editions both show *Flour Mill (Wind & Steam)* but this all disappears from the second editions and, as a speedy example of memorialising this local association, the overall site becomes *Millfield* on the one-inch Third Edition.

[4] Doing rather better than Bentleymill (see above note), this mill survived onto the second edition 1:2500 but only as a shaded circular ground plan incorporated into a composite rectangular ground plan, with no annotation being supplied. The same familiar sort of descent into anonymity can also be inferred on this sheet for mills at Billericay, the last survivor from the trio at Chadwell Heath, Kelvedon and Orsett.

[5] Here it is not so obvious that steam power would eventually win the day. The first editions of the six-inch and the 1:2500 both provide the annotation *East Horndon Mills (Corn)* but there is also a clear circular ground plan. On the second editions for both scales the same annotation appears; the circular ground plan is still present but there is now a second separate annotation of *Windmill* to accompany it. This new found fame was short-lived. This mill was almost certainly gone by 1905.

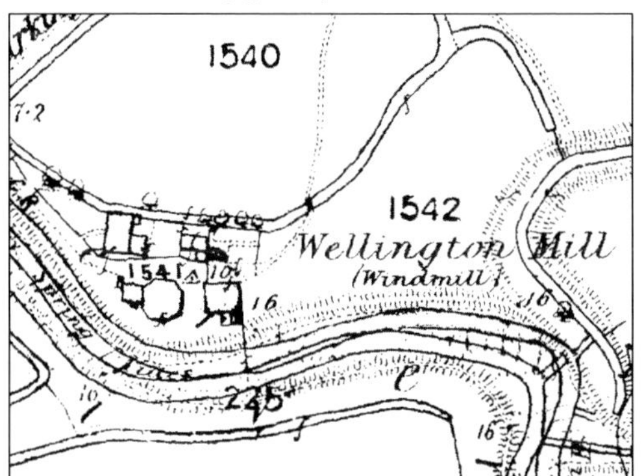

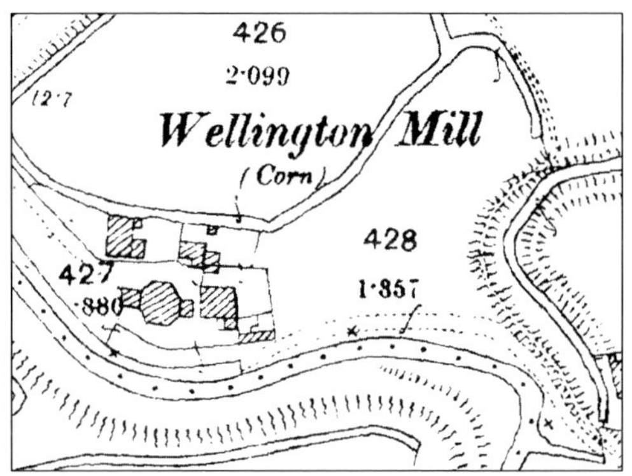

Barking Mill as it appeared at 1:2500 on *(above left)* the first edition and *(above right)* the second edition.

Above: Barking Mill remained an impressively large structure even after the loss of its sails, as this picture taken *circa* 1908 demonstrates.

Right: Horndon on the Hill Mill *circa* 1910.

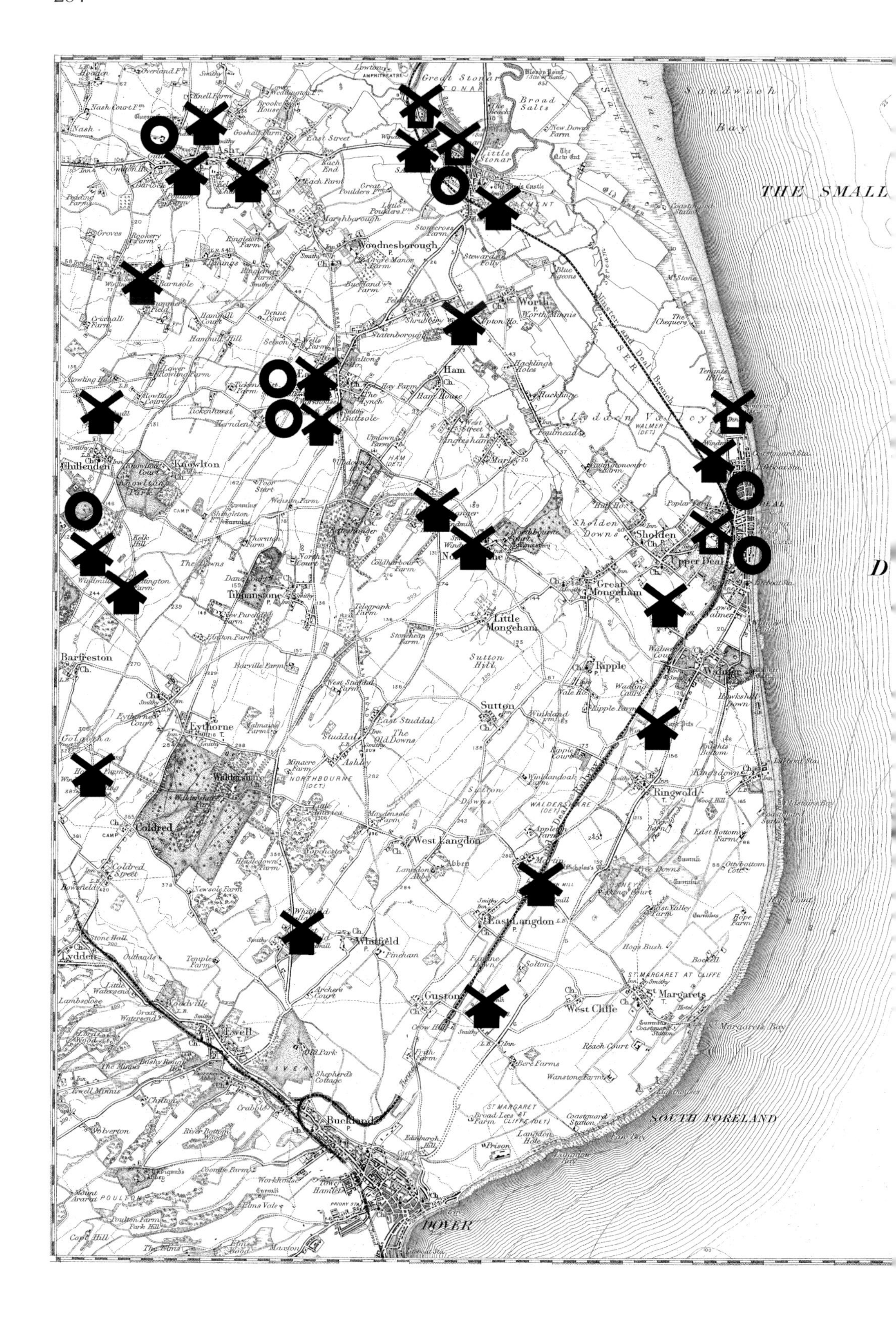

Sheet 290 *Dover*

(see page 284)

Windmill locations as recorded by the Revised New Series *circa* 1895 and/or the first edition 1:2500	19th century recognition			20th century recognition				
	Old Series sheet state *ca* 1865	First edition six-inch	Second edition 1:2500	New Series Third Edition	Popular Edition	National Grid Reference	Seventh Series	Current 1:25,000
Ash	×	✓	✓	✓	×	TR287588	×	×
Ash	✓	✓	✓	✓	✓*OW*	TR282582	×	×
Ash [1]	✓	✓	✓	✓	✓	TR292581	×	×
Chillenden	✓	✓	✓	✓	✓	TR269542	✓S	✓
Deal	✓	✓	✓	×	×	TR373536	×	×
Deal	✓	✓	×	×	×	TR377540	×	×
Deal [2]	✓	✓	×	×	×	TR371522	×	×
Eastry	✓	✓	✓	✓	✓	TR304546	✓	✓
Eastry	✓	✓	✓	✓	✓	TR304544	×	×
Guston	×	✓	✓	✓	✓	TR333444	✓	✓
Hasling Dane	×	✓	✓	✓	✓*OW*	TR266483	×	×
Kettington Farm	×	✓	✓	✓	✓*OW*	TR269517	×	×
Kettington Farm	×	✓	✓	✓	×	TR270517	×	×
Martin	✓	✓	✓	✓	✓	TR342465	×	×
Northbourne	✓	✓	✓	✓	✓	TR326525	×	×
Northbourne	×	✓	✓	✓	×	TR331521	✓	✓
Ripple	✓	✓	✓	✓	✓	TR362490	✓	✓
Sandwich	✓	✓	✓	✓	✓	TR322586	✓	✓
Sandwich	✓	✓	×	×	×	TR324588	×	×
Sandwich	✓	×	×	×	×	TR325588	×	×
Sandwich	✓	✓	×	×	×	TR333579	×	×
Staple	×	✓	✓	✓	×	TR276563	×	×
Upper Deal	✓	✓	✓	✓	✓	TR363510	×	×
Whitfield	✓	✓	✓	✓	×	TR302457	×	×
Worth [3]	✓	✓	✓	✓	×	TR330557	×	×

[1] In a change of depiction that is normally the other way around, the first editions of the large scales show only an unannotated circular ground plan for this mill but on the second edition 1:2500 an annotation of *Windmill (Corn)* is applied. This has then been reduced to just *Windmill* for the second edition of the six-inch.

[2] The depictions of this windmill on the first editions of the large scales are instructive. Anyone searching for the windmills of this area using 1:2500 sheets is condemned to disappointment in the case of this mill. No discernible ground plan appears here at this scale and the annotation for the site is simply *Corn Mill.* Unexpectedly, and rather disturbing for anyone who thinks that using the 1:2500 should not only be more than adequate to discover all mills but also that it represents the best possible way of doing so, is in for a surprise here. With no clear indication of a windmill on the 1:2500, the six-inch on the other hand supplies an annotation of *Corn Mill (Wind & Steam).* By the time of the second editions all trace of this windmill has vanished.

[3] Perhaps to demonstrate that this sheet is not completely anomalous in its 1:2500 windmill depictions, this mill at Worth conforms to the pattern used for most mills on most of the sheets reviewed in this Part Two. A standard annotation of *Windmill (Corn)* is used for this mill on the first editions of both 1:2500 and six-inch scales. On the second edition 1:2500 the circular ground plan is now shaded and the annotation has become *Worth Mill (Corn).*

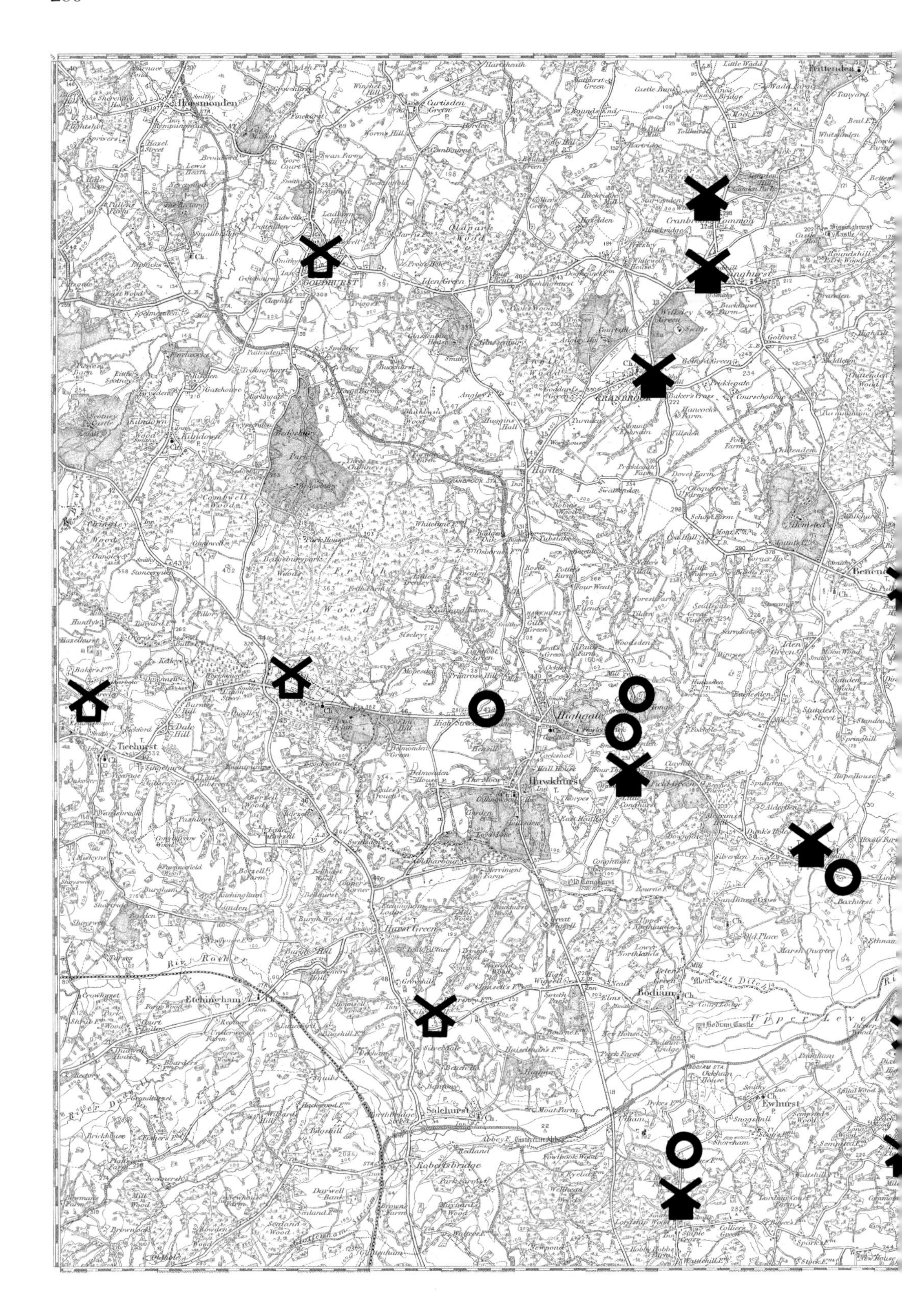
Horsmonden
Goudhurst
Iden Green
Cranbrook
Hartley
Hawkhurst
Highgate
Ticehurst
Etchingham
Hurst Green
Bodiam
Bodiam Castle
Salehurst
Robertsbridge
Ewhurst
Frith Wood
Riv. Rother
Kent Ditch

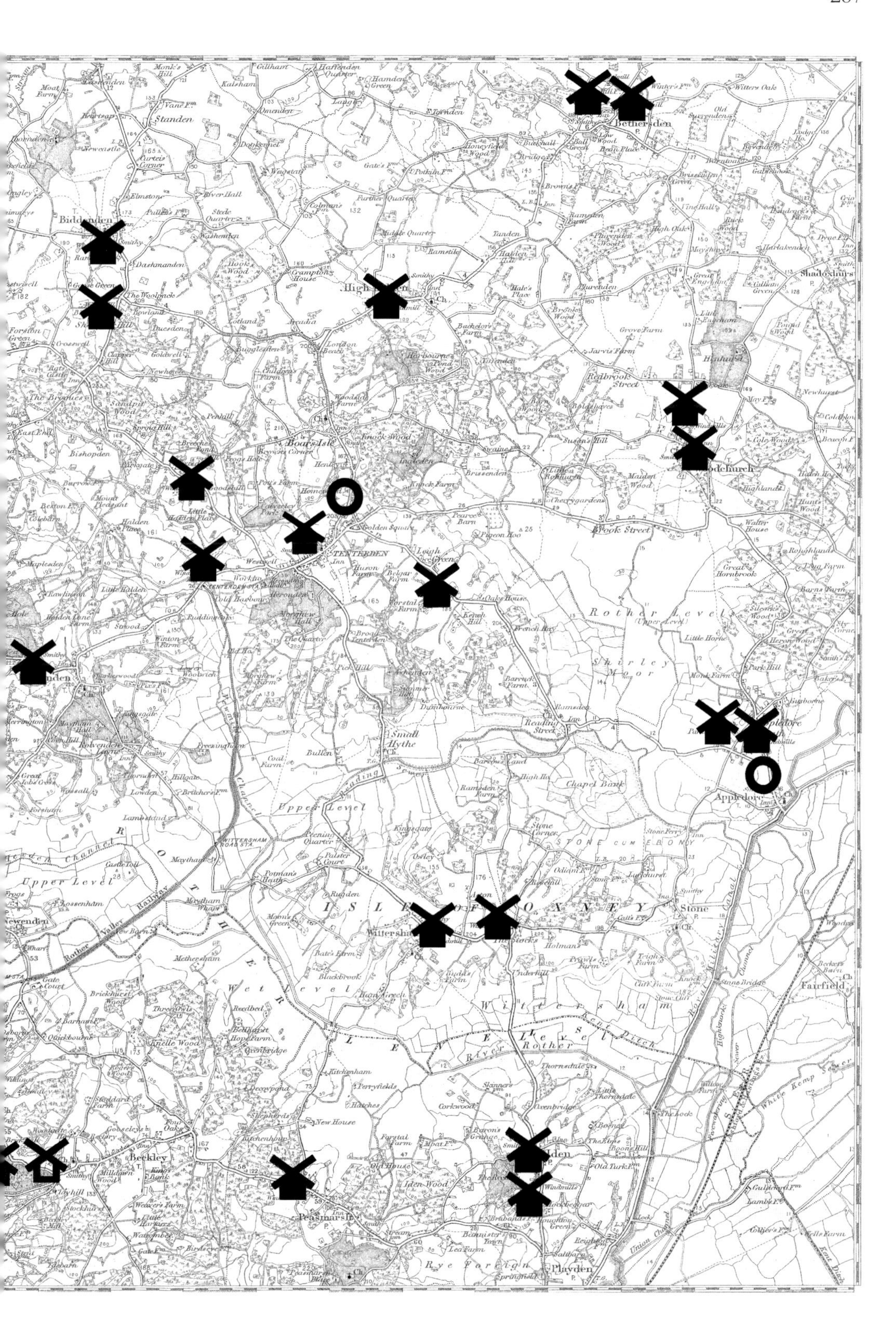
Biddenden
Bethersden
Standen
Tenterden
Boar's Isle
Knock Wood
Woodchurch
Brook Street
Rother Levels
(Upper Level)
Shirley Moor
Reading Street
Small Hythe
Appledore
Chapel Bank
Upper Level
Isle of Oxney
Wittersham
Stone
Wet Level
Rother Levels
River Rother
Fairfield
Peasmarsh
Beckley
Iden
Playden
Rye Foreign
Newenden
Rolvenden
Royal Military Canal
Shadoxhurst
Rother Valley Railway
Union Channel

Sheet 304 *Tenterden*

(see page 286 & 287)

Windmill locations as recorded by the Revised New Series *circa* 1895 and/or the first edition 1:2500	19th century recognition			20th century recognition				
	Old Series mixed sheet states [1]	First edition six-inch	Second edition 1:2500	New Series Third Edition	Popular Edition	National Grid Reference	Seventh Series	Current 1:25,000
Appledore [2]	×	✓	✓	×	×	TQ952304	×	×
Appledore	×	✓	✓	×	×	TQ953304	×	×
Beckley	✓	✓	×	×	×	TQ839235	×	×
Benenden	✓	✓	✓	✓	✓	TQ821324	✓	✓
Bethersden	×	×	✓	✓	✓*OW*	TQ930407	×	×
Bethersden	✓	✓	✓	✓	✓	TQ935404	×	×
Biddenden	×	✓	✓	✓	✓*OW*	TQ851382	×	×
Biddenden	×	✓	✓	✓	×	TQ850371	×	×
Cranbrook	×	✓	✓	×	×	TQ789387	×	×
Cranbrook [3]	×	✓	✓	✓	✓	TQ779359	✓S	✓
Flimwell	×	✓	✓	×	×	TQ718312	×	×
Goudhurst	✓	✓	×	×	×	TQ725379	×	×
High Halden	✓	✓	✓	✓	✓	TQ896373	×	×
Iden	×	✓	✓	✓*Old*	×	TQ917234	×	×
Northiam	✓	✓	✓	✓	✓	TQ823252	×	×
Northiam	×	✓	×	✓*Old*	×	TQ823234	×	×
Northiam	×	✓	×	×	✓	TQ824234	×	×
Peasmarsh	✓	✓	✓	✓	✓	TQ879231	×	×
Playden	✓	✓	×	✓*Old*	×	TQ917232	×	×
Rolvenden	✓	✓	✓	✓	✓*OW*	TQ838315	✓S	✓
Salehurst	✓	✓	×	×	×	TQ743258	×	×
Sandhurst	×	✓	✓	✓	✓*OW*	TQ775295	×	×
Sandhurst	×	✓	✓	✓	✓	TQ804284	×	✓
Sissinghurst	✓	×	✓	✓	✓	TQ789376	×	×
Staple Cross [4]	×	✓	✓	✓	✓	TQ782225	×	×
Tenterden	✓	✓	×	×	×	TQ863344	×	×
Tenterden	✓	✓	✓	✓	×	TQ866330	×	×
Tenterden [5]	×	×	P	×	Pump	TQ882335	×	×
Tenterden	×	✓	✓	✓	×	TQ905326	×	×
Ticehurst	×	✓	×	×	×	TQ684306	×	×
Wittersham [6]	✓	✓	✓	✓	✓	TQ902273	×	×
Wittersham	✓	✓	✓	✓	✓	TQ913273	✓	✓
Woodchurch	✓	✓	✓	✓	✓*OW*	TQ943353	×	×
Woodchurch	×	✓	✓	✓	✓	TQ943352	✓	✓

[1] Old Series sheets 3,4,5, and 6 are needed here. The copy of sheet 3 used was Messenger (see page 233) state 13 (of 19) *circa* 1860; that of sheet 4 was state 13 (of 15) *circa* 1871; that of sheet 5 was state 12 (of 16) *circa* 1864 and, lastly, that of sheet 6 was state 9 (of 21) *circa* 1845.

[2] The difference here between the large scales, one not uncommonly found, is that separate annotations for these two mills are provided at 1:2500 but then at the six-inch scale only one shared annotation is provided.

[3] This fine smock mill still survives having benefitted from more than one phase of restoration. It is one of a small handful of mills that gained early prominence in mill-restoration circles, and so also in the windmill literature. Since then, in times of ever-dwindling numbers of mills, it has maintained its status of being an outstanding smock mill.

[4] The depiction of this mill on the first edition 1:2500 is an example of one that is extremely easy to overlook.

[5] This structure was only ever a windpump. It does not appear on the first edition 1:2500. The second edition has a dot for a ground plan together with an annotation of just *P[ump]*. The third edition finally expands its annotation to *Windpump* meaning that the 1:2500 scale only fully caught up with the one-inch Revised New Series sheet of 1896, which had shown an unannotated symbol, after more than thirty years.

[6] This post mill had a substantial mill mound. The mill itself disappeared shortly after the First World War but its mound survives and is very prominent on the modern 1:25,000 map.

The area shown on this sheet had, in common with those on other sheets, a mix of windmill types but, unlike the other selected sheets, the mix here was really of post mills and smock mills rather than post mills and tower mills. If one associates the tower mills of, say, Lincolnshire or the post mills of Suffolk with excellence in windmills, then it is to Kent that one turns in order to see the same for smock mills.

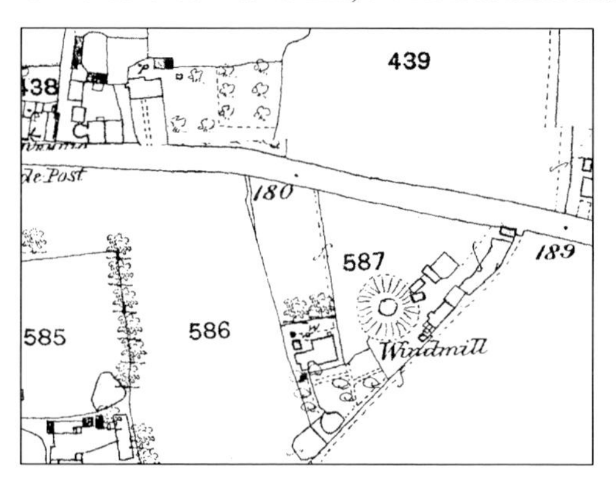

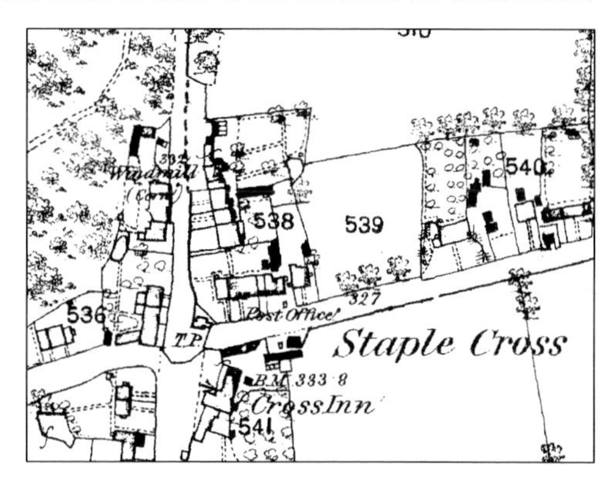

Depictions on the first edition 1:2500 for windmills at Wittersham *(above left)* and Staple Cross *(above right)*.

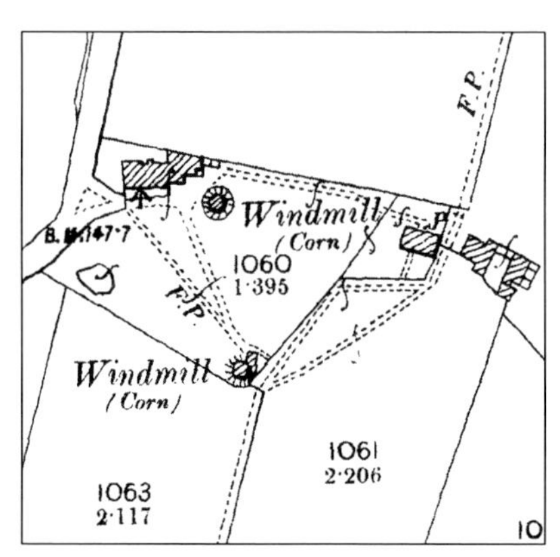

The two smock mills at Woodchurch, the so-called White and Black Mills. Viewed here from the north *(above right)*, the nearer mill sits on or, more correctly, in a large mound. This image is dated *circa* 1912 and this mill is clearly unable to work by sail. The far mill seems to be in working order even if some shutters from its spring sails are missing – this could, however, have been a deliberate way of lessening the mill's speed. The depiction of the first edition 1:2500 for these mills *(above left)* seems to imply that this mill also had a mound (which we can see not to have really been the case), or this may have been an attempt to depict each of the vertical supports of the sail-reefing gallery.

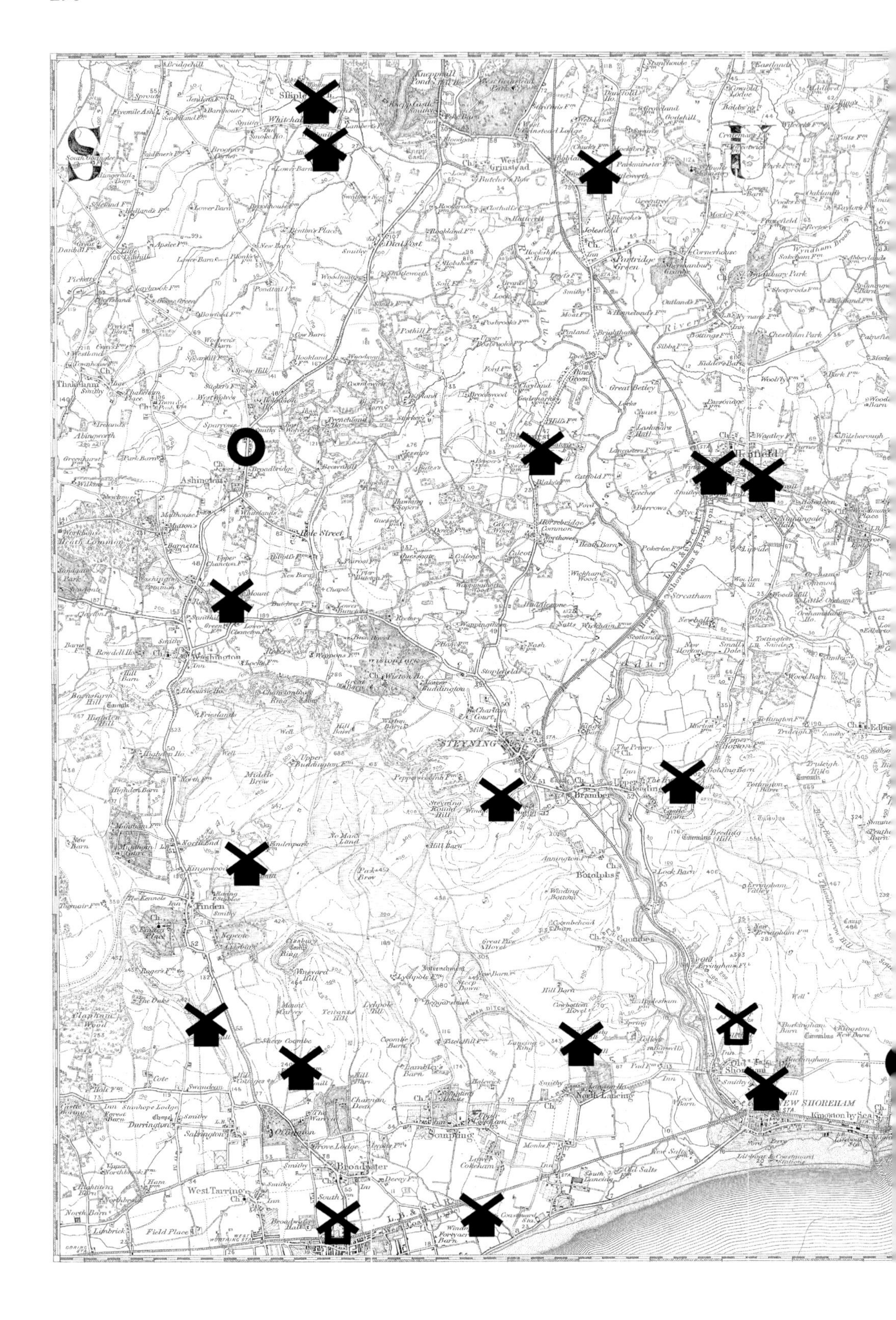

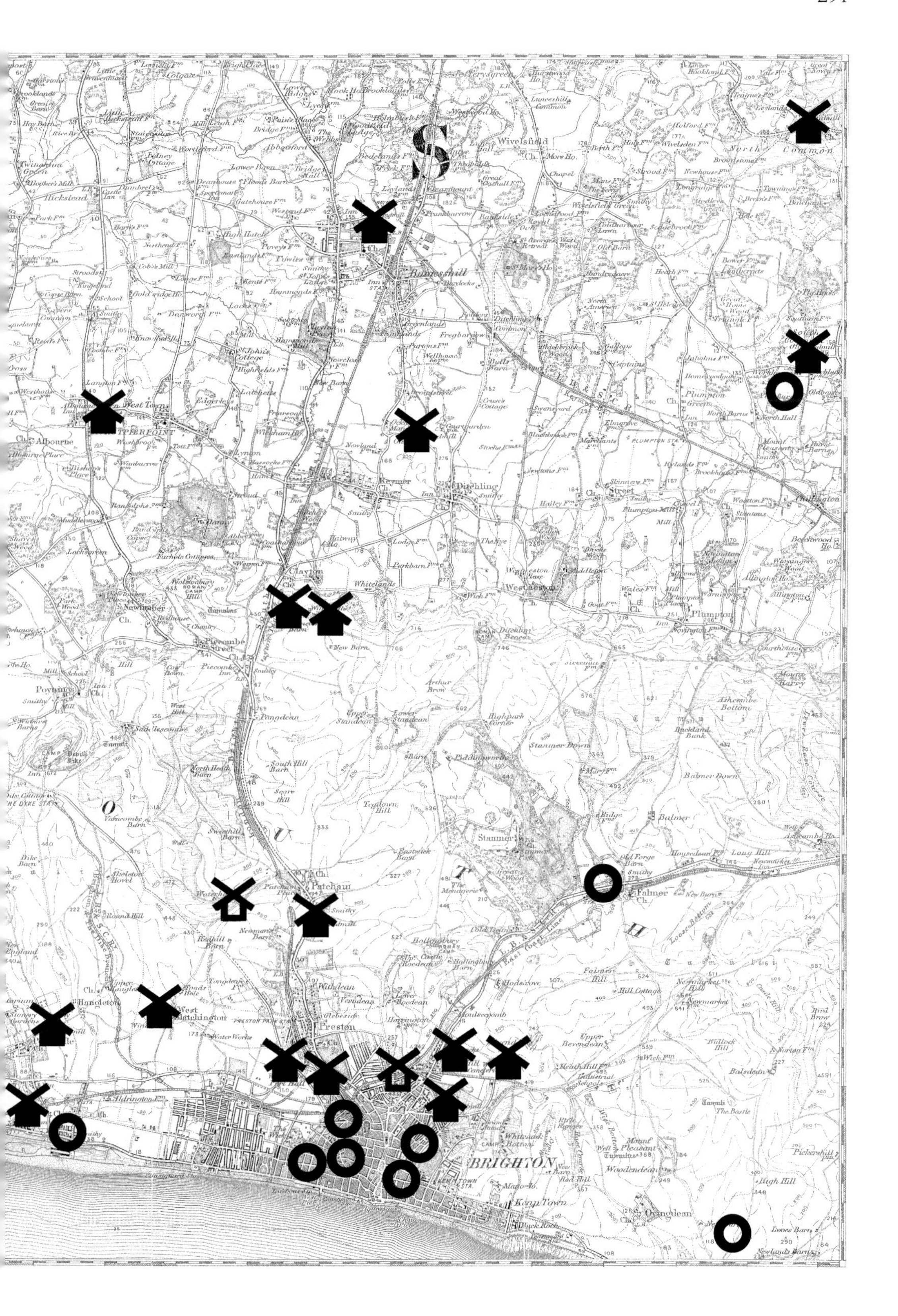
Wivelsfield
Wivelsfield Green
North Common
Hickstead
Twineham Green
Burgesshill
West Town
Albourne
Keymer
Ditchling
Street
Plumpton
Westmeston
Clayton
Newtimber
Pyecombe Street
Poynings
Ditchling Beacon
Stanmer Down
Stanmer
Falmer
Patcham
Withdean
Preston
Hangleton
West Blatchington
Aldrington
BRIGHTON
Kemp Town
Ovingdean
Woodendean
Balsdean
Newmarket Hill
Hollingbury
Old Forge Barn
Pickershill

Sheet 318 *Brighton*

(see pages 290 & 291)

Windmill locations as recorded by the Revised New Series *circa* 1895 and/or the first edition 1:2500	19th century recognition			20th century recognition				
	Old Series sheet states *ca* 1862	First edition six-inch	Second edition 1:2500	New Series Third Edition	Popular Edition	National Grid Reference	Seventh Series	Current 1:25,000
Ashurst	✓	✓	✓	✓	✓*OW*	TQ181160	×	×
Brighton	✓	✓	×	×	×	TQ301060	×	×
Brighton	✓	✓	×	×	×	TQ303057	×	×
Brighton	✓	✓	✓	×	×	TQ316057	×	×
Brighton	✓	✓	×	×	×	TQ325053	×	×
Brighton	✓	✓	×	×	×	TQ326060	×	×
Brighton	×	✓	✓	Pump	×	TQ336058	×	×
Broadwater	✓	✓	✓	✓	✓*OW*	TQ139060	×	×
Burgesshill	×	✓	✓	✓*Old*	✓*OW*	TQ314196	×	×
Clayton	✓	✓	✓	✓	✓*OW*	TQ303134	✓S	✓
Clayton	×	✓	✓	✓	✓*OW*	TQ304134	✓	✓
Durrington	✓	✓	✓	✓	✓*OW*	TQ123067	✓	✓
Finden	×	✓	×	×	×	TQ131092	×	×
Fortyacre Barn [1]	×	×	P	✓*Old*	×	TQ169034	×	×
Henfield	✓	✓	×	×	×	TQ210156	×	×
Henfield	×	✓	✓	✓*Old*	✓*OW*	TQ217154	×	×
Hurstpierpoint	✓	✓	×	×	×	TQ269166	×	×
Jolesfield	✓	✓	✓	✓	✓*OW*	TQ191205	✓S	×
Keymer	×	✓	✓	✓	✓	TQ321162	✓	✓
Leylands Farm	✓	✓	✓	✓	✓	TQ387214	✓	✓
New Shoreham [2]	✓	✓	×	✓*Old*	×	TQ217055	×	×
North Lancing [3]	✓	✓	×	✓*Old*	×	TQ185062	×	×
Old Shoreham	✓	✓	✓	×	×	TQ213066	×	×
Patcham	×	×	✓	✓	✓	TQ291086	✓	✓
Patcham	✓	✓	✓	×	×	TQ304084	×	×
Portslade	✓	✓	✓	✓*Old*	×	TQ260065	×	×
Shipley	×	×	✓	✓	✓	TQ143219	✓	✓
Shipley	×	✓	×	×	×	TQ145213	×	×
South Common	×	✓	✓	✓*Old*	✓*OW*	TQ387175	×	×
Southwick [4]	✓	✓	✓	✓	✓	TQ256052	×	×
Steyning	✓	✓	×	×	×	TQ174103	×	×
Upper Beeding	✓	✓	×	×	×	TQ204106	×	×
Washington	×	✓	✓	✓	✓*OW*	TQ128137	✓S	✓
West Blatchington	×	✓	✓	✓*Old*	✓	TQ279068	✓S	✓
Worthing Station	×	✓	×	×	×	TQ144033	×	×

[1] An unannotatated symbol on the Revised New Series is not matched by any first edition large-scale depictions and then only by the second editions revealing this to be a pump. (The nearby Navarino mills are on sheet 333.)

[2] This mill provides an interesting illustration of the difficulties in acknowledgement. The only hint that a remnant of it might still have existed at the time of the second edition 1:2500 – after having definitely appeared on the first edition – is the retention of the triangulation station symbol associated earlier with the mill. This has been deemed insufficient for acknowledging that this mill appeared on this second edition 1:2500.

[3] In the same vein of being useful to the Survey because of its prominence, as with the above mill, this windmill is shown on the first edition 1:2500 as a circular ground plan annotated as *Windmill (Flour)* but to the plan has been added an arrow indicating the whereabouts of a benchmark together with an additional annotation of *B.M.209.8.*

[4] This mill is ignored by the Revised New Series probably due to lack of map space but has the unusual annotation of *Windmill (Cement)* on the first editions of the large scales. This later becomes an unannotated open circle ground plan on the second edition 1:2500.

Right: Durrington (High Salvington) Mill seen on the first edition 1:2500. One might reasonably think that the depiction of this mill and its context is not nearly as clear as that of many other mills. The boundary line of the parish is particularly badly obscured.

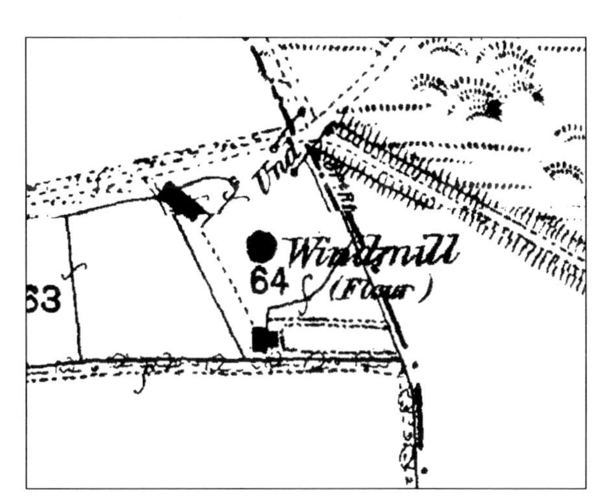

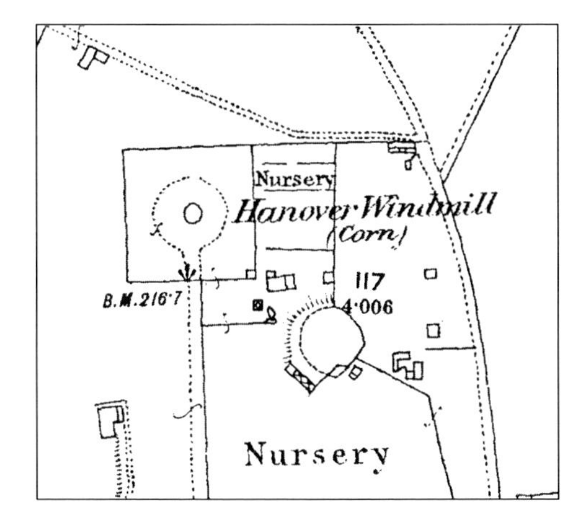

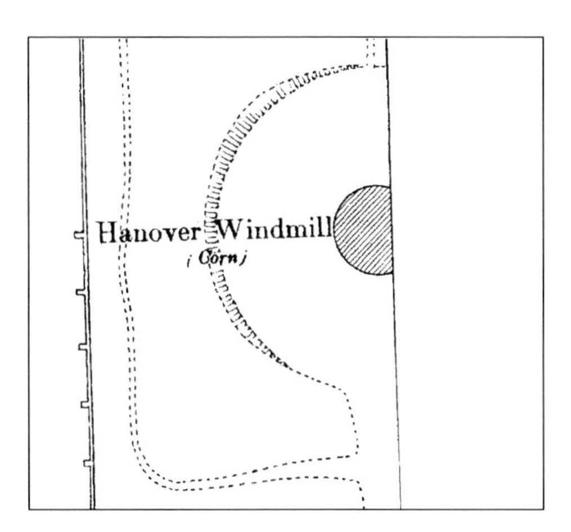

The usefulness of the largest OS map scales to the work of the mill historian is made obvious here by two depictions of Hanover Mill, a mill that could once be found just half a mile east from the centre of Brighton. This mill's depiction on the first edition 1:2500 *(above left)* is consistent with most depictions of windmills at this scale in terms of its reasonable size, lack of ambiguity and capacity for illustrating the spatial context of a mill. The very much larger town scale of 1:500, by offering ground plans whose linear dimensions were five times as great as their 1:2500 equivalents, in practice gave very little extra in the way of information. Additionally, town plans only cover a small part of what, by today's judgment, would comprise most towns. The 1:500 depiction of Hanover Mill *(above right)* upholds these remarks. Thankfully in the case of this windmill, the edge of the 1:500 plan of Brighton was suitably well chosen so that this mill could be shown on it, or at least slightly more than half of it could be! If Hanover Mill had been sited twenty feet further east than it was, it would have been excluded from this scale.

Select Bibliography

Maps and mapping

Andrews, J.H., *A paper landscape: the Ordnance Survey in nineteenth-century Ireland*, Oxford University Press, 1975; second edition, Dublin: Four Courts Press, 2002.

Beech, Geraldine and Mitchell, Rose, *Maps for family and local history: the records of the Tithe, Valuation Office and National Farm Surveys of England and Wales, 1836-1943*, London: The National Archives, 2004.

Bentley, John, 'Research in Somerset using the Ordnance Survey', *Sheetlines* 32 (1992), 9-16.

Bignell, Bill, 'Conventional signs and the Ordnance Survey: the case of mills and the New Series', *Sheetlines* 35 (1993), 10-13.

Black, Jeremy, *Maps and politics*, London: Reaktion Books, 1997.

Booth, J.R.S., *Public boundaries and Ordnance Survey, 1840-1980*, Southampton: Ordnance Survey, 1980.

Catalogue of maps and other publications of the Ordnance Survey, Southampton: Ordnance Survey, 1924. Reprinted with introduction, Kerry: David Archer, 1991.

Clarke, Robin V., 'The use of watermarks in dating Old Series one-inch Ordnance Survey maps', *Cartographic Journal* 6 (1969), 114-129.

Close, Colonel Sir Charles, *The early years of the Ordnance Survey*, Chatham: Institution of Royal Engineers, 1926. Reprinted with introduction by J. B. Harley, Newton Abbot: David & Charles, 1969.

Cole, John, 'Problems of the overhaul - revision of the 1:2500 map', *Sheetlines* 53 (1998), 30-33.

Cole, John, 'Survey diagrams – 1:10,560', *Sheetlines* 56 (1999), 18-21.

Cole, John, 'An introduction to the old 'Town Series' maps', *Sheetlines* 67 (2001), 26-31.

Cooper, Jim, 'One-inch revision in the 1960s', *Sheetlines* 52 (1998), 30-40.

Delano-Smith, Catherine and Kain, Roger J.P., *English maps: a history*, London: British Library, 1999.

Farquharson, Sir John, 'Twelve years' work of the Ordnance Survey, 1887 to 1899', *Geographical Journal* 15 (1900), 565-98.

Fletcher, David, 'The Ordnance Survey's nineteenth century boundary survey: context, characteristics and impact', *Imago Mundi* 51 (1999), 131-46.

Harley, J.B. and Philips, C.W., *The historian's guide to Ordnance Survey maps*, London: The Standing Conference for Social History, 1964.

Harley, J.B., 'Error and revision in early Ordnance Survey maps', *Cartographic Journal* 5 (1968), 115-24.

Harley, J.B., *Ordnance Survey maps: a descriptive manual*, Southampton: Ordnance Survey, 1975.

Hellyer, Roger, 'One-inch engraved maps with hills: some notes on double printing', *Sheetlines* 44 (1995), 11-20.

Hellyer, Roger, *Ordnance Survey small-scale maps, indexes: 1801-1998*, Kerry: David Archer, 1999.

Hellyer, Roger, *A guide to the Ordnance Survey 1:25,000 First Series* (with introductory essay by Richard Oliver), London: Charles Close Society, 2003.

Hellyer, Roger and Higley, Chris, (eds), *Projections and Origins: collected writings of Brian Adams*, London: Charles Close Society, 2006.

Hellyer, Roger and Oliver, Richard, *A guide to the Ordnance Survey one-inch Third Edition maps in colour*, London: Charles Close Society, 2004.

Hellyer, Roger and Oliver, Richard, *Military maps, the one-inch series of Great Britain and Ireland*, London: Charles Close Society, 2004.

Hellyer, Roger and Oliver, Richard, *One-inch engraved maps of the Ordnance Survey from 1847*, London: Charles Close Society, 2009.

Hewitt, Rachel, *Map of a nation: a biography of the Ordnance Survey*, London: Granta, 2010.

Higley, Chris, 'Supplement: Ordnance Survey index diagrams, part 1', *Sheetlines* 85 (2009), 27-38.

Higley, Chris, *Old Series to Explorer: a field guide to the Ordnance map*, London: Charles Close Society, 2011.

Hodson, Yolande, *Ordnance Surveyors' Drawings 1789-c.1840* (with an introduction, summary listing and indexes by Tony Campbell), Reading: Research Publications, 1989.

Hodson, Yolande, *'An inch to the mile': The Ordnance Survey one-inch map 1805-1974*, London: Charles Close Society, 1991.

Hodson, Yolande, *Map making in the Tower of London*, Southampton: Ordnance Survey, 1991.

Hodson, Yolande, *Popular maps, the Ordnance Survey Popular Edition one-inch map of England and Wales 1919-1926*, London: Charles Close Society, 1999.

Instructions for the revision of the one-inch map in the field, Southampton: Ordnance Survey, 1896, in Southampton Circulars, Book 3, ff 127-32. Reprinted in Oliver, Richard, 2003.

James, Sir Henry, (ed), *Account of the methods and processes adopted for the production of the maps of the Ordnance Survey of the United Kingdom*, London: HMSO, 1875.

Johnston, Duncan A., (ed), *Account of the methods and processes adopted...of the United Kingdom*, London: HMSO, 1902.

Kain, Roger J.P., Chapman, John and Oliver, Richard, *The enclosure maps of England and Wales 1595-1918*, Cambridge University Press, 2004.

Kain, Roger J.P. and Oliver, Richard, *The tithe maps of mid-nineteenth century England and Wales*, Cambridge University Press, 1995.

Kain, Roger J.P. and Oliver, Richard, *Historic parishes of England & Wales: electronic map – gazetteer – metadata*, Colchester: History data Service, 2001.

Kain, Roger J.P. and Prince, Hugh C., *The tithe surveys of England and Wales*, Cambridge University Press, 1985.

Keates, J.S., *Understanding maps*, second edition, Harlow: Longman, 1996.

MacEachren, Alan M., *How maps work: representation, visualisation and design*, London: The Guilford Press, 1995.

Margary, Harry, *The Old Series Ordnance Survey maps of England and Wales*, with introductory essays and cartobibliographies by J.B.Harley *et al.*, volumes I to VIII, Lympne Castle: Harry Margary, 1975-92.

Messenger, Guy, *The sheet histories of the Ordnance Survey one-inch Old Series maps of Devon and Cornwall: a cartobibliographic account*, London: Charles Close Society, 1991.

Messenger, Guy, *The sheet histories of the Ordnance Survey one-inch Old Series maps of Essex and Kent: a cartobibliographic account*, London: Charles Close Society, 1991.

Moore, J.N., 'The Ordnance Survey 1:500 town plan of Glasgow: a study of large-scale mapping, departmental policy and local opinion', *Cartographic Journal* 32 (1995), 24-32.

Nicholson, Tim, *The birth of the modern Ordnance Survey small-scale map: the revised New Series colour printed one-inch map of England and Wales 1897-1914*, London: Charles Close Society, 2002.

Oliver, Richard, 'What's what with the New Series', *Sheetlines* 5 (1982), 3-8.

Oliver, Richard, 'New light on the New Series', *Sheetlines* 12 (1985), 7-11.

Oliver, Richard, 'The Ordnance Survey in Great Britain 1835-1870', unpublished University of Sussex D.Phil. thesis, 1986.

Oliver, Richard, 'The Battle of the Scales: contemporary opinions and modern reconsiderations', *Sheetlines* 31 (1991), 59-64.

Oliver, Richard, 'The 'unpopular' one-inch Fourth Edition: an insight into early twentieth century Ordnance Survey small-scale revision policy', *Sheetlines* 32 (1992), 38-46.

Oliver, Richard, 'One-inch Old Series map design in the early 1850s', *Sheetlines* 38 (1994), 19-26.

Oliver, Richard, 'Ordnance Survey one-inch Old Series sheets: some notes on development and dating', *Sheetlines* 50 (1997), 11-31.

Oliver, Richard, *A guide to the Ordnance Survey one-inch New Popular Edition*, London: Charles Close Society, 2000.

Oliver, Richard, *A guide to the Ordnance Survey one-inch Fifth Edition*, London: Charles Close Society, 2000.

Oliver, Richard, 'Design and content changes on one-inch mapping of Britain, 1870-1914', *Sheetlines* 62 (2001), 6-23.

Oliver, Richard, 'The one-inch revision instructions of 1896', *Sheetlines* 66 (2003), 11-25.

Oliver, Richard, *A guide to the Ordnance Survey one-inch Seventh Series*, second edition, London: Charles Close Society, 2004.

Oliver, Richard, *Ordnance Survey maps, a concise guide for historians*, second edition, London: Charles Close Society, 2005.

Oliver, Richard, 'The sheet sizes and Delamere sheet lines of the one-inch Old Series', *Sheetlines* 77 (2006), 27-51.

Oliver, Richard, "Edition codes' and identifications on Ordnance Survey maps', *Sheetlines* 83 (2008), 27-34.

Oliver, Richard, 'The 'shading' of buildings on the 1:2500, 1893-1912: its 'meaning'', *Sheetlines* 83 (2008), 37-8.

Oliver, Richard, 'Photozincography and heliozincography', *Sheetlines* 90 (2011), 41-4.

Oliver, Richard, *The Ordnance Survey in the nineteenth century: maps, money and growth of government*, London: Charles Close Society, (in preparation).

Ordnance Survey of Great Britain – England and Wales – Indexes to the 1/2500 and 6-inch scale maps, (dating from c.1900 but continuously reprinted to c.1935). Reprinted with introduction, Kerry: David Archer, 2002.

Ordnance Survey operational manuals held by the British Library Map Library (BLML) and/or the National Archives (TNA/PRO) include:

Instructions to one-inch field revisers (two parts), 1901. In BLML Maps 207.d.14; Maps 207.b.34 and PRO OS 45/2.

Instructions to draftsmen and plan examiners, 1903. In BLML Maps 207.d.11; Maps 207.b.34.

Instructions to examiners and revisers, 1905. In PRO OS 45/3.

Instructions to draftsmen and plan examiners, provisional edition, 1906. In PRO OS 45/4.

Instructions to draftsmen and plan examiners, 1908. In PRO OS 45/7.

Instructions to surveyors, 1908. In PRO OS 45/8.

Instructions for the revision of the 1-inch map, 1909. In PRO OS 45/9.

Instructions to field examiners and revisers, 1912. In PRO OS 45/9.

Instructions for revision of the small scale maps (provisional): section dealing with drawing and examination, 1914. In PRO OS 45/12.

Owen, Tim and Pilbeam, Elaine, *Ordnance Survey: map makers to Britain since 1791*, London: HMSO and Southampton: Ordnance Survey, 1992.

Rodger, Elizabeth M., *The large scale county maps of the British Isles 1596-1850: a union list*, second edition, Oxford: Bodleian Library, 1972.

Sankey, H. Riall (Capt. R.E.), *The maps of the Ordnance Survey*, originally published by parts in *Engineering*, 1888. Reprinted with an introduction by Ian Mumford, London: Charles Close Society, 1995.

Seymour, W.A., (ed), *A history of the Ordnance Survey*, Folkestone: Dawson, 1980.

Townley, C.H.A., '1:10,560 Ordnance Survey maps – Lancashire sheets 85, 92, 93, 94, 95, 102, 103: notes on first edition', *Sheetlines* 42 (1995), 10-19.

Wheeler, R.C., 'Topographical accuracy of the Old Series one-inch map: artistic licence in the drawing office', *Sheetlines* 28 (1990), 9-10.

Wheeler, Rob, 'Triangulation points – primary, secondary and sacred', *Sheetlines* 46 (1996), 5-8.

Wheeler, Rob, 'Construction lines on some OSDs', *Sheetlines* 51 (1998), 35-45.

Willis, J.C.T., *An outline of the history and revision of 25-inch Ordnance Survey plans*, London: HMSO, 1932.

Winchester, Angus, *Discovering parish boundaries*, Princes Risborough: Shire Publications, 1990.

Winterbotham, Brigadier H.St.J.L., *The National Plans: the ten-foot, five-foot, twenty-five inch and six-inch scales*, (Ordnance Survey Professional Papers, New Series, No 16), London: HMSO, 1934.

Report of Committee on a military map of the United Kingdom, printed at the War Office, 1892 [A.237]. (Unpublished: copies in Royal Geographical Society, Z.72/4 and in TNA PRO WO 33/52, p.639) (abbreviated herein as Baker).

Report of the Departmental Committee appointed by the Board of Agriculture to inquire into the present condition of the Ordnance Survey, British Parliamentary Papers (House of Commons series) 1893-94 [C.6895], LXXII, 305 (abbreviated herein as Dorington).

Sheetlines, the Journal of the Charles Close Society for the study of Ordnance Survey maps has been published three times yearly since 1981.

Windmills

The generally recognised bibliographical guide for secondary windmill sources (Blythman, 2008) extends to more than eight hundred serials, so forcing any compilation of sources here to be rigorous in the choices it makes. This part of the bibliography – which focuses only on published texts for windmills – thus has to keep within a strict set of parameters for deciding what is listed. Accordingly, because of the emphasis placed by this book on knowing about long-vanished mills so as to be able to assess OS maps, a prime concern has been to offer one or more of the county-based sources for windmills for each county where mills were once present in reasonable numbers. For an overview of the English windmill, a few of the general texts for the country's windmills are sensibly included. But much less well represented are texts dealing with aspects of milling such as, for example, the machinery of windmills or their modes of operation. The reader is also reminded that many of the county listings for windmills, whether in short articles or substantial books, were originally based in very large measure on the evidence of different OS map series.

Apling, Harry, *Norfolk corn windmills*, volume one, Norfolk Windmills Trust, 1984.

Baker, P.H.J. and Wailes, Rex, 'The windmills of Derbyshire, Leicestershire and Nottinghamshire, Part 1: post mills', *Transactions Newcomen Society* 33 (1960-61), 113-28.

Baker, P.H.J. and Wailes, Rex, 'The windmills of Derbyshire, Leicestershire and Nottinghamshire, Part 2: tower mills', *Transactions Newcomen Society* 34 (1961-2), 89-104.

Blythman, G.J., *Watermills & windmills of Middlesex*, Baron Birch for Quotes Ltd, 1996.

Blythman, Guy, *British windmills: a bibliographical guide*, published by author, 2008.

Blythman, Guy, *English windmills: a photographic register*, (manuscript copy deposited in the Science Museum library), 2009.

Brunnarius, Martin, *The windmills of Sussex*, Chichester: Phillimore, 1979.

Buckinghamshire County Museum Archaeological Group, 'Buckinghamshire windmills', *Record of Bucks* 20 Part 4 (1978), 516-24.

Coles Finch, William, *Watermills and windmills*, London: C.W. Daniel, 1933. Reprinted in facsimile, Sheerness: Arthur Cassell, 1976. (Deals exclusively with mills in Kent.)

Coulthard, Alfred J. and Watts, Martin, *Windmills of Somerset and the men who worked them*, London: Research Publishing Company, 1978.

De Little, Rodney, *The windmills of England*, third edition, Jolesfield: Colwood, 1997.

Dolman, Peter, *Windmills in Suffolk: a contemporary survey*, Suffolk Mills Group, 1978.

Dolman, Peter, *Lincolnshire windmills: a contemporary survey*, Lincolnshire County Council, 1986.

Douch, H.L., *Cornish windmills*, Truro: Oscar Blackford, nd.

Elliott, J. Steele, 'The windmills of Bedfordshire: past and present', *Bedfordshire Historical Record Society* 14 (1931), 3-50.

Ellis, Monica, (ed), *Water and wind mills in Hampshire and the Isle of Wight*, Southampton University Industrial Archaeology Group, 1978.

Farries, K. G. and Mason, M.T., *The windmills of Surrey and inner London*, London: Charles Skilton, 1966.

Farries, Kenneth G., *Essex windmills, millers & millwrights*, volumes 1 to 5, London: Charles Skilton, 1981-88.

Filby, Peter, 'Fenland drainage windmills of Cambridgeshire and Huntingdonshire', in *Proceedings of the twelfth mill research conference*, Manningtree: Mills Research Group, 1995, 31-44.

Flint, Brian, *Suffolk windmills*, Woodbridge: Boydell, 1979.

Foreman, Wilfred, *Oxfordshire mills*, Chichester: Phillimore, 1983.

Freese, Stanley, *Windmills and millwrighting*, Cambridge University Press, 1957. Reprinted, Newton Abbot: David & Charles, 1971.

Gifford, Alan, *Derbyshire windmills*, Midland Wind & Water Mills Group, 1995.

Gregory, Roy, *East Yorkshire windmills*, Cheddar: Charles Skilton, 1985.

Gregory, Roy, *The industrial windmill in Britain*, Chichester: Phillimore, 2005.
Gregory, Roy and Turner, Laurence, *Windmills of Yorkshire*, Catrine: Stenlake, 2009.
Harrison, John K., *Eight centuries of milling in north east Yorkshire*, North York Moors National Park Authority, 2001.
Hemming, Peter, *Windmills in Sussex*, London: C.W.Daniel, 1936.
Hills, Richard L., *Power from wind: a history of windmill technology*, Cambridge University Press, 1994.
Howes, Hugh, *The windmills and watermills of Bedfordshire*, Copt Hewick: Book Castle, 2009.
Hughes, H.C., 'Windmills in Cambridgeshire and the Isle of Ely', *Proceedings Cambridge Antiquarian Society* 31 (1931), 17-29.
Hughes, J., 'Cumberland windmills', *Transactions Cumberland and Westmorland Antiquarian and Archaeological Society* 72 (1972), 112-41.
Job, Barry, *Staffordshire windmills*, Midland Wind & Water Mills Group, 1985.
Moon, Nigel, *Leicestershire and Rutland windmills*, Wymondham: Sycamore, 1981.
Moore, Cyril, *Hertfordshire windmills & windmillers*, Sawbridgeworth: Windsup, 1999.
Reynolds, John, *Windmills and watermills*, second edition, London: Hugh Evelyn, 1974.
Sass, Jon, *Windmills of Lincolnshire*, Catrine: Stenlake, 2012.
Seaby, Wilfred A., *Warwickshire windmills*, Warwick County Museum, Abstract 1, 1979.
Seaby, Wilfred A. and Smith, Arthur C., *Windmills in Shropshire, Hereford and Worcester: a contemporary survey*, Stevenage Museum, 1984.
Shaw, Tony, *Windmills of Nottinghamshire*, Nottinghamshire County Council, 1995.
Smith, Arthur C., *Windmills in Cambridgeshire: a contemporary survey*, Stevenage Museum, 1975.
Smith, Arthur C., *Windmills in Buckinghamshire and Oxfordshire: a contemporary survey*, Stevenage Museum, 1976.
Smith, Arthur C., *Windmills in Huntingdon and Peterborough: a contemporary survey*, Stevenage Museum, 1977.
Smith, Arthur C., *Drainage windmills of the Norfolk marshes: a contemporary survey*, Stevenage Museum, 1978.
Stainwright, Trevor L., *Windmills of Northamptonshire and the Soke of Peterborough*, Wellingborough: Wharton, 1991.
Starmer, Geoffrey H., *A check list of Northamptonshire windmills*, ca.1970, 9 pages, copy in Northampton Library.
Tebbutt, C.F., 'Huntingdonshire windmills', (in four parts), *Transactions Cambridgeshire and Huntingdonshire Archaeological Society* 5 (1937), 433-8; 6 (1947), 29-33, 62-5, 103-04.
Wailes, Rex, 'Suffolk windmills, Part 1: post mills', *Transactions Newcomen Society* 22 (1941-42), 41-63.
Wailes, Rex, 'Suffolk windmills, Part 2: tower mills', *Transactions Newcomen Society* 23 (1942-43), 37-54.
Wailes, Rex, 'Norfolk windmills, Part 1: corn mills', *Transactions Newcomen Society* 26 (1947-49), 231-58.
Wailes, Rex, 'The windmills of Cambridgeshire: including those of the Isle of Ely, the Soke of Peterborough and Huntingdonshire', *Transactions Newcomen Society* 27 (1949-51), 97-119.
Wailes, Rex, 'Lincolnshire windmills, Part 1: post mills', *Transactions Newcomen Society* 28 (1951-53), 245-53.

Wailes, Rex, *The English Windmill*, London: Routledge & Kegan Paul, 1954.
Wailes, Rex, 'Lincolnshire windmills, Part 2: tower mills', *Transactions Newcomen Society* 29 (1953-55), 103-22.
Wailes, Rex and Russell, John, 'Windmills in Kent', *Transactions Newcomen Society* 29 (1953-55), 221-39.
Wailes, Rex, 'Norfolk windmills, Part 2: drainage and pumping mills including those of Suffolk', *Transactions Newcomen Society* 30 (1955-57), 157-77.
Wailes, Rex, 'Essex windmills', *Transactions Newcomen Society* 31 (1957-59), 153-80.
Watts, Martin, *Wiltshire windmills*, Wiltshire Library & Museum Service, 1980.
Watts, Martin, *The archaeology of mills and milling*, Stroud: Tempus, 2001.
Whitworth, Alan, *Tyke Towers: Yorkshire's windmills*, Blackpool: Landy, 2002.

General

Beresford, Maurice, *History on the ground*, London: Lutterworth, 1957. Reprinted, Stroud: Alan Sutton Publishing, 1984.
Cadava, Eduardo, *Words of light: theses on the photography of history*, Princeton University Press, 1997.
Carter, Paul and Thompson, Kate, *Sources for local historians*, Chichester: Phillimore, 2005.
Cosgrove, Denis and Daniels, Stephen, *The iconography of landscape*, Cambridge University Press, 1988.
Cannadine, David and Reeder, David (eds), *Exploring the urban past: essays in urban history by H.J. Dyos*, Cambridge University Press, 1982.
Grigg, David, *The dynamics of agricultural change*, London: Hutchinson, 1982.
Hamilton, Paul, *Historicism*, London: Routledge, 1996.
Kammen, Carol (ed), *The pursuit of local history: readings on theory and practice*, London: Sage, 1996.
Lilley, Keith, 'J. Brian Harley', in Hubbard, Phil, Kitchin, Rob and Valentine, Gill (eds), *Key thinkers on space and place*, London: Sage, 2004, 174-80.
Palmer, Marilyn and Neaverson, Peter, *Industry in the landscape, 1700-1900*, London: Routledge, 1994.
Ogilvie, Sheilagh C. and Cerman, Markus (eds), *European proto-industrialisation*, Cambridge University Press, 1996.
Rackham, Oliver, *The history of the countryside*, London: Dent, 1986.
Schama, Simon, *Landscape & memory*, London: HarperCollins, 1995.
Short, Brian, *Land and society in Edwardian Britain*, Cambridge University Press, 1997.
Stanford, Michael, *The nature of historical knowledge*, Oxford: Blackwell, 1986.
Waller, P.J., *Town, city & nation: England 1850-1914*, Oxford University Press, 1983.
Walsh, Kevin, *The representation of the past: museums and heritage in the post-modern world*, London: Routledge, 1992.
Weaver, Mike (ed), *British photography in the nineteenth century: the fine art tradition*, Cambridge University Press, 1989.
Wright, Patrick, *On living in an old country: the national past in contemporary Britain*, London: Verso, 1985.

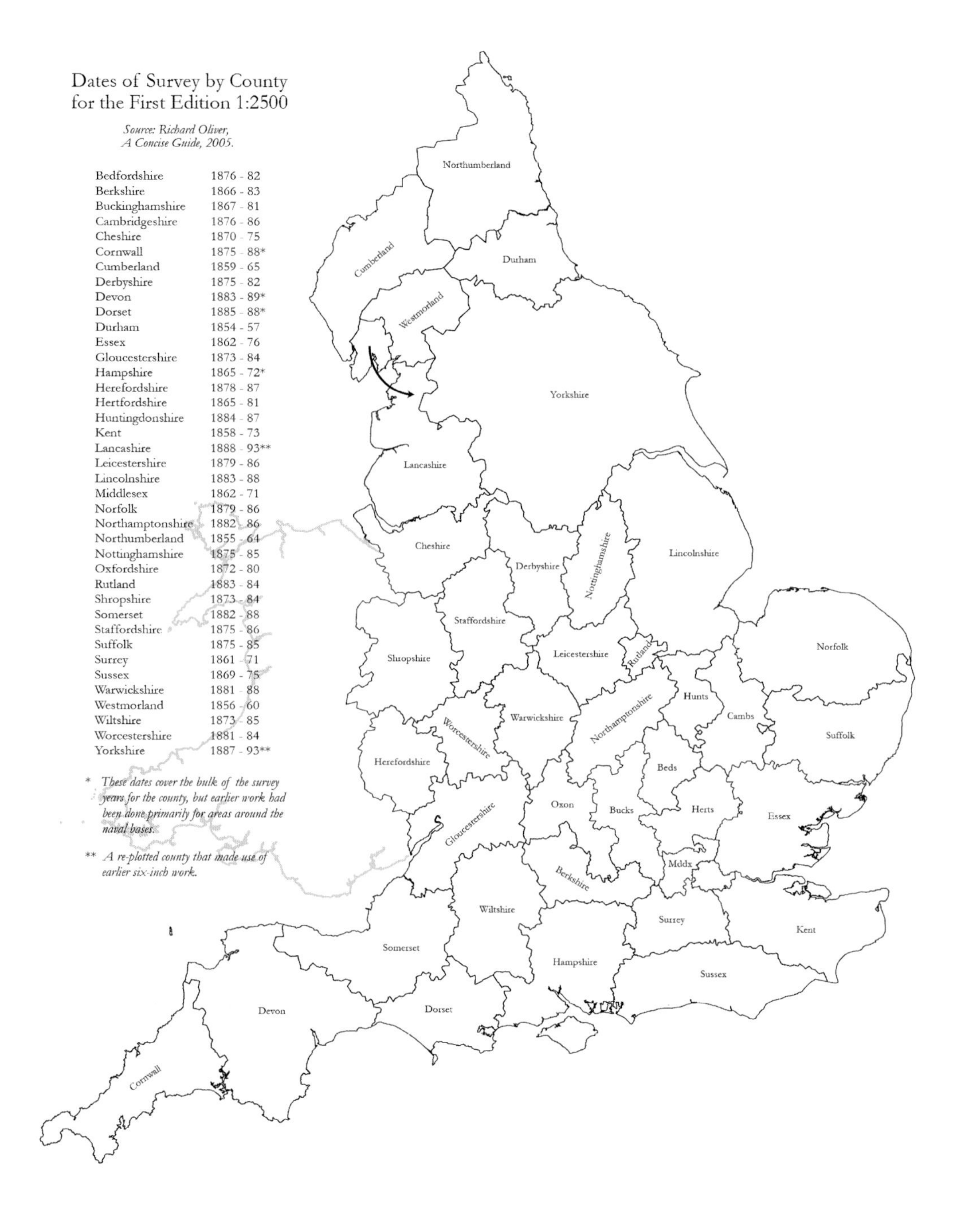

Dates of Survey by County for the First Edition 1:2500

Source: Richard Oliver, A Concise Guide, 2005.

County	Dates
Bedfordshire	1876 - 82
Berkshire	1866 - 83
Buckinghamshire	1867 - 81
Cambridgeshire	1876 - 86
Cheshire	1870 - 75
Cornwall	1875 - 88*
Cumberland	1859 - 65
Derbyshire	1875 - 82
Devon	1883 - 89*
Dorset	1885 - 88*
Durham	1854 - 57
Essex	1862 - 76
Gloucestershire	1873 - 84
Hampshire	1865 - 72*
Herefordshire	1878 - 87
Hertfordshire	1865 - 81
Huntingdonshire	1884 - 87
Kent	1858 - 73
Lancashire	1888 - 93**
Leicestershire	1879 - 86
Lincolnshire	1883 - 88
Middlesex	1862 - 71
Norfolk	1879 - 86
Northamptonshire	1882 - 86
Northumberland	1855 - 64
Nottinghamshire	1875 - 85
Oxfordshire	1872 - 80
Rutland	1883 - 84
Shropshire	1873 - 84
Somerset	1882 - 88
Staffordshire	1875 - 86
Suffolk	1875 - 85
Surrey	1861 - 71
Sussex	1869 - 75
Warwickshire	1881 - 88
Westmorland	1856 - 60
Wiltshire	1873 - 85
Worcestershire	1881 - 84
Yorkshire	1887 - 93**

* *These dates cover the bulk of the survey years for the county, but earlier work had been done primarily for areas around the naval bases.*

** *A re-plotted county that made use of earlier six-inch work.*

Index